Cytokines in Human Health

METHODS IN PHARMACOLOGY AND TOXICOLOGY

Cytokines in Human Health

Immunotoxicology, Pathology, and Therapeutic Applications

Edited by

Robert V. House

DyPort Vaccine Company, LLC
Frederick, MD

Jacques Descotes

Poison Center and Pharmacovigilance Unit
Lyon Cedex, France

HUMANA PRESS ✴ TOTOWA, NEW JERSEY

© 2007 Humana Press Inc.
999 Riverview Drive, Suite 208
Totowa, NJ 07512

www.humanapress.com

Production Editor: Amy Thau

Cover design by Donna Niethe

For additional copies, pricing for bulk purchases, and/or information about other Humana titles, contact Humana at the above address or at any of the following numbers: Tel.: 973-256-1699; Fax: 973-256-8341; E-mail: orders@humanapr.com or visit our website: http://humanapress.com

This publication is printed on acid-free paper. ∞
ANSI Z39.48-1984 (American National Standards Institute) Permanence of Paper for Printed Library Materials.

Printed in the United States of America. 10 9 8 7 6 5 4 3 2 1
eISBN 10-digit: 1-59745-350-1
eISBN 13-digit: 978-1-59745-350-9
Library of Congress Cataloging-in-Publication Data
Cytokines in human health : immunotoxicology, pathology, and
   therapeutic applications / edited by Robert V. House, Jacques
   Descotes.
     p. ; cm. -- (Methods in pharmacology and toxicology)
   Includes bibliographical references and index.
   ISBN 10-digit: 1-58829-467-6 (alk. paper)
   ISBN 13-digit: 978-1-58829-467-8 (alk. paper)
   1. Cytokines. 2. Immunotoxicology.   I. House, Robert V.
II. Descotes, Jacques, 1925-   . III. Series.
   [DNLM: 1. Cytokines--immunology. 2. Cytokines--therapeutic use.
QW 568 C997 2007]
QR185.8.C95C992 2007
616.07'9--dc22
                                        2006026776

Preface

Over the past three decades the field of immunotoxicology, the study of the effects of exposure to drugs, chemicals, or physical/environmental agents on the structure and function of the immune system has benefited from an increasingly detailed understanding of the cellular and molecular basis of innate and acquired immunity. Arguably one of the most promising areas of investigation centers on the intensely complex network of biochemical regulators, namely the cytokines (and, to a lesser degree, their biochemical cousins the chemokines). Cytokines facilitate the initial recognition of foreignness that launches innate host defenses, they form the bridge that allows a nonspecific response to mature into an antigen-specific acquired immune response, and they maintain this response for the life of the individual. Clearly, any event that affects the biology of these important molecules is likely to have significant effects on the overall immune competence of the host.

In *Cytokines in Human Health: Immunotoxicology, Pathology, and Therapeutic Applications*, experts of cytokine biology share their knowledge on various aspects of how modulation of cytokines can affect human health. First, the basic biology of cytokines is reviewed, particularly as this relates to the ability of external influences (whether inadvertent or deliberate) to modify the expression, production, and activity of these molecules. Various chapters describe methodology for measuring cytokine activity, basic description of how differential cytokine modulation determines the type of immune response generated, and how microbes can act as agents of immunotoxicity by directly modulating the cytokine cascade.

This sets the stage for practical preclinical understanding of the effect of cytokines, including how they function in chemical allergy, lung toxicity, and drug abuse. In addition, methods are described for assessing cytokine immunotoxicity. This group of chapters provides the reader with a broad understanding of the range of cytokine activity in human disease.

The final chapters deal with both the desirable as well as the adverse effects of cytokines used as therapeutic agents. Although cytokines show great promise as therapeutic agents, they exhibit a surprising level of toxicity, manifested by flu-like symptoms, vascular leak syndrome, and even the possibility of the induction of autoimmune disease. Even in the absence of therapeutic cytokine-related pathology, their use in humans has been shown in some cases to lead to the induction of anti-cytokine antibodies; this finding is fraught with its own specific consequences.

It spite of the diverse and wide-ranging nature of *Cytokines in Human Health: Immunotoxicology, Pathology, and Therapeutic Applications*, it is our sincere hope that this collection of expert reviews will serve as both a primer as well as a starting point for a more detailed investigation by the reader of the role these fascinating biological regulators play in human health and disease.

Robert V. House
Jacques Descotes

Contents

Contributors

Roxana G. Baluna • *Department of Radiation Oncology, Cancer Therapy and Research Center, San Antonio, TX*

Melissa S. Beck-Westermeyer • *Covance Laboratories, Inc., Madison, WI*

Rebecca J. Dearman • *Syngenta Central Toxicology Laboratory, Alderley Park, Macclesfield, Cheshire, UK*

Jacques G. Descotes • *Poison Center and Pharmacovigilance Unit, Claude Bernard University, Lyon Cedex 03, France*

Robert V. House • *DynPort Vaccine Company LLC, Frederick, MD*

Victor J. Johnson • *Toxicology and Molecular Biology Branch, Health Effects Laboratory Division, National Institute for Occupational Safety and Health, Morgantown, WV*

Michael L. Kashon • *Biostatistics and Epidemiology Branch, Health Effects Laboratory Division, National Institute for Occupational Safety and Health, Morgantown, WV*

Ian Kimber • *Syngenta Central Toxicology Laboratory, Alderley Park, Macclesfield, Cheshire, UK*

Howard M. Kipen • *Department of Pharmacology and Toxicology, Rutgers University, Piscataway, NJ; Environmental and Occupational Health Sciences Institute, UMDNJ-Robert Wood Johnson Medical School, Piscataway, NJ*

Debra L. Laskin • *Department of Pharmacology and Toxicology, Rutgers University, Piscataway, NJ; Environmental and Occupational Health Sciences Institute, UMDNJ-Robert Wood Johnson Medical School, Piscataway, NJ*

Robert J. Laumbach • *Department of Pharmacology and Toxicology, Rutgers University, Piscataway, NJ; and Environmental and Occupational Health Sciences Institute, UMDNJ-Robert Wood Johnson Medical School, Piscataway, NJ*

Hervé Lebrec • *3M Pathology Toxicology, 3M Pharmaceuticals, Saint Paul, MN*

Michael I. Luster • *Toxicology and Molecular Biology Branch, Health Effects Laboratory Division, National Institute for Occupational Safety and Health, Morgantown, WV*

Pierre Miossec • *Department of Immunology and Rheumatology, Hôpital Edouard Herriot, Lyon Cedex 03, France*

James E. Pease • *Leukocyte Biology Section, Biomedical Sciences Division, Faculty of Medicine, Imperial College of Science, Technology, and Medicine, South Kensington Campus, London, UK*

Stephen B. Pruett • *Department of Cellular Biology and Anatomy, LSUHSC, Shreveport, LA*

Kenneth W. Renton • *Department of Pharmacology, Dalhousie University, Halifax, Nova Scotia, Canada*

Michael J. Robertson • *Department of Medicine, Indiana University Medical Center, Indianapolis, IN*

Vasanthi R. Sunil • *Department of Pharmacology and Toxicology, Rutgers University, Piscataway, NJ*

Steven Swanson • *Department of Clinical Immunology, Amgen, Inc., Thousand Oaks, CA*

Peter T. Thomas • *Program Management Services, Covance Laboratories, Inc., Madison, WI*

Rob J. Vandebriel • *Laboratory for Toxicology, Pathology and Genetics, National Institute of Public Health and the Environment, Bilthoven, the Netherlands*

John Vasilakos • *3M Pathology Toxicology, 3M Pharmaceuticals, Saint Paul, MN*

Thierry Vial • *Poison Center and Pharmacovigilance Unit, Lyon Cedex 03, France*

Theresa L. Whiteside • *University of Pittsburgh Cancer Institute, Research Pavilion at the Hillman Cancer Center, Pittsburgh, PA*

Berran Yucesoy • *Toxicology and Molecular Biology Branch, Health Effects Laboratory Division, National Institute for Occupational Safety and Health, Morgantown, WV*

1

Introduction to Cytokines as Targets for Immunomodulation

Theresa L. Whiteside

SUMMARY

Cytokines play a key role in modulation of immune responses. Cytokine networks regulate lymphocyte turnover, differentiation, and activation. Many different cell types, in addition to immune cells, produce cytokines and express receptors for cytokines. Cell-to-cell communication (cellular "crosstalk") is maintained via cytokine networks. In disease, these networks undergo imbalance. By measuring amplification or downregulation of cytokine signaling cascades in response to pathological insults or therapeutic interventions, it might be possible to evaluate disease progression or regression. Multiplex formats for cytokine profiling are now available and appear to be especially useful in monitoring cytokine profile alterations in a variety of human diseases.

Key Words: Immune response; cytokine profiles; cytokines in disease; cytokines in therapy; cytokine crosstalk.

1. INTRODUCTION

In the last decade or so, cytokines, which are low molecular weight glycoproteins produced by immune as well as nonimmune cells, have captured the attention of the scientific and clinical communities. This unprecedented attention to cytokines can be, in part, attributed to their availability in a recombinant form, allowing for the use of individual cytokines in various experimental and clinical settings. Largely, however, it is because of the realization that complex cellular "soups" of the past (also referred to as "lymphokines") contain mixtures of defined soluble factors, cytokines, each of which plays a key role in orchestrating cellular interactions. The field of

From: *Methods in Pharmacology and Toxicology: Cytokines in Human Health:*
Immunotoxicology, Pathology, and Therapeutic Applications
Edited by: R. V. House and J. Descotes © Humana Press Inc., Totowa, NJ

cytokine biology has emerged based on molecular, biochemical, and immunological insights into their structure, function, importance in health and
disease, as well as therapeutic potential (reviewed in ref. *1*) The number of
cytokine genes cloned and individual cytokines identified has grown rapidly to include several distinct families of factors that include lymphokines,
interferons, and hematopoietic and nonhematopoietic growth factors and
appear to be necessary for cell-to-cell communication and signaling. Methods for the identification and measurements of the functional competence of
cytokines have been developed that provide means for ascertaining their
presence and following their levels of activity in body fluids and at tissues
sites. The more recent discovery of chemokines, which are related to
cytokines but remain structurally and functionally distinct *(2)*, has further
expanded the scope of soluble factors involved in the regulation of immunological responses, hematopoietic development, cell-to-cell communication,
as well as host responses to infectious agents and inflammatory stimuli *(3–6)*.
To be able to fulfill this challenging assembly of biological activities,
cytokines and chemokines are endowed with special biological characteristics that allow for effective functioning in the tissue microenvironment.

2. BIOLOGICAL CHARACTERISTICS OF CYTOKINES

The biology of cytokines is complex and, although new information
emerges almost daily, much still remains to be learned. Cytokines share
many characteristics with growth factors and hormones, such as platelet-
derived growth factor, endothelial growth factor, or transforming growth
factor-α. Although classical polypeptide hormones are produced by special-
ized cells, cytokines tend to be secreted by less-specialized cells, and often
many different cell types produce the same cytokine. All cytokines are *pleiotropic*, i.e., they have the ability to interact not with one but a variety of
cellular targets, usually via the specific receptors expressed on the cell surface. Expression of the receptor qualifies the cell as a target for one or more
cytokines. Binding of cytokines to *cell surface receptors* (kDa in the range
of 10^{-9} to 10^{-12} *M*) initiates signaling and new ribonucleic acid (RNA) and
protein synthesis. Frequently, cytokine receptors are composed of several
membrane-associated subunits, and one or more of these subunits might be
shared by distinct cytokine receptors. Also, the presence or absence of individual subunits often determines the affinity of the cytokine receptor for its
ligand as, for example, in the case of receptor for interleukin (IL)-2, in which
the $\alpha\beta\gamma$ subunits are needed for high-affinity binding of IL-2, the $\beta\gamma$ sub-

units function as an intermediate affinity receptor and the α subunit is known as a low-affinity receptor for IL-2 *(7)*. In addition, cytokines can modulate the level of receptor expression for another cytokine or growth factor, a phenomenon referred to as *receptor transmodulation*. For example, tumor necrosis factor (TNF)-α, IL-1, or IL-6 can affect IL-2 receptor expression *(8–10)*, and interferon (IFN)-γ can modulate expression of TNF receptors on many different cell lines *(11)*.

Cytokines are *redundant* in that the same biological function can be executed by several distinct cytokines. This overlapping of cytokine activities, whereby a single cell target shows seemingly identical responses to multiple cytokines, has been referred to as *cytokine crosstalk*. Cytokine redundancy is a biologicalally important feature because it provides a measure of safety for the process of regulation of normal cellular functions. Should one cytokine be absent or its level limited, an option for the substitution by another cytokine exists. Mechanistically, redundancy might be implemented via the shared receptor subunits or via the ability of different cytokine receptors to mediate similar signals, targeting the same molecular pathway(s). Physiologically, cytokine redundancy assures that the vital communication between cells participating in functions essential for survival continues uninterrupted regardless of the absence of one or another cytokine.

Cytokines and chemokines are known to function as *networks* or *cascades* of interacting factors, which are able to induce each other. This implies that cytokines *regulate their own production* via autocrine, juxtacrine, or paracrine pathways in response to microenvironmental stimuli. Cascades of cytokines are able to amplify the functional impact of one component and are related to functional redundancy. Thus, even when a redundant signal is weak, there is the possibility for its amplification via inducing the cascade of functionally related soluble factors. It is also essential to remember that local interactions and the state of activation of cells in the microenvironment are likely to have powerful effects on the cytokine production *in situ*.

Cytokines work in *pharmacological doses* and, once released by the producing cell, have a *short half-life,* often not exceeding a few minutes. This means that tiny quantities of cytokines present in the microenvironment for a short period of time are sufficient to induce functional changes in responding cells. Thus, cytokines are powerful physiological mediators with a potential for inducing *tissue damage*. Cytokines are *local mediators* meant to exert biological effects in their local microenvironment. Although in normal tissues, cytokine-mediated tissue damages do not occur, largely because of carefully regulated cytokine secretion, in pathological conditions, excessive cytokine release from activated cells not infrequently results in a local tissue injury. Further, s*ystemic effects of cytokines* often are quite different and more dramatic than those mediated locally *(12)*.

A division of labor exists among cytokines, in that the T-helper 1 (Th1)-type cytokines (e.g., IL-2, IFN-γ, IL-12, TNF-α) are considered to be responsible for elimination of intracellular infections, notably those caused by viruses or parasites, modulation of organ-specific autoimmune diseases, or mediating allograft rejection and recurrent abortions *(13,14)*. Other human diseases are associated with Th2-type cytokine responses (e.g., IL-3, IL-4, IL-5, IL-13), including atopic disorders, chronic graft-versus host disease, progression to AIDS, cancer metastasis, or the hypereosinophilic syndrome *(14)*. In general, the Th2-type cytokines are involved in promoting antibody-mediated responses. The third category of cytokines, Th3, includes regulatory mediators, such as IL-10 or TGF-β, which are responsible for maintaining a balance in the host microenvironment *(15)*. This division of labor is a reflection of *cytokine polarization* and appears to be useful in the classification of certain human diseases into the Th1 or Th2 categories. It also forms a basis for a concept of the "cytokine profile." Because cytokine polarization occurs in many pathological conditions and cytokine cascades as well as cytokine crosstalk often are characteristically encountered in disease, the profiling of cytokines might prove to be more useful than measuring only one representative cytokine.

Cytokines are *polymorphic*, which means that genetic variability exists among individuals in terms of the ability to produce cytokines in response to a stimulus. Cytokine levels produced as a result of activation can vary widely between individuals, and this trend may be genetically controlled. However, most important, the existing levels of cytokines might determine susceptibility or resistance to disease *(16)*. TNF-α and IL-10 levels, for example, appear to influence responses to and outcome of infectious diseases. Experimental evidence supports the concept that differences in susceptibility to and severity of diseases might be related to genetically defined differences among individuals to produce important cytokines. Thus, low TNF production was associated with a 10-fold risk of fatal outcome from menigococcal meningitis in humans, whereas high IL-10 levels were associated with a 20-fold greater risk of fatality *(17,18)*. These observations have led to extensive examination of cytokine polymorphism and its disease associations and a search for genetic markers that best define such polymorphism *(19)*.

Cytokines mediate interactions between a broad variety of cell types, and their dysregulated production might contribute to the disease pathogenesis. Under normal circumstances, no or only low levels of cytokines are detectable in body fluids or tissues. Therefore, their presence at elevated levels of expression indicates activation of cytokine pathways associated with inflammation or disease progression. Cytokines associated with acute diseases are distinct from those detectable in chronic conditions. Certain cytokines may be present early, others late in the disease. Consistently observed associations of increased cytokine levels with a disease process or disease stage suggest that anticytokine therapy might be effective, at least in some cases. Therefore, an assessment of cytokine profiles in body fluids, tissues, or cells of patients with various diseases is important, particularly if cytokines themselves are used as therapeutic agents.

3. CYTOKINE FAMILIES

Structural differences among cytokines have made it possible to group them within "families" because primary sequences of certain cytokines show structural homology. For example, all members of the IFN-α/β family show at least 30% homology in their amino acid sequences *(20)*. Such common structural features of cytokines permit their groupings into families, for instance, the IL-2/IL-4 family (IL-2, IL-4, IL-5, granulocyte-macrophage colony-stimulating factor) or the TNF family (TNF-α, lymphotoxin-α, LT-β, FasL, CD40L, TRAIL, LIGHT, and others) or the IL-1 family (IL-1α, IL-1β, IL-1 receptor antagonist, IL-18). Structural features of cytokine receptors form the basis for further classification of cytokines into distinct groups, such as the "hematopoietin family receptors" or class I cytokine receptors and "interferon/IL-10 family receptors" or class II cytokine receptors *(1)*. Cytokines signal via their receptors, and the understanding of the receptor structure, expression, and signal transduction is at the heart of cytokine biology. The unique features of the signaling pathways engaged by different cytokines and often associated with JAK and STAT pathways form the basis for cytokine crosstalk.

4. CYTOKINE PROFILES IN HUMAN DISEASE

Activation of a cytokine cascade by a physiological or pathological stimulus usually leads to the production and release of several biologicalally related cytokines. Thus, as indicated previously, IL-2, IFN-γ, and TNF-α production characterizes Th1-type immune responses, whereas IL-4, IL-5, and IL-10 are associated with Th2-type responses *(13,14)*. In an immune response, the cytokine cascade is initiated by specific recognition of the antigenic stimulus by T-cell receptor, subsequent T-cell activation, and secretion of the cytokine elicited by the antigen. Amplification of this initiating signal may then result in a local cytokine cascade. An inflammatory event induces a coordinated release of several proinflammatory cytokines *(3)*. After induction, cytokine concentrations may rapidly and differentially peak and then decrease to undetectable levels.

Today, a wide range of cytokine assays is available, providing researchers with an opportunity to evaluate the biology of cytokines and to establish their therapeutic potential (reviewed in ref. *21*). Moreover, an improved understanding of cytokine interactions in the increasing numbers of pathological conditions has led to a consensus that the simultaneous assessment of many cytokines in a single biological sample, that is, "cytokine profiling," is a preferred approach *(22)*. This change of paradigm in cytokine measurements was brought about by recent technical developments, and it is apparent that cytokine assays that best lend themselves to multiplex-type formats of the future slowly are replacing the traditional biological and immunological cytokine assays of the past *(23)*.

Cytokine measurements in tissue or in the peripheral circulation have been an important part of the process of defining the role various cytokines play in health and disease. New multiplex formats for cytokine assays allow for precise quantification of many different cytokines in a single small volume of body fluid. It has been suggested long ago that local cytokine levels and activity are of considerably greater value for monitoring of pathological events in a target tissue than are systemic cytokine levels *(24)*, with the notable exception of cytokine pharmacokinetics in subjects receiving cytokine therapies. However, tissue cytokine profiles remain largely unexplored, probably because of considerable difficulties in obtaining interstitial fluid. In this respect, a technique of microdialysis, using a commercially available biocatheter inserted into accessible tissue sites, proved to be an excellent approach to the recovery of multiple cytokines in a volume of approx 50 μL of interstitial fluid *(25)*. In our hands, microdialysis of oral mucosa in patients with HIV infection proved to be a powerful strategy to define the cytokine/chemokine profile in the milieu of viral entry, replication, and persistence (T. L. Whiteside and A. Rosenbloom, unpublished data). Combined with *in situ* immunostaining of mucosal biopsies to define intracellular expression of cytokines and cell surface expression of cytokine receptors, this strategy might prove to be more informative and more prognostically important than profiling of serum cytokines. In any event, a comparison between oral mucosa and peripheral blood cytokine profiles in normal volunteers and individuals with HIV at various stages of the disease is in progress to determine the extent of correlations or differences and their relationship to clinical end points.

Although significant progress has been made in linking certain cytokines to pathological changes in disease, interpretations of the results of cytokine profiling are not a trivial matter. Under normal conditions, basal levels of cytokines are low. In diseases such as cancer, chronic infectious diseases, transplantation, or allergy, profiles of cytokines are beginning to emerge that might correlate to clinical end points. In cancer, advanced metastatic disease is associated with the Th2 cytokine profile. To characterize tumor-specific CD4+ T-cell responses in patients with cancer, single-cell assays, such as the enzyme-linked immunosorbent spot, can be used. Upon stimulation of peripheral blood CD4+ T-cells obtained from the peripheral circulation of patients with renal cell carcinoma with major histocompatability class II-restricted peptides (e.g., HLA-DR4-restricted MAGE $6_{121\text{-}144}$ peptide), a Th2-polarized response (IL-5 but not IFN-γ production) is elicited in patients with active disease *(26)*. In contrast, disease-free patients (NED) respond by Th1-type cytokine production (IFN-γ and not IL-5). Furthermore, successful therapy in cancer patients is accompanied by a shift from the Th2 to Th1 cytokine polarization *(26)*. Thus, it might be now possible to follow the frequency of peptide-reactive CD4+ T-cells in the patients' circulation before and after therapy to observe changes in their polarization profile and predict the success or failure of a therapeutic intervention based on the direction of this change. In patients with HIV-1, whose sera often contain viral cytokines and cytokine-receptor homologs that can bind to cellular receptors or to circulating cytokines, respectively, counteracting their biological function, the interpretation of cytokine assays may be very difficult *(27)*. Nevertheless, significant shifts in the cytokine profiles in the course of infection and upon therapy with highly active antiretroviral therapy have been observed in patients with HIV *(28)*. The newly emerging associations between the cytokine profiles and allergic diseases provide an example of how cytokine determinations contribute to a better understanding of disease pathogenesis *(14,29,30)*. For example, T-cells and dendritic cells are considered to be central in the pathogenesis of psoriasis. These cells create a type 1 inflammatory pathway in the skin with an abundant release of IFN-γ, IL-23, and IL-12, which help sustain activation of this pathway *(31)*. TNF-α and IFN-γ, which play a pivotal role in host defense against infections and in the development of Th1 responses, also are involved in the pathogenesis of autoimmune diseases such as rheumatoid arthritis, multiple sclerosis, and type 1 diabetes *(32)*. These and other observations suggest that serial monitoring of cytokine profiles during the course of disease or therapy might be clinically useful. At present, all these correlations are not quite robust enough to be able to consider cytokines or cytokine profiles as "biomarkers" of disease or disease activity. It is expected, however, that the application of genomics and proteomics to cytokine measurements will further facilitate efforts to monitor patients' plasma and/or tissues for levels of cytokines and to evaluate cytokine profiles for their utility as surrogate markers of disease.

5. CYTOKINES AND THE IMMUNE SYSTEM

Immune cells both depend on cytokines for development and function and, upon activation, produce a variety of cytokines. For these reasons, the ability of various lymphocyte or monocyte subsets to respond to and to produce cytokines in response to specific stimuli has been important in the understanding of the immune system. Lymphocyte homeostasis involves continuous maturation and differentiation of antigen-responsive cells and their turnover between the circulating pool and tissues *(33)*. Lymphocyte survival depends on the growth factors and cytokines produced in various niches to which they migrate or home *(34)*. Consequently, cytokine networks determine the lymphocyte turnover and orchestrate their differentiation into functional effector cells. Likewise, cytokines regulate differentiation of dendritic cell precursors into functional antigen-presenting cells *(35)*. The fundamental operation of the immune system is thus cytokine driven, and thymic development of T-cells *(36)*, B-cell differentiation into antibody-producing plasma cells *(37)*, or maturation and activation of natural-killer cells *(38)* are all coordinated by distinct networks of cytokines. Once matured in the designated tissue microenvironment, immune effector cells expressing the required cytokine/chemokine receptor(s) migrate under the direction of chemotactic gradients and thus guided arrive at the target tissue to deliver appropriate signals *(39)*. The latter are entirely dependent on the situation in the targeted microenvironment and can be amplified or inhibited by cytokines produced *in situ*, either by infiltrating immune cells themselves or tissue cells or most likely both cell types. This entire process can be initiated by a sudden alteration in the tissue microenvironment, such as a bacterial or viral infection, injury or death of cells. The remarkable sensitivity, rapidity and regulated amplification of all cytokine-mediated immune responses to tissue injury indicate that the cytokine network functions in tandem with or under control of the neuroendocrine system. Indeed, although cytokines have been thought of as hormones of the immune system, it is necessary to remember that nonimmune cells express cytokine receptors and produce cytokines. Thus, cytokines are ubiquitous and appear to be able to maintain communications between cells of different origin and different systems. For example, it is now well documented that interactions among the central nervous system (CNS), endocrine networks, and immune cells determine the overall immune response. The CNS influences immune responses through the release of humoral factors such as cortisol or epinephrine from the hypothalamic–pituitary–adrenal axis. Lymphoid organs, e.g., lymph nodes, are innervated, and local synthesis of neuropeptides or opioids in these nerve terminals has been documented. Immune cells express surface receptors for neurotransmitters, and nerve fibers have receptors for cytokines *(40)*. Thus, a bidirectional traffic of biochemical mediators, cytokines, and neuropeptides between the CNS and immune cells regulates their interactions. When glucocorticoids are released in response to systemic stress, they downregulate immune responses and cytokine production. Migration inhibitory factor, a cytokine produced by the pituitary, counteracts inhibitory effects of glucocorticoids and restores the immune balance *(41)*. Hence, cytokines play a key role in regulating the neuroendocrine–immune system interface. The mechanisms or molecular pathways involved in these interactions are now being elucidated. The field of neuroimmunology has grasped the likely importance of cytokines in the CNS responses, and much current work is devoted to studies of these interactions in animal models of CNS injury or stress as well as in human models of acute or chronic stress *(42)*.

6. IMMUNOMODULATION BY CYTOKINES IN THERAPY OF HUMAN DISEASES

Ever since recombinant cytokines became available more than 20 years ago, numerous phase I and phase II clinical trials have been performed with these agents worldwide. By and large, these trials have capitalized on preclinical in vitro and in vivo animal model studies demonstrating immunomodulatory effects of cytokines on the host immune system and its components. To date, IL-2 and IFN-α are the only cytokines approved by Food and Drug Administration as therapies for renal cell carcinoma and melanoma, respectively. IFN-α is also approved for therapy of hepatitis C and hairy cell leukemia *(43)*. Nevertheless, the therapeutic potential of many other cytokines is being evaluated in experimental clinical settings.

From the start, two opposite strategies for cytokine-based therapy targeting immune cells have emerged and still exist today. In one, a therapeutic cytokine of choice is systemically administered at very high doses to a subject with defects in the immune system. The objective is to provide the exogenous cytokine in excess to "force" immune recovery and induce a clinical response. The high-dose systemic delivery of cytokines is invariably accompanied by severe toxicity, which may be life threatening (e.g., resulting in a vascular leak syndrome upon high-dose IL-2 therapy) and usually is accompanied by a variety of side effects *(44)*. Although this form of cytokine therapy appears to be nonphysiological and not in agreement with basic principles of the cytokine biology, surprisingly, it has been successful in a limited number of cases *(45)*. The molecular underpinnings of such high-dose IL-2 or IFN-α therapies are not clear, but it has been suggested that cascades of other cytokine (e.g, TNF-α or IL-1β) are responsible for toxicity as well as restoration of immune cell numbers or functions reported in many instances *(46)*. The second approach depends on locoregional delivery of the cytokine to the targeted organ or site (e.g., tumor site, infection site) with the hope of increasing immune cell numbers and augmenting effector cell functions *in situ*. Locoregional delivery of cytokines does not induce severe toxicities, it is usually well tolerated, and it has been shown to alter immune cell infiltration and activation *in situ (47)*. This form of cytokine therapy is based on a rationale that cytokines are local mediators meant to function at short distances in tissue, and although it is appropriate for therapy of local diseases, its usefulness in treatment of the systemic disease might be limited. Still, it should be noted that local, for instance, peritumoral or perilymphatic, delivery of cytokines has been shown to induce local as well systemic changes in immune cell activation *(47)*. Numerous and clever ways of cytokine delivery to target tissues have been introduced, including microspheres, liposomes, viral and nonviral vectors, or plasmids. These strategies, many of which are discussed in this volume, are meant to assure continuous release of the mediator in hope of inducing persistent activation of immune cells in the microenvironment. In animal models of disease, these strategies have been successful in inducing long-lasting and protective changes in the host immune system *(48)*. Their application to therapy of human disease is in progress, pending safety and toxicity considerations. At present, experimental preclinical and clinical studies using either systemic or local cytokine delivery are ongoing in various institutions, and expectations are that immune monitoring and clinical results will help in defining mechanisms responsible for documented objective responses in each of these settings.

A considerable controversy exists in regard to therapeutic delivery of single recombinant cytokines vs natural mixtures of cytokines. Again, two widely held views conflict in that one favors therapy with well-defined, precisely dosed single cytokines or possibly a combination of selected recombinant cytokines, whereas the other insists that naturally derived cytokine mixtures, for instance, partially purified supernatants of mitogen-activated lymphocytes, are more effective in upregulating a variety of immune effector cells and thus providing a broader therapeutic index *(49)*. Although there are advantages and disadvantages to each of these approaches, the controversy emphasizes the need for further studies of cytokines as immunomodulators that potentially are capable of correcting pathological conditions.

Finally, excessive production of cytokines has been associated with a spectrum of human diseases, including posttransplant rejection episodes. In all these conditions, the control of cytokine cascades or inhibition of individual cytokines are desirable. Therapeutic strategies that attempt to restore the cytokine balance disturbed by pathological overproduction of cytokines include systemic or local delivery of inhibitory cytokines, antibodies to cytokines, antibodies or agents blocking cytokine receptors, and *in situ* modifications of cytokine genes, for example, with inhibitory RNA among others. The difficulties associated with these approaches relate to the fact that little is known about the regulatory mechanisms of cytokine production in vivo and that in vitro models of these molecular pathways in cell lines do not adequately reflect cellular controls operative in tissues.

Overall, cytokine-based therapies alone or in combination with other drug modalities are considered to be promising and are continued to be evaluated for treatment of various human diseases. Many novel therapeutic strategies involving cytokine delivery and monitoring of cytokine responses in vivo have been introduced in recent years. The use of fusokines or multifunctional cytokines representing fusion proteins linking two cytokines and expressing them in an expression system convenient for in vivo delivery represent one such strategy *(50)*. Immunocytokines are fusion proteins combining a cytokine and an antibody *(51)*. They have been shown to have potent antitumor effects and are under intensive experimental scrutiny at this time. There are reasons to believe that, in years to come, cytokine-based therapies might fulfill their promise and become a part of the widely applicable therapeutic armamentarium for many human diseases.

7. CONCLUSIONS

The role of cytokines and chemokines in immunomodulation has been well established and generally is considered to be the basis for their biological and therapeutic effects. Various immune cell subsets express receptors for cytokines, produce cytokines upon activation and are functionally dependent on cytokines. However, immunomodulatory effects of cytokines/ chemokines are superceded by their role in networking functions of immune as well as non-immune cells. Cytokines bridge and regulate activities of many different cell types. Their participation in functions of the CNS, endocrine, reproductive, hematopoietic, and immune systems places cytokines at the interface and indicates that they are ubiquitous and essential for survival. Their biological characteristics and therapeutic potential further support this notion. A rapidly expanding body of recent evidence indicates that cytokines are essential for cell-to-cell communication and for orchestrating local cellular events in a variety of tissues. As such, they subserve a central role of linking and integrating signals delivered by different systems. For this reason, the understanding of cytokine crosstalk and their signaling at the molecular level are special challenges to be faced by investigators in the near future. Only through such an understanding are we likely to be able to harness cytokines/chemokines as useful pharmacological and therapeutic agents for human diseases.

ACKNOWLEDGMENTS

Supported in part by grants PO1-DE12321, RO1-CA82016 and RO1-DE13918 from National Institutes of Health.

REFERENCES

1. Vilcek J. The cytokines: an overview. In: Thomson AW, Lotze MT, eds. The Cytokine Handbook. 4th ed, Boston, Academic Press, 2003, pp. 3–18.
2. Zlotnik A, Osamu Y. Chemokines: a new classification system and their role in immunity. Immunity 2000;12:121–127.
3. Dinarello CA, Kluger MJ, Powanda MC. The Physiologicalal and Pathologicalal Effects of Cytokines. New York: Liss; 1990.
4. Oppenheim JJ. Clinical Applications of Cytokines. Role in Pathogenesis, Diagnosis and Therapy. New York: Oxford University Press; 1993.
5. Aggarwal BB, Puri RK, eds. Human Cytokines: Their Role in Disease and Therapy. Cambridge, MA: Blackwell Science; 1995.
6. Meager T. The Molecule Biology of Cytokines. Chichester: John Wiley & Sons; 1998.
7. Taniguchi T, Minami Y. The IL-2/IL-2 receptor system: a current overview. Cell 1993;11:245–267.
8. Smith KA. Interleukin-2: inception, impact, and implications. Science 1988;240:1169–1176.
9. Hatakeyama M, Tsudo M, Minamoto S, et al. Interleukin-2 receptor β chain gene: generation of three receptor forms by cloned human α and β chain cDNAs. Science 1989;244:551–556.
10. Lowenthal JW, Ballard DW, Bogerd H, Bohnlein E, Greene WC. Tumor necrosis factor-α activation of the IL-2 receptor-α gene involves the induction of κB-specific DNA binding proteins. J Immunol 1989;142:3121–3128.
11. Aggarwal BB, Eessalu TE, Hass PE. Characterization of receptors for human tumour necrosis factor and their regulation by γ-interferon. Nature 1985;318: 665–667.
12. Dinarello CA, Wolff SM. Mechanisms of disease: the role of interleukin 1 in disease. N Engl J Med 1993;328:106–113.
13. Romagnani, S. The Th1/Th2 paradigm. Immunol Today 1997;18:263–266.
14. Romagnani, S. The Th1/Th2 Paradigm in Disease. Austin: R.G. Landes Co.; 1997.
15. Mosmann TR, Li L, Sad S. Functions of CD8 T-cell subsets secreting different cytokine patterns. Semin Immunol 1997;9:87–92.

16. Sorensen TT, Nielsen GG, Andersen PK, Teasdale TW. Genetic and environmental influences on premature death in adult adoptees. N Engl J Med 1988; 318:727–732.
17. Westendorp RG, Langermans JA, Huizinga TW, Verweij CL, Sturk A. Genetic influence on cytokine production in meningococcal disease. Lancet 1997;349:1912–1913.
18. van Dissel JT, van Langevelde P, Westendorp RG, et al. Anti-inflammatory cytokine profile and mortality in febrile patients. Lancet 1998;351:950–953.
19. Gallagher G, Eskdale J, Bidwell JL. Cytokine genetics – polymorphisms, functional variations and disease associations. In: Thomson AW, Lotze MT, eds. The Cytokine Handbook, 4th ed. Boston: Academic Press; –2003, pp. 19–55.
20. Langer JA, Pestka S. Interferon receptors. Immunol Today 1989;12:393–400.
21. Whiteside TL. Assays for Ccytokines. In: Thomson AW, Lotze MT, eds. The Cytokine Handbook, 4th ed. Boston: Academic Press; 2003, pp. 1375–1396.
22. De Jager W, te Velthuis H, Prakken BJ, Kuis W, Rijkers GT. Simultaneous detection of 15 human cytokines in a single sample of stimulated peripheral blood mononuclear cells. CDLI 2003;10:133–139.
23. Whiteside TL. Cytokine assays. Biotechniques/Cytokines 2002;33:4–15.
24. Mathey E, Pollard J, Armati P. *In situ* hybridization for cytokines in human tissue biopsies. Methods Mol Biol 2003;204:57–66.
25. Kjellstrom S, Appels N, Ohlrogge M, Laurell T, Marko-Varga G. Microdialysis—a membrane based sampling technique for quantitative determination of proteins. J Chromatogr A 1999;50:539–546.
26. Tatsumi T, Kierstead LS, Ranieri E, et al. Disease-associated bias in T helper type 1 (Th1)/Th2 CD4+ T-cell responses against MAGE-6 in HLA-DRB1*0401+ patients with renal cell carcinoma or melanoma. J Exp Med 2002;196:619–628.
27. Nicholas J, Ruvolo VR, Burns WH, et al. Kaposi's sarcoma-associated human herpesvirus-8 encodes homologues of macrophage inflammatory protein-1 and interleukin-6. Nat Med 1997;3:287–292.
28. Connor RI, Sheridan KE, Ceradini D, Choe S, Landau N. Change in coreceptor use correlates with disease progression in HIV-1 infected individuals. J Exp Med 1997;185:621–628.
29. Lucey DR, Clerici M, Shearer GM. Type 1 and type 2 cytokine dysregulation in human infections, neoplastic and inflammatory diseases. Clin Microbiol Rev 1996;9:532–562.
30. Grewe M, Bruijnzeel-Koomen CA, Schopf E, et al. A role for Th1 and Th2 cells in the immunopatheogenesis of atopic dermatitis. Immunol Today 1998;19:539–361.
31. Lew W, Bowcock AM, Krueger JG. Psoriasis vulgaris: cutaneous lymphoid tissue supports T-cell activation and "Type 1" inflammatory gene expression. Trends Immunol 2004;25:295–305.
32. O'Shea JJ, Ma A, Lipsky P. Cytokines and autoimmunity. Nat Rev Immunol 2002;2:37–45.

33. Surh CD, Sprent J. Regulation of naïve and memory T cell homeostasis. Microbes Infect 2000;4:51–56.
34. Khaled A, Durum SK. The role of cytokines in lymphocyte homeostasis. Biotechniques 2002;33:540–545.
35. Sallusto F, Lanzavecchia A. Efficient presentation of soluble antigen by cultured human dendritic cells is maintained by granulocyte/macrophage colony stimulating factor plus interleukin-4 and down-regulated by tumor nicrosis factor alpha. J Exp Med 1994;179:1109–1118.
36. Schluns KS, Lefrancois L. Cytokine control of memory T-cell development and survival. Nat Rev 2003;3:269–279.
37. Kishimoto T. Factors affecting B cell growth and differentiation. Annu Rev Immunol 1985;3:133–157.
38. Biron C. Activation of natural killer cell responses during viral infections. Curr Opin Immunol 1997;9:24–34.
39. Moser B, Loetscher P. Lymphocyte traffic control by chemokines. Nat Immunol 2001;2:123–128.
40. Haddad JJ, Saade NE, Safieh-Garabedian B. Cytokines and neuro-immune-endocrine interactions: a role for the hypothalamic- pituitary- adrenal revolving axis. J Neuroimmunol 2002;133:1–19.
41. Calandra T, Bucala R. Macrophage migration inhibition factors MIF: a glucocorticoid counter regulator within the immune system. Crit Rev Immunol 1997;17:77–88.
42. Straub RH, Westermann J, Scholmerich J, Falk W. Dialogue between the CNS and the immune system in lymphoid organs. Immunol Today 1998;19:409–413.
43. Foon KA, Maluish AE, Abrams PG, et al. Recombinant leukocyte A interferon therapy for advanced hairy cell leukemia: therapeutic and immunological results. Am J Med 1986;80:351–356.
44. Baluna R, Vitetta ES. Vascular leak syndrome: a side effect of immunotherapy. Immunopharmacology 1997;37:117–132.
45. Rosenberg SA, Lotze MT, Yang JC, et al. Experience with the use of high dose interleukin-2 in the treatment of 652 cancer patients. Ann Surg 1989;210:474–485.
46. Kirkwood JM, Richards T, Zarour HM, et al. Immunomodulatory effects of high-dose and low-dose interferon α2b in patients with high-risk resected melanoma. Cancer 2002;95:1101–1112.
47. Whiteside TL, Letessier E, Hirabayashi H, et al. Evidence for local and systemic activation of immune cells by peritumoral injections of interleukin 2 in patients with advanced squamous cell carcinoma of the head and neck. Cancer Res 1993;53:5654–5662.
48. Pardoll DM. Paracrine cytokine adjuvants in cancer immunotherapy. Annu Rev Immunol 1995;13:399–415.
49. Meneses A, Verastegui E, Barrera JL, DeGarz J, Hadden JW. Lymph node histology in head and neck cancer: impact of immunotherapy with IRX-2. Int Immunopharmacol 2003;3:1083–1091.

50. Gillies SD, Lan Y, Brunkhorst B, Wong WK, Li Y, Lo KM. Bifunctional cytokine fusion proteins for gene therapy and antibody-targeted treatment of cancer. Cancer Immunol Immunother 2002;51:449–460.
51. Lode HN, Xiang R, Becker JC, Gillies SD, Reisfeld RA. Immunocytokines: a promising approach to cancer immunotherapy. Pharmacol Ther 1998;80:277–292.

2

Cytokine Measurement Tools for Immunotoxicology

Rob J. Vandebriel

SUMMARY

Since the early 1990s, the measurement of cytokines has become an important tool to understand mechanisms of immunotoxicity as well as to identify and classify xenobiotics. During the past decade, several major scientific and technical developments have greatly increased the efficiency of measurement. Most of the current attention is focused on methods that allow the simultaneous measurement of many cytokines, be at the messenger ribonucleic acid level (microarrays) or at the protein level (Luminex-based assays). Next, during the last decade quantitative polymerase chain reaction methods have gained widespread use. Finally, intracellular cytokine staining has provided detailed and invaluable information on the immune responses. This chapter describes the most widespread techniques for measuring cytokines. In addition, it tries to suggest which technique may be the optimal choice for specific purposes.

Key Words: Northern blotting; ribonuclease protection assay; polymerase chain reaction; microarray; ELISA; Luminex; ELISPOT; intracellular; FACS; flow cytometry.

1. INTRODUCTION

Cytokines are critical in the regulation of many immune processes, such as host resistance, memory formation, inflammation, apoptosis, and hematopoiesis. The measurement of cytokines has become an important tool in immunotoxicology (1–3). In this chapter, the measurement of cytokines is the predominant focus, although the measurement of cytokine receptor expression is also of major value in immunotoxicology (4).

From: *Methods in Pharmacology and Toxicology: Cytokines in Human Health:*
Immunotoxicology, Pathology, and Therapeutic Applications
Edited by: R. V. House and J. Descotes © Humana Press Inc., Totowa, NJ

Table 1
Advantages and Disadvantages
of Various Cytokine Measurement Techniques

Assay	Advantages	Disadvantages
Bioassay	Only assay measuring biological activity, sensitive	Complex, different for each cytokine, set-up and assays time-consuming
Northern blotting	Controls RNA integrity, specific	Not measuring protein, insensitive, requires high amount of input, set-up time-consuming
RPA	Sensitive, multiplex up to ≈10	Not measuring protein, set-up time consuming
qPCR	Very sensitive, readily extendable with additional cytokines, multiplex up to ≈10	Not measuring protein, set-up time-consuming
Microarray	Sensitive, extreme multiplex capability	Not measuring protein, expensive, set-up very time-consuming
ELISA	Sensitive, simple set-up	No multiplex capability
Luminex-based	Sensitive, substantial multiplex capability	Set-up time-consuming
ELISPOT	Single-cell assay	No phenotyping
Intracellular staining	Single-cell assay, simultaneous phenotyping	Set-up very time-consuming

RPA, ribonuclease protection assay; qPCR, quantitative polymerase chain reaction; ELISA, enzyme-linked immunosorbent assay; ELISPOT, enzyme-linked immunoSPOT.

Basically, three methods of cytokine measurement exist, in historical order biological activity, messenger ribonucleic acid (mRNA) abundance, and production (concentration). Each of the different techniques has specific advantages and disadvantages (Table 1). The eventual choice will depend on specific circumstances, such as the machinery already available, the type of samples that will be measured, the number of samples that will be measured, and the flexibility that is required.

Regardless of the method that is used, it is important to carefully identify which cytokines are to be measured to obtain the most useful information, especially when measuring a limited number of cytokines. Consequently, techniques with higher multiplex capability (i.e., the possibility to measure a number of cytokines simultaneously) have the advantage of allowing extraction of the most useful information afterward. The pleiotropic and redundant nature of cytokines is an additional reason for a comprehensive measurement of cytokines.

2. BIOASSAYS

The bioassay traditionally has been the first method to measure cytokine activity *(5)*. However, its disadvantage is that the assay is relatively labor-intensive and that it is unsuitable for multiplex analysis, resulting in the fact that, nowadays, it is hardly ever used. Nonetheless, one major advantage still is that it can measure cytokine activity instead of its mere presence (protein, measured in immunoassays) or even its expression (mRNA, measured in molecular assays). In addition, it is relatively cheap in terms of consumable cost (no or little requirement for monoclonal antibodies [mAbs]). Its current use is limited to evaluating whether under certain experimental conditions the presence of a certain cytokine as established by immunoassay is linked with biological activity.

3. MOLECULAR ASSAYS

Until recently, cytokine genes were cloned one by one, often on the basis of some structural knowledge of the purified protein(s). The completion of the human, mouse, and rat genome sequences, together with recent developments in bioinformatics, have enabled identification of putative cytokine genes solely on the basis of sequence information. The genome is searched for certain previously identified sequence features, allowing faster discovery of novel cytokine genes.

In immunotoxicology, rats traditionally are the experimental animals of use whereas in the field of immunology, the use of the mouse is much more widespread. Hence, of major importance to immunotoxicology is the recent completion of the rat genome sequence, ensuring the end of an ongoing lag phase in the availability of sequence information between human and mouse on the one hand, and rat on the other.

Molecular assays most often are based on measuring the abundance of mRNA encoding a certain cytokine (subunit), whereas other methods rely on measuring the transcription rate of a certain cytokine. A wide range of techniques exists that is able to measure transcript abundance.

Advantages of molecular assays over immunoassays have previously been discussed *(1)*. In short, molecular assays are able to detect exposure effects earlier than immunoassays. In addition, molecular assays measure at single time points whereas immunoassays measure the average level over a larger time period, its length depending on the half-life of the cytokine investigated. In ex vivo studies immunoassays measure an average between accumulation and decay.

3.1. Northern Blotting

Northern blotting is the classical method to measure transcript abundance. Total RNA or total mRNA is separated by electrophoresis and blotted onto a filter. A labeled DNA or sometimes RNA molecule (probe) is then hybridized to the mRNA molecule of interest by virtue of its sequence being complementary. The probe can be labeled radioactively but, increasingly, chemiluminescent labels are used. The staining intensity is proportional to the number of transcript molecules and can be measured using a PhosphoImager (for radioactive labels) or a FluoImager (for chemiluminescent and fluorescent labels). The size of the transcript can be checked, ensuring that the transcript is intact and the hybridization specific.

A technique derived from Northern blotting is dot or slot blotting. In this application, the RNA or mRNA is directly spotted onto a filter without prior separation on a gel and transfer to a filter, which allows more samples to be tested simultaneously, but lacks the control for RNA integrity and specificity of hybridization. Generally speaking, Northern blotting is performed to check for specificity ("gold standard"), whereas dot and spot blotting have been replaced by the polymerase chain reaction (PCR; *see* Subheading 3.3.).

3.2. RNase Protection Assay

The RNase protection assay (RPA) is a method to simultaneously analyze the abundance of some 10 different transcripts *(6,7)*. The mRNA of interest is hybridized to a labeled antisense RNA probe. This probe is generated by in vitro transcription. The RNase that is used degrades single-stranded (ss) but not double-stranded (ds) RNA. Thus, RNase treatment degrades the ss mRNA as well as excess (ssRNA) probe, resulting in a dsRNA molecule consisting of a fragment of the mRNA molecule hybridized to the probe. Both radioactive and nonradioactive methods have been developed. In the radioactive method, the "protected" probe is subjected to denaturing polyacrylamide gel electrophoresis, after which the radioactivity is visualized by autoradiography. Nonradioactive methods use a biotinylated probe. After denaturing polyacrylamide gel electrophoresis, the "protected" probe is transferred to a membrane, after which biotin is visualized by chemiluminescence using enzyme-coupled streptavidine.

Instead of using RPA for measuring the presence of single transcripts, multiprobe RPA is now in general use. A range of antisense RNA probes, each for a different gene and of different size, allows the simultaneous measurement of some 10 genes. Multiprobe RPA sets are commercially available for human and mouse cytokines, with a possibility of customizing (BD Pharmacia, PharMingen).

The sensitivity of RPA is 10- to 100-fold higher compared with Northern blotting, an important reason being that in RPA the hybridization occurs in solution compared with membrane hybridization in Northern blotting. Besides, to detect rare transcripts RPA allows the input of 10-fold more mRNA compared with Northern blots. Because in RPA the probe size is much smaller compared with the entire transcript assayed in Northern blots, the tolerance to partially degraded sample RNA is much larger in RPA. Collectively, advantages of RPA over Northern blots include the possibility to simultaneously measure the presence of some 10 genes, the increased sensitivity, and the tolerance to partially degraded RNA. Similar to Northern blots, however, RPA does not lend itself to measure high numbers of samples.

3.3. Polymerase Chain Reaction

From its introduction in the mid-1980s, the PCR has evolved to a standard technique that is widely used in many laboratories. Among the many virtues of PCR are the ease of the method, the high sensitivity, and the ease to append additional (cytokine) genes. Although for sequence identification, such as detection of microbial organisms, PCR has become the gold standard, the lack of quantification capability has long hindered the technique from being used for measuring transcript levels. Procedural extensions such as using a dilution series of samples, consecutive numbers of cycles, internal standards, and competitive PCR have provided some improvements and are still being widely used. Kits are commercially available that allow the simultaneous measurement of some 10 cytokines without prior set-up and optimization of experiments.

A breakthrough in quantifying transcript levels using PCR was the introduction of real-time PCR (RT-PCR, not to be confused with reverse transcriptase PCR), which also is denoted as quantitative PCR (qPCR). Instead of the endpoint measurements of traditional PCR, qPCR determines the amount of product during the reaction. While for traditional PCR the amount of product is measured on an agarose gel, qPCR measures the strength of the fluorescence signal instead. Of importance, the measurement in real time together with the high sensitivity allows measurement of product formation in the linear phase of amplification. Therefore, a threshold can be set and the number of cycles required to reach the threshold is the way most often used to measure the number of specific mRNA copies present in the sample. The three major companies that market qPCR all use laser technology to measure the amount of PCR product generated. The method used to generate the signal that is produced during PCR is, however, different, between these companies, as outlined here:

1. The Taqman method (PerkinElmer) employs a 20- to 30-nucleotide probe (denoted Taqman probe) that has a fluorescent reporter dye attached to its 5' end and a fluorescent quencher dye attached to its 3' end. During PCR, the probe hybridizes to its complementary ss sequence within the PCR target. During amplification the probe is degraded because of the 5' → 3' exonuclease activity of Taq DNA polymerase, thereby separating the quencher from the reporter during extension. Because of the release of the quenching effect on the reporter, the fluorescence intensity of the reporter dye increases. Therefore, if the Taqman probe does not hybridize to the PCR product, the reporter and quencher both remain attached to the probe, resulting in an absence of fluorescence. In conclusion, this technique combines real-time measurement of the exponential amplification of PCR products with verification of sequence specificity.

2. The LightCycler method (Roche) makes use of another technique in which two fluorescent dyes interact. Two oligonucleotide probes are used that hybridize head to tail. The 5' probe carries fluorescein at its 3' end, and the 3' probe carries LC Red 640 at its 5' end. When the two probes hybridize to the target sequence the two dyes come in close proximity (1–5 nucleotides), thereby generating a fluorescent signal. The mechanism of energy transfer is called fluorescence resonance energy transfer. Similar to the Taqman method, the Light Cycler method combines real-time measurement of the exponential amplification of PCR products with verification of sequence specificity. Although in the Taqman method fluorescence is generated from zero during every cycle, in the LightCycler method additional fluorescent label is incorporated into the PCR product during every cycle. The LightCycler method is compatible with the DNA binding dye SYBR GREEN I. This dye produces a fluorescent signal depending on the amount of DNA but irrespective of the DNA sequence. Although the use of this dye results in a loss of sequence verification, advantages are a lack of requirement for labeled oligonucleotides, thereby reducing cost and enhancing flexibility.

3. The iCycler method (Bio-Rad) has the ability to simultaneously measure four fluorophores (FAM, HEX, Texas Red, and Cy5), allowing monitoring of up to four amplifications simultaneously.

3.4. Microarrays

Tremendous progress in genomics together with major technological innovations has resulted in the development of microarrays. Microarray technology is the simultaneous individual measurement of the mRNA expression level of thousands of genes in a given sample by means of hybridization. The basic principle of microarray technology is the same for different types of array *(8)*.

First, a single oligonucleotide or a few different oligonucleotides are synthesized per gene. These thousands of oligonucleotides require similar hybridization characteristics as well as a lack of cross-hybridization, implicating that they have to be devised by computer. An alternative for oligonucleotides that was rather common in the early years but is now becoming less common is PCR products amplified from a clone collection. This is a collection of bacteria, each containing a plasmid comprising a different copy DNA (cDNA) insert. Using this method, care has to be taken that the individual clones indeed contain the correct insert; verifying clone sets by sequencing the inserts is therefore not uncommon. Usually the oligonucleotides (or PCR products) are spotted onto a glass surface in a regular array. This process is called spotting or arraying, and requires dedicated machinery. Some companies manufacture oligonucleotides in situ, using either photolithography (Affymetrix) or chemical coupling (Agilent). In order to obtain microarrays several possibilities exist: (1) purchase of ready-made arrays (Affymetrix, Agilent); (2) purchase of custom-made arrays (Affymetrix, Agilent); (3) in-house spotting of an oligonucleotide collection (MWG, Operon, Sigma) or a PCR amplified clone collection (Invitrogen); or (4) in-house spotting of an in-house-prepared PCR amplified clone collection.

Other manufacturers of ready-made arrays include Operon, MWG, and Phase-1. The choice for a specific platform is dependent on a number of considerations. Although commercial arrays are more expensive, they are less prone to errors, have better lot-to-lot reproducibility, and require less time to set up. When performing experiments that have to be integrated with studies in other laboratories, commercial arrays are preferred. Alternatively, arrays can be produced at a single laboratory and subsequently distributed.

Second, (m)RNA is isolated from cells or tissues and cDNA synthesized. This cDNA is labeled during or after synthesis using a fluorescent dye. When the amount of input mRNA is (too) low, linear amplification must be performed. The labeled cDNAs are then hybridized to the array. Although Affymetrix uses single-labeled cDNA (Cy3), other platforms employ two labeled cDNA's (Cy3 and Cy5; most often test and control). The array is read using a scanner (with fitted laser[s]) that measures for each spot the fluorescence intensity (intensities). These data are then transferred to a PC for analysis.

Controls are of major concern when performing microarray experiments. First, the quality and quantity of RNA samples can be checked rigorously using a Bioanalyzer (Agilent) and the amount of labeled cDNA using a NanoDrop spectrophotometer. Second, when using arrays produced in-house the shape of the spots on the arrays and the amount of DNA spotted have to be checked (e.g., by hybridization of labeled random hexamers). Third, after hybridization a similar average staining intensity over the entire array is of importance. Fourth, the intensity ratio of both labels (Cy3/Cy5) is plotted against the intensity of the label for the control (Cy5); this ratio should be independent of the intensity for most of the genes. To exclude artifacts caused by differential incorporation of the two labels into the cDNAs, a "dye swab" is useful. Fifth, when the expression of certain genes is affected and these results, for instance, form the basis of follow-up studies, it is common use to verify those effects by qPCR.

Microarray experiments are performed on replicate samples. The *p* value for statistically significant changes in gene expression is often chosen to be less than 0.001, while ratios of test vs control greater than 2 generally are considered biologically significant. If several time points, dose groups or organs are analyzed, more advanced statistics can be done, such as cluster analysis and/or principal component analysis *(9)*. To this end, several algorithms have been written, most of them being freely available on the internet. Commercial software packages have the advantage of easier data handling, compared to the tedious process of uploading data sets to algorithms on the web.

The number of genes to be analyzed is of interest. Microarrays measure the expression of between 1000 and 20,000 genes. For measuring expression of hundred to a few hundred genes, such as a panel of cytokine genes, macroarrays are available. Such arrays consist of oligonucleotides spotted onto a nylon membrane, while expression is measured using chemiluminescence or fluorescence (SuperArray). A different product line employs spotted PCR products, with expression measured using chemiluminescence or ^{32}P. Arrays that cover the area more or less in between macro- and microarrays are, for example, the ones marketed by BD Clontech. Plastic, glass, and nylon arrays measuring 8000, 3800, and 1200 genes, respectively, are available; detection methods use ^{33}P (all types), ^{32}P (nylon), and fluorescence (glass).

Microarrays are particularly well suited to provide a comprehensive understanding of the toxic profile of a compound. In such analysis, effects on cytokine gene expression are integrated and being interpreted in a much wider context of exposure effects *(10)*. For screening purposes such as identification and classification, however, cytokine macroarrays may be of use. To provide a solid foundation for the use of such macroarrays a range of immunotoxic compounds should first be analyzed by microarray analysis *(11,12)*. From these data compound- or class-specific profiles should be deduced, and they may possibly be amenable for analysis by cytokine macroarrays. It has been shown that indeed specific types of toxicity or classes of compounds can be established on the basis of expression profiles *(13,14)*.

3.5. Reporter-Based (Cell Chip)

A new approach to cytokine measurement in immunotoxicology has been the development of the so-called "cell chip." In this approach cell lines that represent the most important cell types of the immune system are transfected with constructs comprising the regulatory regions of a specific cytokine fused to a reporter gene. Exposure to xenobiotics that affect the mRNA expression of a specific cytokine is measured as an effect on the production of the protein enhanced green fluorescent protein. This protein can be sensitively measured by flow cytometry. For each of the cytokines to be measured separate transfected cell lines were generated. Control experiments showed that effects on the amount of protein as well as fluorescence intensity mimic effects on cytokine expression and production. In addition, exposure effects on mRNA expression were similar to the parental cell line *(15)*. Exposure of the T-cell lymphoma cell line EL4 transfected with reporter constructs for IL-2, IFN-γ, IL-4, and IL-10 to a range of compounds gave promising results especially for the identification of immunosuppressants *(16)*. Testing additional cell lines is, however, needed to fully evaluate the potential of the cell chip.

4. IMMUNOASSAYS

Proteins endow biological activity, whereas mRNA generally does not. Given the fact that altered mRNA expression levels may not always be reflected by altered cytokine production, measuring proteins may be of preference to measuring mRNA levels.

All immunoassays rely on the availability of mAbs. Because mAbs are developed by (expected) public demand, they may be unavailable for cytokines that are rather seldom measured, which is especially true in the case of species other than human and mouse. Because rats often are used in immunotoxicology, this lack of availability may pose problems. Molecular assays can be an escape route, however, because primer sequences for almost any known gene are available in literature, or they can be devised from gene sequences using widely available software tools.

4.1. Enzyme-Linked Immunoassay

Enzyme-linked immunoassays (ELISAs) have been used widely during the last decade to detect cytokines in serum and culture supernatants, and body fluids such as bronchoalveolar lavage fluids. Most often they require two mAbs, one for coating the plate and one for detection. The mAbs usually bind to different epitopes of the cytokine. Single or dual plates are available with mAbs and additional reagents premade. They are marketed for

ELISAs that detect newly discovered cytokines. For established cytokines, however, mAbs and reagents are sold in bulk, allowing the use of tens of plates. These bulk packages often result in 10-fold lower cost per plate.

4.2. Solid-Phase Multiplex Protein Assays

Whereas ELISAs are able to measure tens of samples per plate, each additional cytokine to be measured requires setting up and performing a new test. A technique that allows the simultaneous detection of multiple cytokines in a single well, thereby saving sample volume and time, is the solid phase multiplex protein assay *(17,18)*. It relies on spectrally encoded antibody-conjugated beads that can be measured in a Luminex 100 instrument. After incubating the beads with samples (typically serum or culture supernatant) for 2 h and washing, they are incubated with biotinylated mAbs for one hour. After washing, the beads are stained using R-PE conjugated to streptavidin for 30 min. After washing, the beads are loaded into the Luminex 100 instrument. The instrument monitors the spectral properties of the beads while simultaneously measuring the quantity of the fluorophore. The spectral properties of 100 distinct bead regions can be monitored, allowing a potential for measurement of up to 100 different analytes in a single sample. Currently kits are available to measure up to 25 different cytokines (human, mouse, and rat). Antibody bead kits may be combined, provided that the bead region for each analyte is unique. Compatible kits can be found at http://www.luminexcorp. com/.

4.3. Antibody Arrays

Antibody arrays are available in various formats. Novagen markets a microarray format that is able to detect 12 different cytokines in 15 samples (human; mouse becomes available). This format requires 50 µL of sample. It is compatible with current microarray scanners and software.

Ray Biotech markets a membrane format that is able to detect up to 120 (human), 60 (mouse), or 20 (rat) different cytokines in one sample. This format requires 1 mL of culture supernatant, or 1 mL of (undiluted or 1:10 diluted) serum.

5. SINGLE-CELL IMMUNOASSAYS

5.1. ELISPOT

The enzyme-linked immunosorbent spot (ELISPOT) assay is an assay for the analysis of cytokine (or immunoglobulin) production at the single-cell level *(19,20)*. It is particularly useful for analyzing specific immune responses to whole antigens or peptides. In this technique, 96-well ELISPOT microtiter plates are coated with a mAb directed against the cytokine of study. After washing and blocking against nonspecific binding, the cell suspension (such as animal spleen or lymph node cells, or human peripheral blood mononuclear cells [PBMCs]) is added to the chamber and incubated in the presence of a specific stimulus, with nonspecific stimuli (such as mitogens) and nonstimulated cells as positive and negative controls, respectively. The cytokine that is released during culture is captured in the immediate vicinity of the cells. After removing the cells by washing, a second biotinylated mAb against the same cytokine (but a different epitope) is added. After incubation and washing, streptavidine-labeled enzyme (mostly alkaline phosphatase or peroxidase) is added. Finally, substrate for alkaline phosphatase or peroxidase is added, and color formation is allowed until spots emerge at the sites of responsive cells. The number of spots is counted using imaging software and expressed per 10^6 cells.

5.2. Intracellular Cytokine Staining

A major advantage of intracellular cytokine staining over the ELISPOT assay is the ability to analyze multiple parameters per cell *(21)*. Such data can be very valuable to gain insight into immune response regulation and memory formation. This technique can also be used on whole blood without requiring separation of PBMCs; a protocol is as follows. Heparinized blood samples are stimulated with antigen plus CD28 and CD49d in the presence of secretion inhibitor Brefeldin A for 6 h at 37°C. The superantigen staphylococcal enterotoxin B (SEB) is added to some tubes as a positive control for antigen stimulation. Ethylene diamine tetraacetic acid is then added to stop activation and remove adherent cells. Erythrocytes are lyzed and the cells fixed by adding lysing solution for 10 min at room temperature. Cells are then washed and resuspended in a permeabilizing solution. Cells are washed and staining is performed using both cell surface and intracellular staining antibodies for 30 min at room temperature. Three- or four-color staining generally is used, using the fluorochromes fluorescein isothiocyanate (FITC), phycoerythrin (PE), and PerCP-Cy5.5, or FITC, PE, PerCP-Cy5.5, and antigen-presenting cell. As a control, isotype-matched control antibodies should be used to detect nonspecific staining. Finally, the cells are washed and fixed in paraformaldehyde for flow cytometry. Depending on the sensitivity of the epitopes to formaldehyde fixation, it may be required to perform cell surface staining before permeabilization and intracellular cytokine staining. The protocols for intracellular cytokine staining of human PBMCs or animal spleen or lymph node cells are rather similar. It is clear that three- or four-color fluorescence-activated cell sorting analysis requires careful data acquisition and analysis. Therefore, especially when setting up the staining protocols, it is advised to use as many controls as possible such as positive controls (e.g., SEB), negative controls, and negative staining controls. It is advisable to set the gates using a positive control. Because activated cells may down-modulate CD4 and CD8, it is important to include CD4^{dim} and CD8^{dim} cells when setting the gates. Of importance, although CD4^{dim} SSC^{lo} cells are activated T cells, CD4^{dim} SSC^{hi} cells are monocytes that can nonspecifically bind antibodies causing background staining and should thus be excluded. Finally, staining with fluorescent mAbs of a cell type that needs to be excluded may decrease background staining.

A more advanced application is the simultaneous detection of proliferation, cell surface staining, and intracellular cytokine staining *(22)*. Proliferation is measured by the incorporation of bromodeoxyuridine (BrdU), whereas BrdU is detected using a fluorescent labeled anti-BrdU mAb. DNase is added to denature DNA, thereby enhancing the availability of incorporated BrdU to the anti-BrdU mAb. It is clear that this technique is not amenable to large series of samples in routine investigations.

6. CONCLUSIONS

The "data explosion" has not left methods of cytokine measurement untouched. Both for molecular assays and immunoassays the trend is toward getting more data from each sample, at lower expense of time and money. Of course, these developments require major investments in machinery, a trend seen across the whole of biomedical research.

In my view three methods are currently the most appealing, each one being particularly powerful in a specific application:

1. Microarray analysis allows measuring the expression of virtually all (known) genes, allowing a very detailed and comprehensive picture of toxicant effects.
2. Luminex-based assays allow the simultaneous measurement of many cytokines in a way that is amenable for high-throughput processing.
3. Intracellular cytokine staining allows pinpointing to only those cells that underlie the different outcomes of an immune response. Of course, RPA, qPCR, and antibody arrays also are techniques of major value in cytokine measurements.

In conclusion, depending on the number of samples, the type of samples, and the underlying questions (mechanistic, single-cell, high-throughput), different techniques can be considered. It is clearly a challenge to immunotoxicologists to use these methods for better mechanistic understanding, faster screening of compounds on a stronger knowledge background and, hopefully, a reduction in the use of animals.

ACKNOWLEDGMENTS

Prof. H Van Loveren is acknowledged for reviewing the manuscript.

REFERENCES

1. Vandebriel RJ, Van Loveren H, Meredith C. Altered cytokine (receptor) mRNA expression as a tool in immunotoxicology. Toxicology 1998;130:43–67.
2. House RV. Theory and practice of cytokine assessment in immunotoxicology. Methods 1999;19:17–27.
3. House RV. Cytokine measurement techniques for assessing hypersensitivity. Toxicology 2001;158:51–58.
4. Cohen MD, Schook LB, Oppenheim JJ, Freed BM, Rodgers KE. Symposium overview: alterations in cytokine receptors by xenobiotics. Toxicol Sci 1999;48:163–169.
5. House RV. Cytokine bioassays: an overview. Dev Biol Stand 1999;97:13–19.
6. Kinloch RA, Roller RJ, Wassarman PM. Quantitative analysis of specific messenger RNAs by ribonuclease protection. Methods Enzymol 1993;225:294–303.
7. Tymms MJ. Quantitative measurement of mRNA using the RNase protection assay. Methods Mol Biol 1995;37:31–46.
8. Duggan DJ, Bittner M, Chen Y, Meltzer P, Trent JM. Expression profiling using cDNA microarrays. Nat Genet 1999;21S:10–14.
9. Eisen MB, Spellman PT, Brown PO, Botstein D. Cluster analysis and display of genome-wide expression analysis. Proc Natl Acad Sci USA 1998;95: 14,863–14,868.
10. Ezendam J, Staedtler F, Pennings J, Vandebriel RJ, Pieters R, Harleman JH, Vos JG. Toxicogenomics of subchronic hexachlorobenzene exposure in Brown Norway rats. Environ Health Perspect Toxicogenomics 2004;112:782–791.5.
11. Vandebriel RJ. Toxicogenomics (microarray technology). In: Vohr HW, ed. Encyclopedic Reference of Immunotoxicology. Berlin: Springer-Verlag; 2005, pp. 651–654.
12. Baken KA, Vandebriel RJ, Pennings JLA, Kleinjans JC, van Loveren H. Toxicogenomics in the assessment of immunotoxicity. Methods, in press.
13. Hamadeh HK, Bushel PR, Jayadev S, et al. Gene expression analysis reveals chemical-specific profiles. Toxicol Sci 2002;67:219–231.
14. Hamadeh HK, Bushel PR, Jayadev S, et al. Prediction of compound signature using high density gene expression profiling. Toxicol Sci 2002;67:232–240.
15. Ulleräs E, Trzaska D, Arkusz J, et al. Development of the "Cell Chip": a new in vitro alternative technique for immunotoxicity testing. Toxicology 2005;206:245–256.
16. Ringerike T, Ulleräs E, Völker R, et al. Detection of immunotoxicity using T-cell based cytokine reporter cell lines ("Cell Chip"). Toxicology 2005;206:257–272.
17. Carson RT, Vignali DA. Simultaneous quantitation of 15 cytokines using a multiplexed flow cytometric assay. J Immunol Methods 1999;227:41–52.

18. Vignali DA. Multiplexed particle-based flow cytometric assays. J Immunol Methods 2000;243:243–255.
19. Czerkinsky CC, Nilsson LA, Nygren H, Ouchterlony O, Tarkowski A. A solid-phase enzyme-linked immunospot (ELISPOT) assay for enumeration of specific antibody-secreting cells. J Immunol Methods 1983;65:109–121.
20. Hutchings PR, Cambridge G, Tite JP, Meager T, Cooke A. The detection and enumeration of cytokine-secreting cells in mice and man and the clinical application of these assays. J Immunol Methods 1989;120:1–8.
21. Suni MA, Picker LJ, Maino VC. Detection of antigen-specific T cell cytokine expression in whole blood by flow cytometry. J Immunol Methods 1998;212:89–98.
22. Mehta BA, Maino VC. Simultaneous detection of DNA synthesis and cytokine production in staphylococcal enterotoxin B activated CD4+ T lymphocytes by flow cytometry. J Immunol Methods 1997;208:49–59.

3

The Relevance of the T1/T2 Paradigm in Immunotoxicology

Example of Drug Hypersensitivity Reactions

Hervé Lebrec and John Vasilakos

SUMMARY

Early events in an immune response stimulate the production of cytokines that direct the subsequent development of T-cells. The innate immune system initiates T-cell activation by inducing naïve T-cells to differentiate into functional effector or memory T-cells. Effector T-cells produce cytokines and chemokines, which in turn affect the function (differentiation, proliferation, migration, etc.) of innate and adaptive immune cells. The effect of chronic immunization, infection, or disease results in the differentiation of naïve T-cells into polarized subsets of effector T-cells that can be differentiated from each other by the cytokines each T-cell subset produces. The Th1/Th2 paradigm is the best characterized example of T-cell polarization. In simplest terms, T-helper cell type 1 (Th1) and T-helper cell type 2 (Th2) cells are defined as differentiated CD4$^+$ $\alpha\beta$ T-cells that produce predominantly interferon (IFN)-γ or interleukin (IL)-4, respectively *(1)*. Th1 and Th2 cells can be differentiated phenotypically and functionally by parameters other than IFN-γ and IL-4 production, but these two cytokines quintessentially define the Th1/Th2 paradigm. The function of Th1 and Th2 cells is to help or instruct the innate and adaptive immune systems to respond in a specific manner to pathogens and cancer cells. In general, Th1 cells help the immune system respond to intracellular pathogens, and Th2 cells help the immune system respond to extracellular pathogens. Polarized Th1 and Th2 immunity should be considered endpoints of T-cell differentiation in response to chronic immunization or disease. Similar to CD4$^+$ T-cells, CD8$^+$ T-cells can develop into IFN-γ- or IL-4-producing cells, also called T cytotoxic-type 1 (Tc1) and T cytotoxic-type 2 (Tc2) cells, respectively. The nomenclature can be simplified to include both

From: *Methods in Pharmacology and Toxicology: Cytokines in Human Health: Immunotoxicology, Pathology, and Therapeutic Applications*
Edited by: R. V. House and J. Descotes © Humana Press Inc., Totowa, NJ

Th1 and Tc1 cells as T1 cells. Similarly, Th2 and Tc2 cells are grouped together as T2 cells. Finally, uncontrolled T-cell responses can lead to pathologies, such as T1-mediated organ autoimmunity or T2-mediated asthma and allergy.

This chapter will provide an overview of T1/T2 responses to xenobiotics, with a focus on nonclinical evidence of the involvement of the T1/T2 paradigm in hypersensitivity reactions and on the clinical evidence of various T-cell populations involved in drug hypersensitivities. This chapter will then focus on the regulation of T1 and T2 responses, both in humans and in the mouse. Finally, it will provide an overview of methods available to measure T1 and T2 immunity.

Key Words: Immunotoxicology; hypersensitivity; Th1; Th2; drugs.

1. T1/T2 RESPONSES TO XENOBIOTICS

1.1. Nonclinical Models and T1/T2 Responses in Hypersensitivity to Xenobiotics

Many experimental models have been used to try to correlate the diversity of clinical or pathological features of immune-related diseases to the heterogeneity of immune responses involved. Especially, people have focused on the possible correlation between these clinical or pathological features and the aforementioned Th1 and Th2 subpopulations of $CD4^+$ helper T-lymphocytes as well as the Tc1 and Tc2 phenotypes of $CD8^+$ lymphocytes. An important focus has been to try to associate type 1 (Tc1 or Th1) and type 2 (Tc2 or Th2) subsets with different clinical outcomes of xenobiotic-induced hypersensitivity reactions. Historically, the involvement of Th1- or Th2-like cytokines in sensitization mainly has been studied in the mouse with reference contact sensitizers, such as dinitrochlorobenzene (DNCB), dinitrofluorobenzene, trinitrochlorobenzene, and oxazolone, or respiratory sensitizers such as trimelitic anhydride (TMA). All these data clearly demonstrate that strong contact sensitizers activate IFN-γ-producing (Tc1) effector lymphocytes and that IFN-γ plays a central role for the inflammatory cutaneous reaction *(2)*. IFN-γ-producing $CD8^+$ lymphocytes are activated after exposure of mice to DNCB or formaldehyde but not to respiratory sensitizers such as TMA *(3,4)*. In addition, neutralization of IFN-γ at the time of challenge of mice with picryl chloride induces a strong reduction of contact sensitization *(5)*. However, IFN-γ not only is produced by $CD8^+$ T-lymphocytes during contact sensitization and depletion of $CD4^+$ cells from lymph nodes of formaldehyde-treated mice results in a substantial reduction in IFN-γ production *(4)*. It has been described that only respiratory sensitizers such as isocyanates or TMA could trigger IL-10 or IL-4 responses (Th2 responses) in the mouse after cutaneous treatment. However, when using reverse transcription polymerase chain reaction (RT-PCR), IL-4 mRNA can be detected in the lymph node from oxazolone or DNCB-treated mice (2). These nonclinical studies showed a clear involvement of type 1 and type 2 cytokines in xenobiotics-specific immune responses.

2. INVOLVEMENT OF T-CELL IN DRUG-HYPERSENSITIVITY REACTIONS

In recent years, drug-induced hypersensitivity reactions and underlying immunological mechanisms have been extensively studied, providing a better understanding of the implication of T-lymphocytes and various associated cytokines in these adverse reactions.

It is now well established that drug-specific CD4+ and CD8+ T-lymphocytes can be isolated from the peripheral blood or at cutaneous lesional sites of patients exhibiting delayed-type drug hypersensitivity (6–11). The involvement of specific T-cells is not restricted to delayed-type reactions because CD4+ and CD8+ T-lymphocytes have been isolated from the peripheral blood of patients with generalized urticaria and angioedema to β-lactam antibiotics (9). More recently, cellular and molecular features of presentation of drugs to specific T-cells, and T-cell responses to drugs have been a subject of growing interest. The recognition of small chemical entities by T-lymphocytes has been essentially explained as a consequence of these compounds being reactive and behaving as haptens. Haptens bind to peptides or proteins and can then be recognized by T-cells in a major histocompatability complex (MHC)-dependent way. As an example, it has been known for many years that penicillin reacts in vivo with nucleophilic amino acids, especially with lysine residues, to form antigenic determinant groups (12). Benzylpenicylloyl binding sites have then been identified on human serum albumin (13). It has been demonstrated that T-cell responses to β-lactam antibiotics are MHC-restricted and that synthesized peptides or natural proteins (serum albumin) modified by penicillin are recognized by penicillin-specific T-cell clones (9,14). Some drugs are not chemically reactive and need to be metabolized to behave as haptens. Sulfamethoxazole-nitroso is a reactive metabolite of sulfomethoxazole able to bind covalently to proteins and peptides. Werner Pichler (15) described a possible different way for drugs to interact with T-cells in the context of drug-induced hypersensitivity reactions. This possible mechanism is described as a "pharmacological interaction" of drugs with T-cell receptors (the "p-i concept"). It relies on the possibility to activate drug-specific T-cells in the presence of antigen-presenting cells made unable to process antigens while metabolizing capabilities are inhibited, and on the demonstration that drug is bound in a labile way as it can be washed away from the cell surface. Overall, these data clearly indicate the involvement of T-cells in drug hypersensitivity reactions.

3. DIFFERENT DRUG HYPERSENSITIVITY CLINICAL OUTCOMES ARE ASSOCIATED WITH THE INVOLVEMENT OF DIFFERENT LYMPHOCYTE SUBPOPULATIONS AND COMPLEX CYTOKINE NETWORKS

In a recent article, Pichler *(15)* reviewed how type 1 and type 2 cytokines are involved in various types of delayed drug hypersensitivity reactions characterized by different clinical outcomes and how the most recent information fits within the 1968 classification of allergic reactions responsible for clinical hypersensitivity and disease from Gell and Coombs *(16)*. He distinguishes four subtypes of type IV reactions as IVa (defined as Th1 and involving monocyte activation), IVb (defined as Th2 and involving eosinophilic inflammation), IVc (involving perforin and granzyme B expressing cytotoxic $CD4^+$ and $CD8^+$ T-lymphocytes), and IVd (involving IL-8 production, neutrophil recruitment, and activation). Table 1 *(17–24)* summarizes how different clinical outcomes of drug hypersensitivity reactions have been associated with T-cells and associated cytokines and known pathological consequences.

As illustrated in Table 1, drug reactions involve complex responses (multiple cell types and cytokines), and some drugs may induce different type of reactions in different individuals. For example, clinical manifestations of β-lactam antibiotics-induced allergy include immediate IgE-mediated reactions such as urticaria, Quincke edema (0.5–4.5 % of the treatments) and anaphylactic shock (0.01–0.2% *[25,26]*). Other hypersensitivity reactions include maculopapular rashes, which occur in up to 9.5% of the treated patients *(6,27,28)*, allergic contact dermatitis, and skin reaction of other types *(29)*. In addition, a given individual may develop a multiple drug allergy syndrome, that is, a clinical situation characterized by reactions against more than one different class of, both pharmacologically and structurally, unrelated drugs. Scala et al. *(30)* demonstrated the coexistence in the same patient of a delayed hypersensitivity to both penicillin G and β-methasone, driven, respectively, by penicillin G-specific Th2-skewed $CD8^+$ and β-methasone specific Th0 $CD4^+$ cells.

Table 1
Outcomes of Drug Hypersensitivity Reactions and Known Pathological Consequences

Clinical condition	Cytokines and associated cells	Examples of incriminated molecules
Anaphylactic shock, angioedema, Quincke edema, urticaria	IL-4, no IFN-γ from PBMCs (17) IL-4 from Tc2 and IFN-γ from Th1 and Tc1 lymphocytes (18) IL-4, IL-5, IL-13 from CD4$^+$ cells (19)	β-Lactam antibiotics, amiodarone, diclofenac, dipyrone, ibuprofen
Acute generalized exanthematous pustulosis (AGEP; characterized by neutrophilic skin inflammation)	IL-4, IL-5, IFN-γ, RANTES, GM-CSF, IL-8 from T-cells (20) TNF-α, IFN-γ, GM-CSF, CXCL8/IL-8 from CD4$^+$ T-cells (21)	Amoxicillin, sulfamethoxazole, celecoxib
Maculopapular exanthema and drug-induced hypersensitivity syndrome (maculopapular exanthema associated with systemic signs, including eosinophilia)	IFN-γ and IL-5 from CD4$^+$ (predominantly) and CD8$^+$ cells, and RANTES, MCP-3, IL-8, EOTAXIN (22,23)	Amoxicillin, sulfamethoxazole, ceftriaxone, cefazolin, imipenem, metazolon, carbamazepin
Bullous exanthema, toxic epidermal necrolysis (TEN), Steven-Johnson syndrome	Tumor necrosis factor-α, IFN-γ, IL-2 from peripheral blood monuclear cells (24)	Phenytoin, carbamazepin

4. REGULATION OF T1/T2 RESPONSES: THE ROLE OF DENDRITIC CELLS

A number of factors impact the development of Th1 and Th2 cells. All appear to be dependent fundamentally on a stepwise interaction between dendritic cells (DCs) and naïve CD4+ T-cells. DCs are the primary sensors of the innate immune system detecting the presence of pathogens through highly conserved pattern recognition receptors, including the recently identified toll-like receptors (TLRs [31,32]). In humans, there are 10 TLRs identified that act as receptors for a number of ligands from bacteria, viruses, parasites, and fungi. When DCs encounter pathogens, they internalize the pathogens via endocytosis or phagocytosis. The activation of TLRs induces the activation of DCs or maturation, resulting in the expression of chemokine receptors that are important for migration to lymph nodes where naïve T-cells reside, the expression of cell surface co-stimulatory markers that are important for T-cell activation, and the production of cytokines important for naïve T-cell differentiation. During DC maturation, the pathogen's proteins are proteolytically cleaved into short peptides that are coexpressed with MHC class I or MHC class II molecules on the cell surface. Once DCs initiate contact with the naïve T-cells expressing the appropriate T-cell receptor, DC costimulatory markers interact with CD28 on the T-cell surface, culminating in the initiation of naïve T-cell activation. Other DC costimulatory molecules such as OX40L and B7H also interact with T-cells via OX40 and ICOS, respectively, enhancing the activation of naïve T-cells. The cytokines produced by DCs play the most critical role for naïve T-cell differentiation into effector Th1 or Th2 cells (33,34). IL-12 (p70 composed of p40/p35 heterodimers) is the most critical cytokine for inducing Th1 cells and is produced by activated DCs. IL-4 is the most critical cytokine for the induction of Th2 cells, but the source of IL-4 is unknown. IFN-γ and IL-4 enhance the generation of their own Th cell subset and concomitantly inhibit the generation of the opposing Th cell subset. IL-18 produced by DCs during DC–T-cell interaction also modulates Th1 development. Although IL-18 does not appear to directly induce Th1 development alone, IL-18 can strongly augment IL-12-induced Th1 generation by acting after IL-12-induced Th1 development to increase IFN-γ production by differentiated Th1 cells (35). In addition to IL-12 and IL-18, type I IFN can enhance Th1 differentiation by enhancing the expression of IL-12 receptor subunits on naïve T-cells, thereby increasing the ability of T-cell activation by IL-12. Conversely, the absence of IL-12 results in the differentiation of naïve T-cells into IL-4-

producing effector cells. Between IL-4 and IL-12, IL-4 appears to be dominant in that small amounts of IL-4 produced over a short time result in skewing the immune response to a Th2 phenotype *(36)*. For Th1 development, prolonged exposure to IL-12 appears necessary for the development of naïve CD4 cells into Th1 cells.

It should be noted that DCs are heterogenous populations of cells that express different TLRs and therefore respond to different types of pathogens *(37–40)*. For instance, in humans, myeloid DCs express TLR4 and TLR8 but not TLR9 like plasmacytoid DCs. Therefore, myeloid DCs respond to pathogens that contain lipopolysaccharide (TLR4 agonist), a cell wall component of Gram-negative bacteria, and uridine-rich single-stranded ribonucleic acid (RNA; TLR8 agonists) found in various viruses. Plasmacytoid DCs express TLR7 and TLR9 and respond to single-stranded RNA viruses and to unmethylated CpG oligonucleotides (TLR9 agonist) found in many types of bacteria *(41,42)*. Human myeloid DCs appear to be the primary DC subset that produces IL-12, and plasmacytoid DCs are the primary type I IFN-producing DC subset. Each of these DC subsets can influence Th1 or Th2 development on the basis of the cytokines they produce, both seemingly driving Th1 responses.

5. T1/T2 RESPONSES IN MICE AND HUMANS: KEY DIFFERENCES

In general, Th1 and Th2 immunity are similar in both human and mouse systems, but there are a number of notable differences *(43–46)*. In particular, Th1 immunity in mice is highly dependent on IL-12. Whereas Th1 cells generated in humans rely on both IL-12 and IFN-γ. The 70-kDa biologically active IL-12 protein binds to the IL-12 receptor composed of two subunits (IL12Rβ1 and IL-12Rβ2). In humans, IL-12Rβ2 expression is greatly enhanced by IFN-α. In mice, IFN-α does not seem to affect naïve T-cell differentiation; however, in the human system, it can affect both naïve and activated populations of T-cells. Effective IL-12 signaling requires the generation of IFN-α, resulting in upregulation of IL-12Rβ2. Once IL-12 binds to the dimeric IL-12 receptor, Stat-4 activation ensues culminating in the transcription of IFN-γ *(47)*. Another difference between mouse and human systems is that mouse Th2 cells are the predominant IL-10-producing CD4[+] T-cells, whereas both human Th1 and Th2 cells can produce IL-10. Despite a number of differences between human and mouse Th1/Th2 immunity, the essential feature of polarized T-cell subsets based on IFN-γ and IL-4-producing cells remain the same.

5.1. Cytokines That Define the Th1 and Th2 Paradigm

Although IFN-γ and IL-4 production are hallmarks of Th1 and Th2 cells, respectively, other cytokines are produced by Th1 or Th2 cells (Fig. 1).

The differences between Th1 and Th2 cells in terms of cytokine production generally are a matter of magnitude rather than absolute production. For instance, Th1 cells produce greater levels of IL-2 and TNF-β (lymphotoxin) than Th2 cells. Conversely, Th2 cells produce higher levels of IL-5, IL-9, and IL-13 than Th1 cells *(33,48)*. This Th1/Th2 cytokine profile is typical in both human and murine systems. Although individual cytokines are useful for characterizing an immune response as either Th1 or Th2, Th1/Th2 immunity is best characterized as a ratio of Th1 cytokines to Th2 cytokines rather than the absolute production of a given type of cytokine. For example, a Th1 response would have a comparatively high IFN-γ:IL-4 or IFN-γ:IL-5 ratio than a Th2 response, which would have lower levels of IFN-γ relative to IL-4, IL-5, or IL-13.

The role or function of both Th1 and Th2 immunity is to protect against specific types of infectious disease and cancer *(49–52)*. Th1 cells appear most important for protection against intracellular pathogens inducing cell-mediated immunity, whereas the generation of Th2 cells is important for protection against extracellular pathogens inducing humoral immunity. The promotion of cell-mediated immunity or delayed-type hypersensitivity is a hallmark of Th1 immunity and is mainly the result of the cytokines Th1 cells produce. Cell-mediated immunity is characterized by the activation of macrophages, natural-killer cells, and IFN-γ-producing CD8 T-cells. IFN-γ also enhances antigen processing, presentation, and DC costimulatory marker expression. IFN-γ enhances the generation of Th1 cells by enhancing IL-12 production, and IFN-γ inhibits Th2 generation by inhibiting Stat6-induced IL-4 production. IFN-γ enhances the production of a number of chemokines that promote Th1 immunity. IP-10 (CXCL10) and MIG (CXCL9) are chemokines that predominantly recruit Th1 and Tc1 cells *(53,54)*. MIP-1α (CCL3) and MIP-1β (CCL4) also chemoattract activated CD4, CD8, and memory T-cells. MCP-1 (CCL2) chemoattracts monocytes and macrophages, as well as Th1 cells. The chemokines produced by Th1 cells establish an environment that favors cell mediated immunity and enhances the ability to destroy intracellular pathogens via macrophage-mediated mechanisms or cell killing mechanisms by natural-killer or CD8 T-cells.

In contrast, IL-4 production by Th2 cells promotes humoral immunity, which is characterized by the production of neutralizing immunoglobulins (IgG) and antibodies that induce mast cell and eosinophil degranulation (IgE *[55–57]*). Neutralizing IgG is critical for protection against extracellular

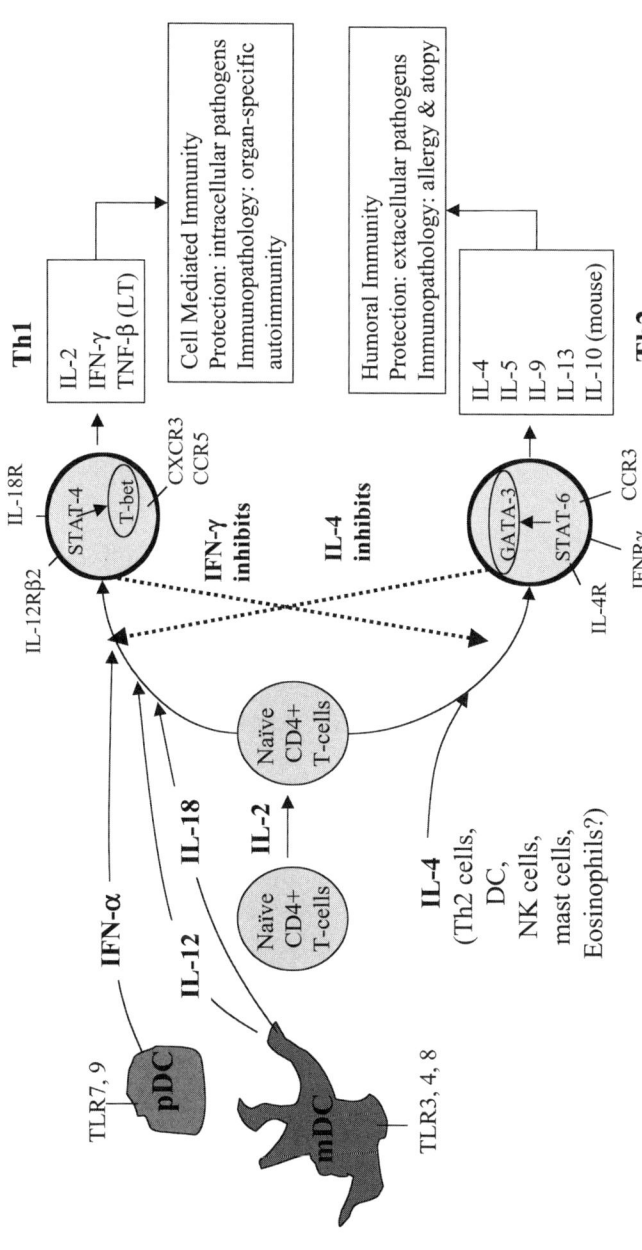

Fig. 1. Generation of human Th1 and Th2 cells. Modified from O'Garra (33), Glimcher (47), and Theofilopoulos (63).

bacteria, and bacterial toxins, as well as protection against disseminating viral infections. IgE is important for protection against mucosal parasites such as helminths. IL-4 enhances the production of the IgE receptor, FcεRII (CD23), on masT-cells, eosinophils, and macrophages *(58)*. IL-4 antagonizes the function of IFN-γ by inhibiting IFN-γ-induced cytolytic antibodies (IgG2a in mice), while concomitantly enhancing the production of opsonizing immunoglobulins (IgG1 in mice). IL-4 secreted by polarized Th2 cells inhibits Th1 generation by inhibiting Stat4-induced IFN-γ production. Therefore, Th1 and Th2 cells reciprocally inhibit each other.

Like CD4$^+$ T-cells, naive CD8$^+$ T-cells can differentiate into effector populations that predominantly produce IFN-γ or IL-4 *(59,60)*. IFN-γ- and IL-4-producing CD8$^+$ T-cells are referred to as T-cytotoxic cell type 1 (Tc1) or T-cytotoxic cell type 2 (Tc2) cells, respectively. Cytotoxic T-cells recognize or target other cells expressing foreign antigens (i.e., pathogen-specific or tumor-specific antigens) that are associated with MHC class I molecules. Unlike MHC class II molecules that are solely expressed on antigen presenting cells such as DCs, macrophages, and B-cells, class I molecules are present on virtually all cell types. Therefore, cytotoxic T-cells have the capability of recognizing and destroying most cells that express antigens associated with MHC class I. Cytotoxic T-cells not only destroy cells infected by intracellular pathogens but also produce a variety of cytokines that can have immune regulatory functions, similar to CD4 T-helper cells. Tc1 and Tc2 are analogous to Th1 and Th2 cells in terms of the cytokines they secrete. However, both Tc1 and Tc2 effector cells appear to kill target cells equally well; therefore, the key differentiating feature between Tc1 and Tc2 cells is the cytokine profile they produce.

A final note on T-cell subsets should be made regarding regulatory T-cells or suppressor T-cells. These heterogeneous subsets of CD4 and CD8 T-cells primarily function as inhibitors of effector T-cell function *(61,62)*. When in contact with effector T-cells, suppressor T-cells inhibit effector T-cell cytokine production and the ability of cytotoxic T-cells to kill. Suppressor T-cells appear critical for preventing or diminishing autoimmunity and chronic inflammatory diseases. Suppressor T-cells also appear to be critical barrier to overcome for successful cancer vaccine therapy because suppressor T-cells help inhibit T-cell effector generation to self antigens, which are indeed the same antigens used to generate anticancer immunity.

6. MEASURING T1 AND T2 IMMUNITY

Because the defining characteristic of T1 and T2 immunity is based on differential cytokine production by T-cell subsets, the most reliable methods used to identify and differentiate T1 from T2 immune response is by measuring cytokines. The most reliable means of assessing T1 and T2 immunity is to measure the proteins that define the paradigm. In general, using cytokine-specific antibodies is the most reliably sensitive, specific, and quantitative way to measure T1 and T2 responses. The most rigorous means used to differentiate T1 and T2 immunity is to measure both T1 and T2 cytokines and determine cytokine ratios. In practice, IFN-γ protein is considerably easier to measure than IL-4 protein mainly because of the actual amount of IFN-γ produced relative to IL-4. Therefore, ratios of IFN-γ to IL-5 or IL-13 (or IL-10 for mice) often are reported. Although legitimate, the strictest definition of T1 and T2 immunity is difference or ratio between IFN-γ and IL-4. Numerous methods can be used to detect the cytokines. Quantitatively, enzyme-linked immunosorbent assays (ELISAs) or modifications of ELISAs are the most reliable tools. Often, appropriate antibodies are not available, and biological assays are used to determine the level or presence of a given cytokine. Despite being extremely sensitive, biological assays can be difficult to interpret because of the pleitropic nature of cytokines. As an example, IL-4 induces naïve T-cell proliferation similar to IL-2. To differentiate between the activities of these two cytokines, neutralizing the activity of one of the cytokines usually is necessary. Therefore, the interpretation of biological assays generally is more difficult than ELISAs. One of the most reliable means of assessing which T-cell subsets produce these cytokines is intracellular flow cytometry. The advantage of intracellular flow cytometric analysis is that one can quantitate the number of cells making a given cytokine. Specific populations of cells within a complex mixture of cells, such as peripheral blood, can be identified to produce the cytokine of interest. For example, the number of CD4+ T-cells that produce IFN-γ relative to those that make IL-4 or IL-5 can be differentiated from each other and from CD8+ T-cells within the same population that make the same cytokines. The disadvantage of flow cytometry is that the technique does not have the high-throughput capacity of ELISAs, and the cytokine-specific antibodies for intracellular staining are not as readily available as those for ELISAs. Although, in general T1 and T2 cytokines are secreted once the mRNA is translated into protein. Additional antibody-mediated methods used to differentiate T1 and T2 cytokines have been evaluated by immunohistochemistry and Western blotting. These techniques can be effective but lack the sensitivity of ELISAs or flow cytometry. The most sensitive technique used to measure T1 and T2 cytokines is RT-PCR, where messenger RNA that encodes for the proteins of the various cytokines can be reliably measured. The advantage of RT-PCR is that small amounts of material are required to measure a given cytokine messenger (m)RNA or transcript compared with other analytical techniques. The disadvantage of RT-PCR is that mRNA is measured, not protein, and proteins are the mediators of biological function. Indeed, not all cytokines are regulated strictly at the transcriptional level. In other words, some cytokines such as IL-1 and tumor necrosis factor-α, require post-translational modifications to be secreted from the cell as biologically active proteins. Hence, induction of cytokine mRNA simply indicates that the cell or tissues contain the transcript of interest, not necessarily the protein. Overall, there are a number of analytical approaches used to differentiate between T1 and T2 responses using cytokine-specific antibodies or RT-PCR.

REFERENCES

1. Mosmann T, Cherwinski H, Bond M, Giedlin M, Coffman R. Two types of murine helper T-cell clone. I. Definition according to profiles of lymphokine activities and secreted proteins. 1986. J Immunol 1986;136:2348–2357.
2. Xu H, Dilulio NA, Fairchild RL. T-cell populations primed by hapten sensitization in contact sensitivity are distinguished by polarized patterns of cytokine production: IFN-γ producing (Tc1) effector CD8+ T and IL-4/IL-10 producing (Th2) negative regulatory CD4+ T-cells. J Exp Med 1996;183:1001–1012.
3. Dearman RJ, Basketter DA, Kimber I. Characterization of chemical allergens as a function of divergent cytokine secretion profiles induced in mice. Toxicol Appl Pharmacol 1996;138:308–316.
4. Dearman RJ, Moussavi A, Kemeny DM, Kimber I. Contribution of CD4+ and CD8+ T lymphocyte subsets to the secretion patterns induced in mice during sensitization to contact and respiratory chemical allergens. Immunology 1996;89:502–510.
5. Dieli F, Asherson GL, Sireci G, Dominici R, Sciré E, Salerno A. Development of IFN-γ-producing CD8+ T lymphocytes during contact sensitivity. J Immunol 1997;158:2567–2575.
6. Warington RJ, Silviu-Dan F, Magro C. Accelerated cell-mediated immune reactions in penicillin allergy. J Allergy Clin Immunol 1993;92:626–628.
7. Hertl M, Bohlen H, Jugert F, Boecker C, Knaup R, Merk HF. Predominance of epidermal CD8+ T lymphocytes in bullous cutaneous reactions caused by β-Lactam antibiotics. J Invest Dermatol 1993;101:794–799.
8. Hertl M, Merk HF. Lymphocyte activation in cutaneous drug reactions. J Invest Dermatol 1995;105:95S–98S.
9. Brander C, Mauri-Hellweg D, Bettens F, Rolli H, Goldman M, Pichler WJ. Heterogenous T-cell responses to β-Lactam-modified self-structures are observed in penicillin-allergic individuals. J Immunol 1995;155:2670–2678.

10. Mauri-Hellweg D, Bettens F, Mauri D, Brander C, Hunziker T, Pichler WJ. Activation of drug-specific CD4+ and CD8+ T-cells in individuals allergic to sulfonamides, phenytoin, and carbamazepine. J Immunol 1995;155:462–472.

11. Hertl M, Geisel J, Boecker C, Merk HF. Selective generation of CD8+ T-cell clones from the peripheral blood of patients with cutaneous reactions to beta-lactam antibiotics. Br J Dermatol 1993;128:619–626.

12. Levine BB, Ovary Z. Studies on the mechanism of the formation of the penicillin antigen. J Exp Med 1961;114:875–904.

13. Yvon M, Anglade P, Wal J-M. Identification of the binding sites of benzyl penicilloyl, the allergenic metabolite of penicillin, on the serum albumin molecule. FEBS Lett 1990;263:237 240.

14. Weltzien HU, Padovan E. Molecular features of penicillin allergy. J Invest Dermatol 1998;110:203–206.

15. Pichler W. Delayed Drug Hypersensitivity Reactions. Ann Intern Med 2003; 139:683–693.

16. Coombs PR, Gell PG. Classification of allergic reactions responsible for clinical hypersensitivity and disease. In: Gell RR, ed. Clinical Aspects of Immunology. Oxford: Oxford Univ Press, 1968, pp. 575–596.

17. Posadas S, Leyva L, Torres M, et al. Subjects with allergic reactions to drug show in vivo polarized patterns of cytokine expression depending on the chronology of the clinical reaction. J Allergy Clin Immunol 2000;106:769–776.

18. Gaspard I, Guinnepain MT, Laurent J, et al. IL-4 and IFN-γ mRNA Induction in peripheral lymphocytes specific for β-lactam antibiotics in immediate or delayed hypersensitivity reactions. J Clin immunol 2000;20:107–116.

19. Brugnolo F, Annunziato F, Sampognaro S, et al. Highly Th2-skewed cytokine profile of β-lactam-specific T-cells from nonatopic subjects with adverse drug reactions. J Immunol 1999;163:1053–1059.

20. Britschgi M, Steiner U, Schmid S, et al. T-cell involvement in drug-induced acute generalized exanthematous pustulosis. J Clin Invest 2001;107:1433–1441.

21. Schaerli P, Britschgi M, Keller M, et al. Characterization of Human T-cells That Regulate Neutrophilic Skin Inflammmation. J Immunol 2004;173:2151–2158.

22. Pichler W, Yawalkar N, Britschgi M, et al. Cellular and molecular pathophysiology of cutaneous drug reactions. Am J Clin Dermatol 2002;3:229–238.

23. Choquet-Kastylevsky G, Intrator L, Chenal C. Increased levels of interleukin 5 are associated with the generation of eosinophilia in drug-induced hypersensitivity syndrome. Br J Dermatol 1998;139:1026–1032.

24. Leyva L, Torres M, Posadas S, et al. Anticonvulsivant-induced toxic epidermal necrolysis: monitoring the immunological response. J Allergy Clin Immunol 2000;105:157–165.

25. Lin RY. A perspective on penicillin allergy. Arch Intern Med 1992;152:930–937.

26. Vega JM, Blanca M, Garcia JJ, et al. Immediate allergic reaction to amoxicillin. Allergy 1994;49:317–322.

27. Boguniewicz M, Leung DYM. Management of the patient with allergic reactions to antibiotics. Pediatr Pulmonol 1992;12:113–122.
28. RomanoA, Di Fonso M, Papa G, et al. Evaluation of adverse cutaneous reactions to aminopenicillins with emphasis on those manifested by maculopapular rashes. Allergy 1995;50:113–118.
29. Barbaud A, Bene MC, Faure G. Immunological physiopathology of cutaneous adverse drug recations. Eur J Dermatol 1997;7:319–323.
30. Scala E, Giani M, Pastore S, et al. Distinct delayed response to β-methasone and penicillin-G in the same patient. Allergy 2003;58:439–444.
31. Pulendran B, Palucka K, Banchereau J. Sensing pathogens and tuning immune responses. Science 2001;293:253–256.
32. Banchereau J, Fay J, Pascual V, Palucka AK. Dendritic cells: controllers of the immune system and a new promise for immunotherapy. 2003;252:226–235.
33. O'Garra A. Cytokines induce the development of functionally heterogenous T helper cell subsets. Immunity 1998;8:275–283.
34. Macatonia S, Hosken N, Litton M, et al. Dendritic cells produce IL-12 and direct the development of Th1 cells from naive CD4+ T-cells. 1995. J Immunol 1995;154:5071–5079.
35. Robinson D, Shibuya K, Mui A, et al. IGIF does not drive Th1 development, but synergizes with IL-12 for IFN-γ production, and activates IRAK and NF-κB. Immunity 1997;7:571–581.
36. Abehsira-Amar O, Gibert M, Joliy M, Theze J, Jankovic D. IL-4 plays a dominant role in the differential development of Tho into Th1 and Th2 cells. J Immunol 1992;148:3820–3829.
37. Barton GM, Medzhitov R. Control of adaptive immune responses by Toll-like receptors. Curr Opin Immunol 2002;14:380–383.
38. Medzhitov R, Janeway C. The Toll receptor family and microbial recognition. Trends Microbiol. 2000;8:452–456.
39. Pasare C, Medzhitov R. Toll-like receptors and acquired immunity. Semin Immunol 2004;16:23–26.
40. Schnare M, Barton GM, Holt AC, Takeda K, Akira S, Medzhitov R. Toll-like receptors control activation of adaptive immune responses. Nat Immunol 2001;2:947–950.
41. Kaisho T, Akira S. Regulation of dendritic cell function through Toll-like receptors. Curr Mol Med 2003;3:373–385.
42. Kadowaki N, Ho S, Antonenko S, Malefyt RW, Kastelein RA, Bazan F, Liu YJ. Subsets of human dendritic cell precursors express different toll-like receptors and respond to different microbial antigens. J Exp Med 2001;194:863–869.
43. Parronchi P, De Carli M, Manetti R, et al. IL-4 and IFN (alpha and gamma) exert opposite regulatory effects on the development of cytolytic potential by Th1 or Th2 human T-cell clones. J Immunol 1992;149:2977–2983.
44. Schandene L, Del Prete GF, Cogan E, et al. Recombinant interferon-alpha selectively inhibits the production of interleukin-5 by human CD4+ T-cells. J Clin Invest 1996;15:309–315.

45. Parronchi P, Mohapatra S, Sampognaro S, et al. Effects of interferon-alpha on cytokine profile, T-cell receptor repertoire and peptide reactivity of human allergen-specific T-cells. 1996. Eur J Immunol 1996;26:697–703.

46. Del Prete G, De Carli M, Almerigogna F, Giudizi M, Biagiotti R, Romagnani S. Human IL-10 is produced by both type 1 helper (Th1) and type 2 helper (Th2) T-cell clones and inhibits their antigen-specific proliferation and cytokine production. 1993. J Immunol 1993;150:353–360.

47. Glimcher LH, Murphy KM. Lineage commitment in the immune system: the T helper lymphocyte grows up. Genes Dev 2000;14:1693–1711.

48. Maggi E, Parronchi P, Manetti R, et al. Reciprocal regulatory effects of IFN-gamma and IL-4 on the in vitro development of human Th1 and Th2 clones. J Immunol 1992;148:2142–2147.

49. Pardoll DM, Topalian SL. The role of CD4+ T-cell responses in antitumor immunity. Curr Opin Immunol 1998;10:588–594.

50. Kemp M, Kurtzhals J, Bendtzen K, et al. Leishmania donovani-reactive Th1- and Th2-like T-cell clones from individuals who have recovered from visceral leishmaniasis. 1993. Infect Immun 1993;61:1069–1073.

51. Miralles G, Stoeckle M, McDermott D, Finkelman F, Murray H. Th1 and Th2 cell-associated cytokines in experimental visceral leishmaniasis. Infect Immun 1994;62:1058–1063.

52. Sjolander A, Baldwin TM, Curtis JM, Handman E. Induction of a Th1 immune response and simultaneous lack of activation of a Th2 response are required for generation of immunity to leishmaniasis. J Immunol 1998;160:3949–3957.

53. Hilkens CM, Schlaak JF, Kerr IM. Differential responses to IFN-alpha subtypes in human T-cells and dendritic cells. J Immunol 2003;171:5255–5263.

54. Padovan E, Spagnoli GC, Ferrantini M, Heberer M. IFN-alpha2a induces IP-10/CXCL10 and MIG/CXCL9 production in monocyte-derived dendritic cells and enhances their capacity to attract and stimulate CD8+ effector T-cells. J Leukoc Biol 2002;71:669–676.

55. Toellner K-M, Luther SA, Sze DMY, et al T Helper 1 (Th1) and Th2 characteristics start to develop during T-cell priming and are associated with an immediate ability to induce immunoglobulin class switching. J Exp Med 1998;187:1193–1204.

56. Del Prete G, De Carli M, Ricci M, Romagnani S. Helper activity for immunoglobulin synthesis of T helper type 1 (Th1) and Th2 human T-cell clones: the help of Th1 clones is limited by their cytolytic capacity. J Exp Med 1991;174:809–813.

57. Finkelman F Katona I, Mosmann T, Coffman R. IFN-gamma regulates the isotypes of Ig secreted during in vivo humoral immune responses. J Immunol 1988;140:1022–1027.

58. Rousset F, Malefijt RW, Slierendregt B, et al. Regulation of Fc receptor for IgE (CD23) and class II MHC antigen expression on Burkitt's lymphoma cell lines by human IL-4 and IFN-gamma. J Immunol 1988;140:2625–2632.

59. Vukmanovic-Stejic M, Vyas B, Gorak-Stolinska P, Noble A, Kemeny DM. Human Tc1 and Tc2/Tc0 CD8 T-cell clones display distincT-cell surface and functional phenotypes. Blood 2000;95:231–240.

60. Halverson DC, Schwartz GN, Carter C, Gress RE, Fowler DH. In vitro generation of allospecific human CD8+ T-cells of Tc1 and Tc2 phenotype. Blood 1997;90:2089–2096.

61. Shevach EM. Regulatory T-cells in autoimmmunity. Annu Rev Immunol 2000;18:423–449.

62. Shevach EM. Regulatory T-cells. Introduction. Semin Immunol 2004;16:69–71.

63. Theofilopoulos AN, Koundouris S, Korio DH, Lawson BR. The role of IFN-gamma in systemic lupus erythematosus: a challenge to the Th1/Th2 paradigm in autoimmunity. Arthritis Res 2001;3:136–141.

4

Microbial Exploitation and Subversion of the Human Chemokine Network

James E. Pease

SUMMARY

The chemokine network, comprising cell surface G protein-coupled receptors and soluble small molecular-weight protein ligands, constitutes a highly evolved system that facilitates leukocyte recruitment in both innate and adaptive immunity. As such, it has attracted attention from the research community as a means of modulating the immune system, a hypothesis that it appears has already been tested rigorously by microbes during coevolution. Several examples exist to support the notion that viruses, protozoa, and helminths have derived strategies of either exploitation or subversion, for example, using the chemokine network to gain cellular entry or to evade host immune surveillance. It is anticipated that, in the coming years, close examination of the mechanisms underlying these processes should provide opportunities for the generation of novel therapeutics. These may be of use to both thwart the microbial defense strategies and also to treat a variety of inflammatory diseases in which the inappropriate or excessive production of chemokines is pathologically implicated.

Key Words: Chemokines; chemokine receptors; microbes; inflammation.

1. INTRODUCTION

Chemokines (*chemo*tactic cyto*kines*) are small peptides that are potent inducers of leukocyte migration and that are involved in a myriad of host processes, including angiogenesis, leukocyte trafficking, and immune responses *(1)*. Chemokines constitute a family of approx 40 members in the human, which are classified according to the position of their amino terminal-conserved cysteine residues. The majority of chemokines reside within two major classes, namely the CC chemokines (in which the two amino terminal cysteine residues are adjacent) and the CXC chemokines (in which the two

From: *Methods in Pharmacology and Toxicology: Cytokines in Human Health:*
Immunotoxicology, Pathology, and Therapeutic Applications
Edited by: R. V. House and J. Descotes © Humana Press Inc., Totowa, NJ

amino terminal cysteine residues are separated by a single amino acid residue). Two minor classes, the CX3C and C chemokines, also exist and have only three members between them *(2–4)*. Chemokines exert their effects by binding to chemokine receptors on the surface of cells, predominantly leukocytes. These are seven transmembrane (7TM) receptors belonging to the family of G protein coupled receptors and, to date, 20 specific chemokine receptors have been identified in humans *(5,6)*.

The ability of chemokines to induce leukocyte recruitment plays a pivotal role in host immunity as they guide leukocytes to both the organs in which they are to reside (e.g., intestine, skin, thymus) and also to sites of microbial infection. The specific task of each chemokine and its receptor in such processes is being gradually teased apart by the generation of mice deficient in either a specific chemokine or receptor and in conjunction with in vitro techniques, the precise role of both receptor and chemokine in selective leukocyte recruitment is steadily being resolved. However, it is humbling to find that we are not the first creatures to take such a close look at the chemokine network. Long before the first chemokine was identified in 1977 *(7)*, microbes were busy manipulating the network at a variety of levels to achieve both evasion of the host defense systems and increased propagation. Indeed, such bombardment of the immune system has been put forward as the driving force behind the generation of host defense protein diversity *(8)*.

Such manipulation of the chemokine system can be conveniently broken down into two main strategies, namely exploitation and subversion (Fig. 1). The former makes use of host chemokines or their receptors to facilitate processes such as cellular entry and microbial proliferation. In contrast, subversion seeks to nullify the host chemokine network by blunting the chemokine:receptor interaction, thereby helping the microbe to evade detection. In this chapter we will look in detail at such strategies and discuss the potential therapeutic modalities for correcting such microbe-induced immunomodulation.

2. EXPLOITATION OF THE CHEMOKINE NETWORK: USE OF HOST CHEMOKINE RECEPTORS TO GAIN CELL ENTRY

2.1. Human Immunodeficiency Virus-1

The use of chemokine receptors by human immunodeficiency virus-1 (HIV-1) to gain cellular entry into leukocytes represents perhaps the most infamous example of exploitation of the chemokine network. In this process, a subunit of the HIV-1 envelope glycoprotein named gp120 forms a heterotrimeric complex with the leukocyte cell surface protein CD4 and a

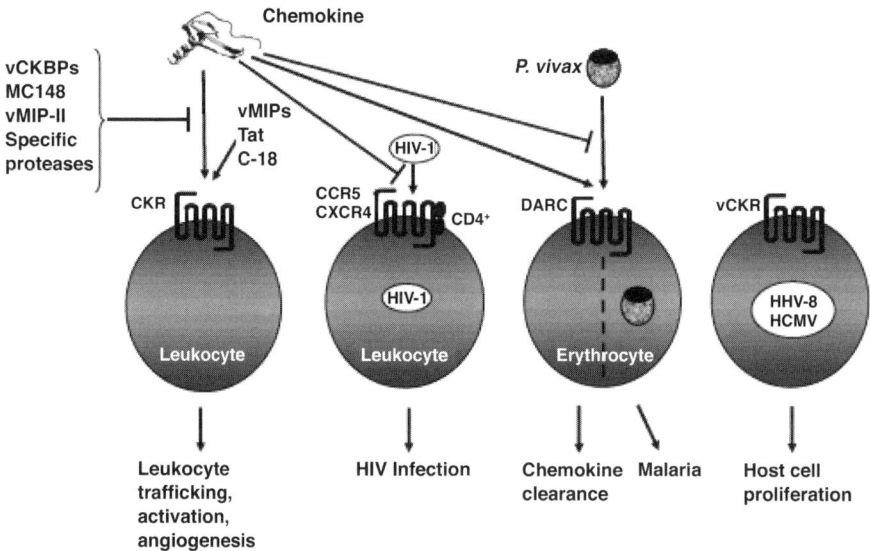

Fig. 1. An overview of microbial exploitation and subversion of the human chemokine network. The figure illustrates the main ways in which microbes have subverted or exploited the chemokine network. Chemokines function by activating G protein-coupled receptors on leukocytes, ultimately resulting in their activation and chemotactic migration (left). This process can be efficiently blocked by viral chemokine binding proteins (vCKPBs), viral chemokine antagonists (such as MC148, vMIP-II), and proteolytic degradation of the chemokine ligand. Similarly, leukocyte recruitment can be activated via the same cell receptors by chemokine mimetics such as the viral macrophage inflammatory proteins (vMIPs), the HIV-1 protein Tat and the *Toxoplasma gondii* protein cyclophillin C-18. Chemokine receptors also can be exploited as cell entry factors by viruses (e.g. HIV-1) and protozoa (e.g., *Plasmodium vivax*). Finally some viruses, notably human herpes virus-8 (HHV-8) and human cytomegalovirus (HMCV) encode homologs of chemokine receptors that are thought to hijack the host cell signaling machinery and induce proliferation of both the infected cell and the virus.

chemokine receptor, often referred to as a coreceptor. For many years, it was known that HIV-1 used the cell surface protein CD4 to enter and infect cells, but cell transfectants engineered to express CD4 were not permissible for viral entry, suggesting the required presence of an unknown cofactor. By means of an elegant molecular cloning approach, Feng and colleagues *(9)* identified an orphan 7TM receptor, which they named "fusin" that in conjunction with CD4 facilitated fusion of the viral and host cell membranes.

Subsequent work identified the chemokine ligand for fusin *(10,11)*, immediately cementing the link between the chemokine field with that of HIV-1 research. In keeping with nomenclature, Fusin was renamed CXCR4 and has been shown to be responsible for the cellular entry of T-lymphocyte (T)–tropic laboratory strains of HIV-1. In contrast, the related chemokine receptor CCR5 allows the entry of macrophage (M)-tropic strains *(12–14)*.

During the entry process, it is thought that the formation of a heterotrimeric complex of envelope protein, CD4, and chemokine receptor unmasks a cryptic portion of the transmembrane subunit of the envelope protein named gp41. This possesses a highly hydrophobic section at its amino terminus, which readily inserts itself into the membrane of the host cell, ultimately leading to fusion of the virion and cell membranes and the release of the viral genome into the cytoplasm of the target cell (reviewed in ref. *15*). Blockade of HIV-1 entry via either coreceptor can be blocked by their specific ligands, a point that explains the earlier identification of CCR5 ligands as HIV-suppressive factors produced by CD8+ T-cells *(16)*.

The importance of CCR5 in HIV-1 infection is highlighted by homozygous inheritance of the naturally occurring CCR5Δ32 mutation, which results in the truncation of the CCR5 protein and, ultimately, an absence of CCR5 on the cell surface *(17,18)*. The mutation renders otherwise perfectly healthy homozygous individuals highly resistant to infection by M-tropic HIV-1 viruses. The origins of such a mutation are still unaccounted for, and a popular theory that it originally provided protection against bubonic plague in people of Northern European ancestry has been discredited by recent experimentation *(19)*. Likewise, the importance of CXCR4 in HIV-1 infection is supported by the finding that infection can be facilitated and maintained by viral strains exclusively utilizing CXCR4 *(20)*. In addition to these heavyweights, other minor coreceptors such as CXCR6 *(21)*, CCR3 *(22,23)*, CCR8 *(24)*, and CX3CR1 *(25)* have been shown to facilitate viral entry in vitro, although a role for them in HIV-1 pathogenesis has yet to be clearly shown.

2.2. Plasmodium vivax

The use of host chemokine receptors as an entry factor by HIV-1 is closely paralleled in the protozoan world, as the causative agent of human malaria, *Plasmodium vivax*, uses the Duffy antigen receptor for chemokines (DARC) to enter erythrocytes *(26,27)*. Duffy was originally defined serologicalally in the 1950s as a minor red blood cell antigen *(28)*, and more than 25 yr elapsed before it was it was identified as a host factor required for vivax malaria by correlation of the absence of the Duffy antigen in most black

Africans with natural resistance *(29)*. The molecular identification of DARC was accelerated by parallel studies from two different groups. The first study identified an erythrocyte receptor for the chemokine CXCL8 *(30)*, and the second study showed that the receptor was absent from Duffy-negative individuals and unable to bind chemokine in the presence of anti-Duffy monoclonal antibodies, suggesting that the Duffy antigen and the erythrocyte chemokine receptor were the same molecule *(31)*. Subsequent cloning of the specific complementary deoxyribonucleic acid (cDNA) suggested that it encoded a 7TM protein with homology to chemokine receptors *(32)*. In addition to its expression on erythrocytes, it has also been identified on the postcapillary venule endothelial cells of several organs *(33)* and on subsets of neurons in the central nervous system *(34)*.

Moreover, there also exists a mutation in the promoter of the DARC gene that confers high-level resistance to infection by *P. vivax*, mirroring the CCR3Δ32 story. The DARC-negative phenotype is the result of an inherited mutation in a GATA site of the DARC promoter, which is thought to control transcription specifically in the erythroid lineage *(35)*. This mutation is thought to be fixed at high levels in Africans, presumably because of positive selective pressure exerted by *P. vivax* malaria; indeed, more than 95% of Africans in endemic regions and 70% of African Americans lack erythroid expression of DARC.

The physiological role of DARC has been more difficult to unravel because, upon chemokine binding, no signal appears to be transduced *(36)*, which is thought to be caused by the lack of the DRY motif in the putative third transmembrane helix, which is present in the majority of signaling chemokine receptors and thought to play a critical role in maintaining receptor conformation *(37)*. Studies of mice in which the DARC gene has been deleted suggest that the receptor functions as a biological "sink" for chemokines, with both an anti-inflammatory role and antiangiogenic role *(38)*. To date, no confirmed association of DARC loss with susceptibility to disease has been found, although it has been postulated to predispose African-American men to a greater incidence of prostate cancer on the basis of their reduced ability to clear angiogenic chemokines *(39)*. Another recent report suggests that erythrocytes infected by *P. falciparum* can adhere to the membrane-bound form of the chemokine CX3CL1, which is expressed on the surface of vascular endothelial cells *(40)*. The identity of the parasitic molecule responsible for facilitating chemokine binding remains unknown, but interference with this interaction also may be of therapeutic benefit in the treatment of malaria.

3. USE OF VIRAL CHEMOKINE RECEPTORS FOR SIGNALING PROCESSES

3.1. Human Cytomegalovirus

Human cytomegalovirus (HCMV) is a species-restricted β-herpesvirus which in vivo infects epithelial, lymphoid, and myeloid cells. Infection of healthy individuals is associated with asymptomatic or, at most, a mild mononucleosis syndrome, whereas in immunocompromized hosts (such as patients with AIDS), severe retinal, pulmonary, and gastrointestinal complications are known to occur. Within the genome of HCMV lies an ORF named US28, which shares 30% identity at the amino acid level with the human chemokine receptor, CCR1. The virus presumably acquired this gene by molecular piracy in a manner akin to retroviral oncogenes *(41)*.

Unlike DARC, US28 has been shown in vitro to transduce a signal both constitutively *(42)* and in response to host chemokines *(43–45)*. Both clinical isolates and laboratory strains of HCMV also can induce production of the chemokine CCL5 after infection, which coincides with the transcription of the US28 ORF *(46)*. Modulation of CCL5 production early during CMV infection might be perceived to have a regulatory effect on viral replication, acting in an autocrine fashion at US28, which would be expressed on the surface of the HCMV infected cell. Although this hypothesis is interesting, supportive in vivo data currently are lacking. Expression of US28 also has been reported to render cells capable of binding to the chemokine CX3CL1 in vitro *(47)*, and because this chemokine is expressed by endothelial cells, it has been postulated to play a role in the dissemination of HCMV *(48)*. US28 also has been shown to act as a scavenger of CC chemokines, including CCL5 *(49,50)* and therefore might be presumed to perturb host recruitment of leukocytes in vivo. A role for US28 in the pathogenesis of atherosclerosis also has been postulated as it can induce the chemokinesis of vascular smooth muscle cells *(51)*. Like CCR5 and CXCR4, US28 also can function as an HIV-1 co-receptor in vitro *(52)*, although its activity appears to be highly cell dependent *(53)*, and little evidence has been forthcoming to suggest it can be used by viral envelopes from primary isolates *(54)*.

It is plausible, however, that HCMV and HIV-1 may establish a symbiotic relationship, whereby HCMV facilitates HIV-1 infection via US28 and HIV-1 facilitates the replication of HCMV by establishing the immunosuppression needed for the emergence of HCMV from latency.

3.2. Human Herpes Virus-8

Human herpes virus-8 (HHV-8; also known as Kaposi's sarcoma herpesvirus or KSHV), is a γ-herpesvirus that infects B-lymphocytes and has been implicated as a co-factor in several proliferative disorders, including multiple myeloma, primary effusion and, most notably, Kaposi's sarcoma *(55–57)*. Within its genome resides homologs of multiple cellular immunoregulatory genes, notably several chemokines and a chemokine receptor *(58)*. The chemokine receptor is encoded by ORF 74 and is most closely related to human CXCR2 *(59)*. Alongside DARC, they are the only receptors described to date that are able to bind chemokines of both the CC and CXC classes *(60)*. ORF 74 is constitutively active, inducing cell proliferation in the absence of ligand when transfected into rat fibroblasts *(61)*. This constitutive activity can be blocked by the chemokine CXCL10, which acts as an inverse agonist *(62,63)*.

Collectively, these biological properties suggest that ORF 74 may act as a viral oncogene in Kaposi's sarcoma. The involvement of ORF 74 in the pathogenesis of HHV-8 infection has been ably demonstrated by the generation of transgenic mice expressing the receptor, which display lesions with all the histological hallmarks of Kaposi's sarcoma *(64,65)*. ORF 74 has also been demonstrated to induce several proinflammatory cytokines, including the chemokines CXCL8 and CCL5 via an NFκB mediated pathway *(66,67)*. Of interest, HHV-8 also encodes three chemokine homologs (*see* Subheading 4.1.2.), which raises the question as to whether or not the function of ORF 74 is regulated in vivo not by host chemokines but by HHV-8-encoded chemokines.

4. SUBVERSION OF THE CHEMOKINE NETWORK

4.1. Microbial Generation of Chemokine Receptor Agonists and Antagonists

4.1.2. HHV-8-Encoded Chemokine Receptor Agonists and Antagonists

Besides a chemokine receptor, the HHV-8 genome also contains three ORFs named K4, K6, and K4.1, which encode CC chemokines, designated vMIP-I, vMIP-II, and vMIP-III on the basis of their homology to the chemokine macrophage inflammatory protein (MIP)-1α/CCL3 *(68,69)*. vMIP-I and vMIP-II share 60% sequence identity, whereas vMIP-III is more distantly related, with approx 37% identity to vMIP-I and vMIPII. The vMIPs are angiogenic, act as chemokine receptor agonists/antagonists, and also function as HIV-1 suppressive factors *(70–72)*. The former activity is consistent with the characteristic angiogenesis of Kaposi's sarcoma and, as mentioned in the previously, raises the possibility that they act via the product of ORF 74.

vMIP-I and vMIP-III are respective agonists for the receptors CCR8 *(73,74)* and CCR4 *(72)*, both of which are specifically expressed by Th2-polarized cells. In contrast, vMIP-II has a much broader spectrum of activity and, alongside its antagonistic activity at CCR1, CCR2, CCR5, and CXCR4 *(71)*, it has been reported to act as an agonist at CCR3 and CCR8, recruiting both human eosinophils and Th2 lymphocytes via these receptors *(70,75,76)*. Immunohistochemical analysis of Kaposi's lesions has correlated vMIP-II expression with an increased numbers of Th2/CCR3+ lymphocytes compared with CXCR3+, CCR5+/Th1 leukocytes, suggesting that vMIP-II polarizes the immune response in vivo away from a Th1 cytotoxic reaction, to an antiparasitic Th2 reaction response, thereby promoting escape from the host defenses *(75)*.

4.1.3. Molluscum contagiosum-*Encoded Chemokine Receptor Antagonists*

The human cutaneous poxvirus *Molluscum contagiosum* virus (MCV) types one and two have been shown to encode CC chemokine orthologues named MC148R1 and MC148R2 respectively, which share 87% identity with each other and approx 25% identity to human CC chemokines *(77,78)*. Both chemokines lack agonist activity but have reported antagonist effects against a variety of chemokine receptors, notably CCR8 *(79,80)*. It is curious that CCR8 should be targeted in this fashion by a virus.

Indeed, apart from the two host ligands CCL1 and CCL16, which are confirmed agonists *(81–83)*, the remaining ligands are virally derived, namely vMIP-I, vMIP-II, and MC148. This interest in CCR8 does not appear to be restricted to MCV and HHV-8. Yaba-like disease virus, an as yet-unclassified member of the yatapoxvirus genus, contains two ORFs with homology to chemokine receptors *(84)*. One of these, named 7L shares 53% identity with CCR8 and can bind and signal in response to CCL1 *(85)*. The reasons for this deliberate targeting/mimicking of CCR8 and its ligands are still unresolved, although it suggests that they may be key players in viral defense.

4.1.4. HIV-1 *and* Toxoplasma gondii-*Encoded Chemokine Receptor Agonists*

The transcription factor Tat is released from HIV-1 infected cells and, in addition to its role in transactivating the transcription of HIV-1 mRNAs, it has been shown to mimic angiogenic growth factors *(86)*, bind integrins *(87)*, and to induce the chemotaxis of neutrophils *(88)*, monocytes *(87)*, and mast cells *(89)*, the latter two activities presumably mediated by CCR2 and CCR3 *(90)*. Activities at either of these receptors may serve to recruit further target cells for infection. In contrast, the reported activity of Tat at CXCR4 is antagonistic, blocking the entry of T-tropic but not M-tropic

HIV-1 strains, which may promote the usage of CCR5 during the early stages of infection *(91,92)*. Although possessing limited homology to chemokines, Tat does possess both a CXC and CC motif, and a synthetic peptide incorporating these motifs has been shown to activate monocytes and macrophages in vitro *(90)*.

A cyclophilin produced by the protozoan *Toxoplasma gondii* (C-18) recently has been reported to act as a CCR5 agonist, stimulating dendritic cells to produce interleukin-12 *(93)*. Like Tat, its potency is all the more remarkable considering it has no obvious sequence homology with known CC chemokine ligands of CCR5. It is unclear what benefit the acquisition of such activity is to the parasite as, logically, the cyclophillin should stimulate cell-mediated immunity and therefore hinder, rather than encourage, parasite infection. The authors postulate that because the transmission of *T. gondii* is dependent upon the feline predation of infected small mammals and birds, induction of a host cell-mediated response against itself may prevent intermediate host mortality, thereby maintaining the life cycle of the parasite. The structural requirements for CCR5 activation by the cyclophilin appear to be analogous to those acquired by HIV-1 for engagement with the same receptor, because HIV-1 infection of CCR5 expressing target cells is sensitive to recombinant C-18 *(94)*.

4.2. Microbial Chemokine Scavengers and Degraders

4.2.1. Production of Chemokine Binding Proteins by Poxviruses and Herpesviruses

In addition to containing soluble cytokine receptors within their genome with which to fox the host defenses, several orthopoxviruses maintain genes encoding for soluble proteins, which they can excrete in large amounts and which can bind chemokine in solution. Such proteins have been termed viral chemokine binding proteins (vCKBPs *[95]*) and, to date, three different types have been described. The myxoma virus vCKBP-I is a soluble IFN-γ receptor that also binds CC, CXC, and C chemokines through a glycosaminoglycan (GAG) binding site and may function by perturbing the haptotactic chemokine gradient in vivo *(96)*. Similarly, the poxvirus vCKBP-II protein binds CC chemokines with high affinity but functions by directly preventing the binding of chemokines to receptors *(97)*. More recently, a third secreted protein, termed vCKBP-II,I has been shown to be produced not by a poxvirus, but by a herpesvirus, namely murine γ-herpesvirus 68 (MHV-68 *[98,99]*). This protein has a broader range of specificity, sequestering chemokines of all known classes and combines the functions of both vCKBP-I and vCKBP-II by impeding binding of the chemokine to both chemokine receptors *(100)* and

GAGs *(101)*. The solution structure of a vCKBP from cowpox virus has been determined *(102)* and is unrelated to the known structures of any other mammalian proteins suggesting that unlike viral chemokine homologues, it was not acquired by the piracy of mammalian genes. A conserved region of acidic residues within the protein is thought to play a role in binding the typically basic chemokines.

4.2.2. Degradation of the Eosinophil Chemoattractant CCL11 by Necator americanus

The chemokine eotaxin-1/CCL11 was originally identified as a potent eosinophil chemoattractant present in bronchoalveolar lavage fluid from allergen-challenged sensitized guinea pigs *(103)*. The cDNA was subsequently identified and using this sequence, CCL11 orthologs have been identified in several species, including humans *(104)*. CCL11 shows great fidelity for the chemokine receptor CCR3, which is expressed on the surface of eosinophils *(105)*, basophils *(106)*, mast cells *(107)*, and a subpopulation of Th2 lymphocytes *(108)*, all of which are important in host defense against helminth infection. Moreover, CCL11 is constitutively expressed at high levels in the small intestine and colon *(109)*, and the deletion of CCR3 in the mouse results in reduced basal trafficking of eosinophils to the intestinal mucosa *(110)*. It is therefore not entirely surprising that the hookworm *Necator americanus* has chosen to exploit the CCR3:CCL11 axis. This parasite is a leading cause of malnutrition in developing countries *(111)*, and within its armory is an as yet-unidentified metalloproteinase which specifically cleaves CCL11 but not related chemokines *(112)*. This presumably serves to perturb eosinophil recruitment in the host, thereby blunting the immune response at the site of helminth infection.

5. EXPLOITING THE EXPLOITERS

Having seen that microbes are keen students of the chemokine network and have developed strategies to subvert and exploit this facet of our host defenses, it seems only proper that we should attempt to turn the tables, both to thwart the microbes themselves and also to exploit the fruits of their labors, turning the microbial chemokine mimetics into useful anti-inflammatory reagents. In terms of blocking the microbial usage of host chemokine receptors, much has already been achieved in vitro. As members of the GPCR superfamily, which currently represent more than 60% of the targets for marketed prescription drugs *(113)*, chemokine receptors are highly "drugable," with several small molecule antagonists of CCR5 and CXCR4 having graduated from the bench to clinic trials *(114)*. A combination of such compounds either alone or in conjunction with currently prescribed inhibitors (such as reverse transcriptase and protease inhibitors) may well prove useful additions to the HIV-1 physician's armory. Likewise, a mutant of the chemokine CXCL1, which fails to activate neutrophils but can readily bind to DARC, has been shown to inhibit the invasion of erythrocytes by *P. knowlesi* in vitro *(115)*. A small molecule with the same biological effects could conceivably be a useful antimalarial drug.

The virally encoded chemokine scavengers and mimetics also hold promise as prototypic anti-inflammatory drugs. The vCKBP of vaccinia virus has been shown to block CCL11-induced infiltration of eosinophils in a guinea pig skin model *(97)*, whereas the administration of the vCKBP of cowpox virus to the respiratory tract had localized effects on allergen induced airways hyperreactivity but no undesired effects on systemic immune responses *(116)*. vMIP-I and vMIP-II have been shown to block HIV-1 infection of peripheral blood mononuclear cells and brain cells in vitro *(70,117)* and vMIP-II to block leukocyte recruitment in rat models of experimental glomerulonephritis *(118)*, spinal cord contusion injury *(119)*, and a murine model of viral infection *(120)*. Additionally, the introduction of plasmids encoding vMIP-II and MC148 into murine cardiac allografts has been shown to result in a reduction in both alloantibody production and the numbers of donor-specific cytotoxic T-lymphocytes, resulting in a prolonging of graft survival *(121)*. vMIP-I and MC148 also have recently found use as tools to both target and attract antigen-presenting cells. Fusion of either viral ORF with that of a single-chain lymphoma-specific Ig resulted in a construct whose protein product could be effectively targeted to antigen-presenting cells in vitro and which alone could be used as DNA vaccines to elicit anti-tumor immunity in vivo *(122)*.

In conclusion, we have much to learn from the mechanisms by which microbes have targeted mammalian chemokine systems during millions of years of coevolution. It is anticipated that during the coming decades "exploitation of the exploiters" may lead to novel therapeutics, useful for the treatment of several clinically important diseases.

REFERENCES

1. Rot A, Von Andrian UH. Chemokines in innate and adaptive host defense: basic chemokinese grammar for immune cells. Annu Rev Immunol 2004;22: 891–928.
2. Bazan JF, Bacon KB, Hardiman G, et al. A new class of membrane-bound chemokine with a CX3C motif. Nature 1997;385:640–644.
3. Kelner GS, Kennedy J, Bacon KB, et al. Lymphotactin: a cytokine that represents a new class of chemokine. Science 1994;266:1395–1399.

4. Yoshida T, Imai T, Takagi S, et al. Structure and expression of two highly related genes encoding SCM-1/human lymphotactin. FEBS Lett 1996;395:82–88.

5. Murphy PM, Baggiolini M, Charo IF, et al. International union of pharmacology. XXII. Nomenclature for chemokine receptors. Pharmacol Rev 2000;52: 145–176.

6. Murphy PM. International Union of Pharmacology. XXX. Update on chemokine receptor nomenclature. Pharmacol Rev 2002;54:227–229.

7. Deuel TF, Keim PS, Farmer M, Heinrikson RL. Amino acid sequence of human platelet factor 4. Proc Natl Acad Sci USA 1977;74:2256–2258.

8. Murphy PM. Molecular mimicry and the generation of host defense protein diversity. Cell 1993;72:823–826.

9. Feng Y, Broder CC, Kennedy PE, Berger EA. HIV-1 entry cofactor: functional cDNA cloning of seven-transmembrane, G protein-coupled receptor. Science 1996;272:872–877.

10. Bleul CC, Farzan M, Choe H, et al. The lymphocyte chemoattractant SDF-1 is a ligand for LESTR/fusin and blocks HIV-1 entry. Nature 1996;382:829–833.

11. Oberlin E, Amara A, Bachelerie F, et al. The CXC chemokine SDF-1 is the ligand for LESTR/fusin and prevents infection by T-cell-line-adapted HIV-1 [published erratum appears in Nature 1996 Nov 21;384(6606):288]. Nature 1996;382:833–835.

12. Alkhatib G, Combadiere C, Broder CC, et al. CC CKR5: a RANTES, MIP-1alpha, MIP-1beta receptor as a fusion cofactor for macrophage-tropic HIV-1. Science 1996;272:1955–1958

13. Deng H, Liu R, Ellmeier W, et al. Identification of a major co-receptor for primary isolates of HIV-1. Nature 1996;381:661–666.

14. Dragic T, Litwin V, Allaway GP, et al. HIV-1 entry into CD4+ cells is mediated by the chemokine receptor CC-CKR-5. Nature 1996;381:667–673.

15. Eckert DM, Kim PS. Mechanisms of viral membrane fusion and its inhibition. Annu Rev Biochem 2001;70:777–810.

16. Cocchi F, DeVico AL, Garzino-Demo A, Arya SK, Gallo RC, Lusso P. Identification of RANTES, MIP-1α, and MIP-1β as the major HIV-suppresive factors produced by CD8+ T cells. Science 1996;270:1811–1815.

17. Samson M, Libert F, Doranz BJ, et al. Resistance to HIV-1 infection in caucasian individuals bearing mutant alleles of the CCR-5 chemokine receptor gene. Nature 1996;382:722–725.

18. Dean M, Carrington M, Winkler C, et al. Genetic restriction of HIV-1 infection and progression to AIDS by a deletion allele of the CKR5 structural gene. Hemophilia Growth and Development Study, Multicenter AIDS Cohort Study, Multicenter Hemophilia Cohort Study, San Francisco City Cohort, ALIVE Study. Science 1996;273:1856–1862.

19. Mecsas J, Franklin G, Kuziel WA, Brubaker RR, Falkow S, Mosier DE. Evolutionary genetics: CCR5 mutation and plague protection. Nature 2004;427:606.

20. Michael NL, Nelson JA, KewalRamani VN, et al. Exclusive and persistent use of the entry coreceptor CXCR4 by human immunodeficiency virus type 1 from a subject homozygous for CCR5 delta32. J Virol 1998;72:6040–6047.
21. Zhang YJ, Zhang L, Ketas T, Korber BT, Moore JP. HIV type 1 molecular clones able to use the Bonzo/STRL-33 coreccptor for virus entry. AIDS Res Hum Retroviruses 2001;17:217–227.
22. Choe H, Farzan M, Sun Y, et al. The β-chemokine receptors CCR3 and CCR5 facilitate infection by primary HIV-1 isolates. Cell 1996;85:1135–1148.
23. Alkhatib G, Berger EA, Murphy PM, Pease JE. Determinants of HIV-1 coreceptor function on CC chemokine receptor 3. Importance of both extracellular and transmembrane/cytoplasmic regions. J Biol Chem 1997;272:20,420–20,426.
24. Lee S, Tiffany HL, King L, Murphy PM, Golding H, Zaitseva MB. CCR8 on human thymocytes functions as a human immunodeficiency virus type 1 coreceptor. J Virol 2000;74:6946–6952.
25. Combadiere C, Salzwedel K, Smith ED, Tiffany HL, Berger EA, Murphy PM. Identification of CX3CR1. A chemotactic receptor for the human CX3C chemokine fractalkine and a fusion coreceptor for HIV-1. J Biol Chem 1998;273:23,799–23,804.
26. Hadley TJ, Peiper SC. From malaria to chemokine receptor: the emerging physiological role of the duffy blood group antigen. Blood 1997;89:3077–3091.
27. Pogo AO, Chaudhuri A. The Duffy protein: a malarial and chemokine receptor. Semin Hematol 2000;37:122–129.
28. Young MD, Eyles DE, Burgess RW, Jeffrey GM. Experimental testing of the immunity of negroes to Plasmodium vivax. J Parasitol 1955;41:315–322.
29. Miller LH, Mason SJ, Dvorak JA, McGinniss MH, Rothman IK. Erythrocyte receptors for (Plasmodium knowlesi) malaria: Duffy blood group determinants. Science 1975;189:561–563.
30. Darbonne WC, Rice GC, Mohler MA, et al. Red blood cells are a sink for interleukin 8, a leukocyte chemotaxin. J Clin Invest 1991;88:1362–1369.
31. Horuk R, Chitnis CE, Darbonne WC, et al. A receptor for the malarial parasite Plasmodium vivax: the erythrocyte chemokine receptor. Science 1993;261:1182–1184.
32. Chaudhuri A, Polyakova J, Zbrzezna V, Williams K, Gulati S, Pogo AO. Cloning of glycoprotein D cDNA, which encodes the major subunit of the Duffy blood group system and the receptor for the Plasmodium vivax malaria parasite. Proc Natl Acad Sci USA 1993;90:10,793–10,797.
33. Peiper SC, Wang ZX, Neote K, et al. The Duffy antigen/receptor for chemokines (DARC) is expressed in endothelial cells of Duffy negative individuals who lack the erythrocyte receptor. J Exp Med 1995;181:1311–1317.
34. Horuk R, Martin AW, Wang Z, et al. Expression of chemokine receptors by subsets of neurons in the central nervous system. J Immunol 1997;158:2882–2890.

35. Tournamille C, Colin Y, Cartron JP, Le Van Kim C. Disruption of a GATA motif in the Duffy gene promoter abolishes erythroid gene expression in Duffynegative individuals. Nat Genet 1995;10:224–228.
36. Neote K, Mak JY, Kolakowski LF Jr., Schall TJ. Functional and biochemical analysis of the cloned Duffy antigen: identity with the red blood cell chemokine receptor. Blood 1994;84:44–52.
37. Auger GA, Pease JE, Shen X, Xanthou G, Barker MD. Alanine scanning mutagenesis of CCR3 reveals that the three intracellular loops are essential for functional receptor expression. Eur J Immunol 2002;32:1052–1058.
38. Dawson TC, Lentsch AB, Wang Z, et al. Exaggerated response to endotoxin in mice lacking the Duffy antigen/receptor for chemokines (DARC). Blood 2000;96:1681–1684.
39. Lentsch AB. The Duffy antigen/receptor for chemokines (DARC) and prostate cancer. A role as clear as black and white? FASEB J 2002;16:1093–1095.
40. Hatabu T, Kawazu S, Aikawa M, Kano S. Binding of plasmodium falciparuminfected erythrocytes to the membrane-bound form of Fractalkine/CX3CL1. Proc Natl Acad Sci USA 2003;100:15,942–15,946.
41. Varmus HE. The molecular genetics of cellular oncogenes. Annu Rev Genet 1984;18:553–612.
42. Casarosa P, Bakker RA, Verzijl D, et al. Constitutive signaling of the human cytomegalovirus-encoded chemokine receptor US28. J Biol Chem 2001;276: 1133–1137.
43. Gao JL, Murphy PM. Human cytomegalovirus open reading frame US28 encodes a functional beta chemokine receptor. J Biol Chem 1994;269:28,539–28,542.
44. Neote K, DiGregorio D, Mak JY, Horuk R, Schall TJ. Molecular cloning, functional expression, and signalling characteristics of a C-C chemokine receptor. Cell 1993;72:415–425.
45. Kuhn DE, Beall CJ, Kolattukudy PE. The cytomegalovirus US28 protein binds multiple CC chemokines with high affinity. Biochem Biophys Res Commun 1995;211:325–330.
46. Michelson S, Dal Monte P, Zipeto D, et al. Modulation of RANTES production by human cytomegalovirus infection of fibroblasts. J Virol 1997;71:6495–6500.
47. Haskell CA, Cleary MD, Charo IF. Unique role of the chemokine domain of fractalkine in cell capture. Kinetics of receptor dissociation correlate with cell adhesion. J Biol Chem 2000;275:34,183–34,189.
48. Beisser PS, Laurent L, Virelizier JL, Michelson S. Human cytomegalovirus chemokine receptor gene US28 is transcribed in latently infected THP-1 monocytes. J Virol 2001;75:5949–5957.
49. Bodaghi B, Jones TR, Zipeto D, et al. Chemokine sequestration by viral chemoreceptors as a novel viral escape strategy: withdrawal of chemokines from the environment of cytomegalovirus-infected cells. J Exp Med 1998;18: 855–866.
50. Randolph-Habecker JR, Rahill B, et al. The expression of the cytomegalovirus chemokine receptor homolog US28 sequesters biologicalally active CC chemokines and alters IL-8 production. Cytokine 2002;19:37–46.

51. Streblow DN, Soderberg-Naucler C, Vieira J, et al. The human cytomegalovirus chemokine receptor US28 mediates vascular smooth muscle cell migration. Cell 1999;99:511–520.

52. Pleskoff O, Treboute C, Brelot A, Heveker N, Seman M, Alizon M. Identification of a chemokine receptor encoded by human cytomegalovirus as a cofactor for HIV-1 entry. Science 1997;276:1874–1878.

53. Ohagen A, Li L, Rosenzweig A, Gabuzda D. Cell-dependent mechanisms restrict the HIV type 1 coreceptor activity of US28, a chemokine receptor homolog encoded by human cytomegalovirus. AIDS Res Hum Retroviruses 2000;16:27–35.

54. Singh A, Besson G, Mobasher A, Collman RG. Patterns of chemokine receptor fusion cofactor utilization by human immunodeficiency virus type 1 variants from the lungs and blood. J Virol 1999;73:6680–6690.

55. Chang Y, Cesarman E, Pessin MS, et al. Identification of herpesvirus-like DNA sequences in AIDS-associated Kaposi's sarcoma. Science 1994;266:1865–1869.

56. Cesarman E, Chang Y, Moore PS, Said JW, Knowles DM. Kaposi's sarcoma associated herpesvirus-like DNA sequences in AIDS-related body-cavity-based lymphomas. N Engl J Med 1995;332:1186–1191.

57. Cesarman E, Knowles DM. Kaposi's sarcoma-associated herpesvirus: a lymphotropic human herpesvirus associated with Kaposi's sarcoma, primary effusion lymphoma, and multicentric Castleman's disease. Semin Diagn Pathol 1997;14:54–66.

58. Cesarman E, Nador RG, Bai F, et al. Kaposi's sarcoma-associated herpesvirus contains G protein-coupled receptor and cyclin D homologs which are expressed in Kaposi's sarcoma and malignant lymphoma. J Virol 1996;70:8218–8223.

59. Russo JJ, Bohenzky RA, Chien MC, et al. Nucleotide sequence of the Kaposi sarcoma-associated herpesvirus (HHV8). Proc Natl Acad Sci USA 1996;93: 14,862–14,867.

60. Horuk R, Colby TJ, Darbonne WC, Schall TJ, Neote K. The human erythrocyte inflammatory peptide (chemokine) receptor. Biochemical characterization, solubilization, and development of a binding assay for the soluble receptor. Biochemistry 1993;32:5733–5738.

61. Arvanitakis L, Geras-Raaka E, Varma A, Gershengorn MC, Cesarman E. Human herpesvirus KSHV encodes a constitutively active G-protein-coupled receptor linked to cell proliferation. Nature 1997;385:347–350.

62. Geras-Raaka E, Varma A, Ho H, Clark-Lewis I, Gershengorn MC. Human interferon-gamma-inducible protein 10 (IP-10) inhibits constitutive signaling of Kaposi's sarcoma-associated herpesvirus G protein-coupled receptor. J Exp Med 1998;188:405–408.

63. Rosenkilde MM, Kledal TN, Brauner-Osborne H, Schwartz TW. Agonists and inverse agonists for the herpesvirus 8-encoded constitutively active seventransmembrane oncogene product, ORF-74. J Biol Chem 1999;274:956–961.

64. Yang TY, Chen SC, Leach MW, et al. Transgenic expression of the chemokine receptor encoded by human herpesvirus 8 induces an angioproliferative disease resembling Kaposi's sarcoma. J Exp Med 2000;191:445–454.

65. Guo HG, Sadowska M, Reid W, Tschachler E, Hayward G, Reitz M. Kaposi's sarcoma-like tumors in a human herpesvirus 8 ORF74 transgenic mouse. J Virol 2003;77:2631–2639.

66. Pati S, Cavrois M, Guo HG, et al M. Activation of NF-kappaB by the human herpesvirus 8 chemokine receptor ORF74: evidence for a paracrine model of Kaposi's sarcoma pathogenesis. J Virol 2001;75:8660–8673.

67. Schwarz M, Murphy PM. Kaposi's sarcoma-associated herpesvirus G protein coupled receptor constitutively activates NF-kappa B and induces proinflammatory cytokine and chemokine production via a C-terminal signaling determinant. J Immunol 2001;167:505–513.

68. Moore PS, Boshoff C, Weiss RA, Chang Y. Molecular mimicry of human cytokine and cytokine response pathway genes by KSHV. Science 1996;274: 1739–1744.

69. Nicholas J, Ruvolo VR, Burns WH, et al. Kaposi's sarcoma-associated human herpesvirus-8 encodes homologues of macrophage inflammatory protein-1 and interleukin-6. Nat Med 1997;3:287–292.

70. Boshoff C, Endo Y, Collins PD, et al. Angiogenic and HIV-inhibitory functions of KSHV-encoded chemokines. Science 1997;278:290–294.

71. Kledal TN, Rosenkilde MM, Coulin F, et al. A broad-spectrum chemokine antagonist encoded by Kaposi's sarcoma- associated herpesvirus. Science 1997;277:1656–1659.

72. Stine JT, Wood C, Hill M, et al. KSHV-encoded CC chemokine vMIP-III is a CCR4 agonist, stimulates angiogenesis, and selectively chemoattracts TH2 cells. Blood 2000;95:1151–1157.

73. Dairaghi DJ, Fan RA, McMaster BE, Hanley MR, Schall TJ. HHV8-encoded vMIP-I selectively engages chemokine receptor CCR8. Agonist and antagonist profiles of viral chemokines. J Biol Chem 1999;274:21,569–21,574.

74. Endres MJ, Garlisi CG, Xiao H, Shan L, Hedrick JA. The Kaposi's sarcoma related herpesvirus (KSHV)-encoded chemokine vMIP-I is a specific agonist for the CC chemokine receptor (CCR)8. J Exp Med 1999;189:1993–1998.

75. Weber KS, Grone HJ, Rocken M, et al. Selective recruitment of Th2-type cells and evasion from a cytotoxic immune response mediated by viral macrophage inhibitory protein-II. Eur J Immunol 2001;31:2458–2466.

76. Sozzani S, Luini W, Bianchi G, et al. The viral chemokine macrophage inflammatory protein-II is a selective Th2 chemoattractant. Blood 1998;92:4036–4039.

77. Senkevich TG, Bugert JJ, Sisler JR, Koonin EV, Darai G, Moss B. Genome sequence of a human tumorigenic poxvirus: prediction of specific host response evasion genes. Science 1996;273:813–816.

78. Krathwohl MD, Hromas R, Brown DR, Broxmeyer HE, Fife KH. Functional characterization of the C–C chemokine-like molecules encoded by molluscum contagiosum virus types 1 and 2. Proc Natl Acad Sci USA 1997;94:9875–9880.

79. Damon I, Murphy PM, Moss B. Broad spectrum chemokine antagonistic activity of a human poxvirus chemokine homolog. Proc Natl Acad Sci USA 1998; 95:6403–6407.

80. Luttichau HR, Stine J, Boesen TP, et al. A highly selective CC chemokine receptor (CCR)8 antagonist encoded by the poxvirus molluscum contagiosum. J Exp Med 2000;191:171–180.
81. Tiffany HL, Lautens LL, Gao JL, et al. Identification of CCR8: a human monocyte and thymus receptor for the CC chemokine I-309. J Exp Med 1997;186: 165–170.
82. Garlisi CG, Xiao H, Tian F, et al. The assignment of chemokine-chemokine receptor pairs: TARC and MIP-1 beta are not ligands for human CC-chemokine receptor 8. Eur J Immunol 1999;29:3210–3215.
83. Howard OM, Dong HF, Shirakawa AK, Oppenheim JJ. LEC induces chemotaxis and adhesion by interacting with CCR1 and CCR8. Blood 2000;96:840–845.
84. Lee HJ, Essani K, Smith GL. The genome sequence of Yaba-like disease virus, a yatapoxvirus. Virology 2001;281:170–192.
85. Najarro P, Lee HJ, Fox J, Pease J, Smith GL. Yaba-like disease virus protein 7L is a cell-surface receptor for chemokine CCL1. J Gen Virol 2003;84:3325–3336.
86. Albini A, Benelli R, Presta M, et al. HIV-tat protein is a heparin-binding angiogenic growth factor. Oncogene 1996;12:289–297.
87. Brake DA, Debouck C, Biesecker G. Identification of an Arg-Gly-Asp (RGD) cell adhesion site in human immunodeficiency virus type 1 transactivation protein, tat. J Cell Biol 1990;111:1275–1281.
88. Benelli R, Barbero A, Ferrini S, et al. Human immunodeficiency virus transactivator protein (Tat) stimulates chemotaxis, calcium mobilization, and activation of human polymorphonuclear leukocytes: implications for Tat-mediated pathogenesis. J Infect Dis 2000;182:1643–1651.
89. de Paulis A, De Palma R, Di Gioia L, et al. Tat protein is an HIV-1-encoded beta-chemokine homolog that promotes migration and up-regulates CCR3 expression on human Fc epsilon RI+ cells. J Immunol 2000;165:7171–7179.
90. Albini A, Ferrini S, Benelli R, et al. HIV-1 Tat protein mimicry of chemokines. Proc Natl Acad Sci USA 1998;95:13,153–13,158.
91. Xiao H, Neuveut C, Tiffany HL, et al. Selective CXCR4 antagonism by Tat: implications for in vivo expansion of coreceptor use by HIV-1. Proc Natl Acad Sci USA 2000;97:11,466–11,471.
92. Ghezzi S, Noonan DM, Aluigi MG, et al. Inhibition of CXCR4-dependent HIV-1 infection by extracellular HIV-1 Tat. Biochem Biophys Res Commun 2000;270:992–996.
93. Aliberti J, Valenzuela JG, Carruthers VB, et al. Molecular mimicry of a CCR5 binding-domain in the microbial activation of dendritic cells. Nat Immunol 2003;4:485–490.
94. Golding H, Aliberti J, King LR, et al. Inhibition of HIV-1 infection by a CCR5-binding cyclophilin from *Toxoplasma gondii*. Blood 2003;102:3280–3286.
95. Alcami A, Koszinowski UH. Viral mechanisms of immune evasion. Trends Microbiol 2000;8(9):410–418.
96. Lalani AS, Graham K, Mossman K, et al. The purified myxoma virus gamma interferon receptor homolog M-T7 interacts with the heparin-binding domains of chemokines. J Virol 1997;71:4356–4363.

97. Alcami A, Symons JA, Collins PD, Williams TJ, Smith GL. Blockade of chemokine activity by a soluble chemokine binding protein from vaccinia virus. J Immunol 1998;160:624–633.

98. Parry CM, Simas JP, Smith VP, et al. A broad spectrum secreted chemokine binding protein encoded by a herpesvirus. J Exp Med 2000;191:573–578.

99. van Berkel V, Barrett J, Tiffany HL, et al. Identification of a gamma herpesvirus selective chemokine binding protein that inhibits chemokine action. J Virol 2000;74:6741–6747.

100. Webb LM, Clark-Lewis I, Alcami A. The gammaherpesvirus chemokine binding protein binds to the N terminus of CXCL8. J Virol 2003;77:8588–8592.

101. Webb LM, Smith VP, Alcami A. The gammaherpesvirus chemokine binding protein can inhibit the interaction of chemokines with glycosaminoglycans. FASEB J 2004;18:571–573.

102. Carfi A, Smith CA, Smolak PJ, McGrew J, Wiley DC. Structure of a soluble secreted chemokine inhibitor vCCI (p35) from cowpox virus. Proc Natl Acad Sci USA 1999;96:12,379–12,383.

103. Jose PJ, Griffiths-Johnson DA, Collins PD, et al. Eotaxin: A potent eosinophil chemoattractant cytokine detected in a guinea-pig model of allergic airways inflammation. J Exp Med 1994;179:881–887.

104. Ponath PD, Qin S, Ringler DJ, et al. Cloning of the human eosinophil chemoattractant, eotaxin. Expression, receptor binding and functional properties suggest a mechanism for the selective recruitment of eosinophils. J Clin Invest 1996;97:604–612.

105. Ponath PD, Qin S, Post TW, et al. Molecular cloning and characterization of a human eotaxin receptor expressed selectively on eosinophils. J Exp Med 1996;183:2437–2448.

106. Uguccioni M, Mackay CR, Ochensberger B, et al. High expression of the chemokine receptor CCR3 in human blood basophils. Role in activation by eotaxin, MCP-4, and other chemokines. J Clin Invest 1997;100:1137–1143.

107. Ochi H, Hirani WM, Yuan Q, Friend DS, Austen KF, Boyce JA. T helper cell type 2 cytokine-mediated comitogenic responses and CCR3 expression during differentiation of human mast cells in vitro. J Exp Med 1999;190:267–280.

108. Sallusto F, Mackay CR, Lanzavecchia A. Selective expression of the eotaxin receptor CCR3 by human T helper 2 cells. Science 1997;277:2005–2007.

109. Kitaura M, Nakajima T, Imai T, et al. Molecular cloning of human eotaxin, an eosinophil-selective CC chemokine, and identification of a specific eosinophil eotaxin receptor, CC chemokine receptor 3. J Biol Chem 1996;271:7725–7730.

110. Humbles AA, Lu B, Friend DS, et al. The murine CCR3 receptor regulates both the role of eosinophils and mast cells in allergen-induced airway inflammation and hyperresponsiveness. Proc Natl Acad Sci USA 2002;99:1479–1484.

111. Chan MS, Medley GF, Jamison D, Bundy DA. The evaluation of potential global morbidity attributable to intestinal nematode infections. Parasitology 1994;109:373–387.

112. Culley FJ, Brown A, Conroy DM, Sabroe I, Pritchard DI, Williams TJ. Eotaxin is specifically cleaved by hookworm metalloproteases preventing its action in vitro and in vivo. J Immunol 2000;165:6447–6453.

113. Gurrath M. Peptide-binding G protein-coupled receptors: new opportunities for drug design. Curr Med Chem 2001;8:1605–1648.

114. Schols D. HIV co-receptors as targets for antiviral therapy. Curr Top Med Chem 2004;4:883–893.

115. Hesselgesser J, Chitnis CE, Miller LH, et al. A mutant of melanoma growth stimulating activity does not activate neutrophils but blocks erythrocyte invasion by malaria. J Biol Chem 1995;270:11,472–11,476.

116. Dabbagh K, Xiao Y, Smith C, et al. Local blockade of allergic airway hyperreactivity and inflammation by the poxvirus-derived pan-CC-chemokine inhibitor vCCI. J Immunol 2000;165:3418–3422.

117. Hibbitts S, Reeves JD, Simmons G, et al. Coreceptor ligand inhibition of fetal brain cell infection by HIV type 1. AIDS Res Hum Retroviruses 1999;15:989–1000.

118. Chen S, Bacon KB, Li L, et al. In vivo inhibition of CC and CX3C chemokine induced leukocyte infiltration and attenuation of glomerulonephritis in Wistar-Kyoto (WKY) rats by vMIP-II. J Exp Med 1998;188:193–198.

119. Ghirnikar RS, Lee YL, Eng LF. Chemokine antagonist infusion attenuates cellular infiltration following spinal cord contusion injury in rat. J Neurosci Res 2000;59:63–73.

120. Lindow M, Nansen A, Bartholdy C, et al. The virus-encoded chemokine vMIPII inhibits virus-induced Tc1-driven inflammation. J Virol 2003;77:7393–7400.

121. DeBruyne LA, Li K, Bishop DK, Bromberg JS. Gene transfer of virally encoded chemokine antagonists vMIP-II and MC148 prolongs cardiac allograft survival and inhibits donor-specific immunity. Gene Ther 2000;7:575–582.

122. Ruffini PA, Biragyn A, Coscia M, et al. Genetic fusions with viral chemokines target delivery of nonimmunogenic antigen to trigger antitumor immunity independent of chemotaxis. J Leukoc Biol 2004;76:77–85.

5

The Use of Cytokines in the Identification and Characterization of Chemical Allergens

Rebecca J. Dearman and Ian Kimber

SUMMARY

Many chemicals cause skin sensitization, resulting in allergic contact dermatitis, a delayed type hypersensitivity reaction. Chemical respiratory allergy is also an important occupational health problem, but there are currently available no validated methods for hazard identification, partly because of the fact that the relevant cellular and molecular mechanisms of sensitization of the respiratory tract have been unclear, with particular controversy regarding an obligatory role for IgE. Increasing evidence now exists that respiratory sensitization is associated with the preferential activation of type 2 T-lymphocytes and the expression of type 2 cytokines interleukin (IL)-4, IL-5, IL-10, and IL-13. Type 2 cell products favor immediate type hypersensitivity reactions, serving as growth and differentiation factors for mast cells and eosinophils, the cellular effectors of the clinical manifestations of the allergic responses, and promoting IgE antibody production. In contrast, chemical contact allergens induce type 1 cytokine expression profiles. There has been considerable interest in the application of cytokine profiling for the characterization of chemical allergens, with cytokine phenotypes analyzed in freshly isolated tissue, or after culture in the presence or absence of mitogen at the level of protein secretion or messenger ribonucleic acid expression. The most common configuration of cytokine profiling, measurement of induced cytokine secretion patterns in the absence of restimulation, shows considerable promise as an approach for the identification and characterization of chemical respiratory allergens.

Key Words: Chemical respiratory allergy; skin sensitization; hazard identification; cytokine fingerprinting; cytokines; interleukins.

From: *Methods in Pharmacology and Toxicology: Cytokines in Human Health:
Immunotoxicology, Pathology, and Therapeutic Applications*
Edited by: R. V. House and J. Descotes © Humana Press Inc., Totowa, NJ

1. INTRODUCTION

Many chemicals encountered in the workplace are known to cause allergic contact dermatitis, a form of delayed type hypersensitivity (DTH *[1]*). Allergic hypersensitivity of the respiratory tract and occupational asthma is also an important occupational health problem, although fewer chemicals have been implicated as causative agents *(2)*. A number of methods for the prospective identification of chemicals with the potential to cause contact allergy are available, including assays that rely on measurement of challenge-induced elicitation responses, such as the guinea pig maximization test *(3)* and the occluded patch test of Buehler *(4)*, and a method developed more recently, the murine local lymph node assay (LLNA), in which contact allergenic potential is assessed as a function of proliferative responses provoked in the induction phase of contact allergy *(5)*. In contrast, there are no widely accepted or fully validated test methods for the prospective identification of chemicals with the potential to cause sensitization of the respiratory tract *(6)*. Respiratory allergenic potential has been measured as a function of inhalation challenge-induced changes in respiratory parameters in previously sensitized animals, with guinea pigs usually being the species of choice *(7–9)*. An alternative approach that does not rely upon the elicitation of clinical manifestations of respiratory distress is to determine respiratory sensitizing activity by characterization of the immune response in rodents provoked by chemical respiratory allergens *(10,11)*. However, a major constraint to the development of such tests has been continuing uncertainty regarding the mechanisms of chemical respiratory hypersensitivity. Although it is generally acknowledged that immediate-type hypersensitivity responses and asthma or rhinitis induced by protein allergens are dependent on IgE-mediated mechanisms, there is no such consensus regarding chemical respiratory hypersensitivity *(12)*. In recent years, however, an increased understanding of the immunobiology of chemical respiratory allergy has resulted in new opportunities for the development of tests for the identification of potential respiratory allergens. This article considers the application of one such method, cytokine profiling, for the characterization and identification of chemical allergens.

2. IMMUNOLOGY OF ALLERGIC RESPONSES

Adaptive immune responses, including the development of allergic responses, are determined largely by the activity of functional subpopulations of $CD4^+$ T-helper (Th) and $CD8^+$ T-cytotoxic (Tc) cells. The most polarized forms of these develop from common precursors and are designated Th1 and Th2, and Tc1 and Tc2, respectively *(13–15)*. They have been identified in rodents and humans and are characterized by their differential cytokine repertoires *(13,16–19)*. Th1 and Th2 cells produce some cytokines (interleukin [IL] 3 and granulocyte/macrophage colony-stimulating factor) in common, but only Th1 cells express interferon γ (IFN-γ), IL-2, and tumor necrosis factor-α, and only Th2 cells express IL-4, IL-5, IL-6, IL-10, and IL-13 *(17,18)*. Tc1 and Tc2 cells display selective cytokine secretion patterns analogous with Th1 and Th2 cells, respectively *(15)*. Although these cytokine profiles discriminate between the main subpopulations of T-lymphocytes that differentiate during the course of immune responses, other phenotypes also have been identified. In addition, individual T-cells may display heterogeneity with respect to co-ordinate cytokine expression within populations that overall have type 1 or type 2 characteristics *(15,20,21)*.

The ability of the immune system to tailor the quality of the response to challenge appropriately according to the nature of the insult is the result of the existence of these functional subpopulations of T-cells. Activation of type 1 cells results in the provision of cell-mediated immunity, whereas humoral immune function and the costimulation and differentiation of B lymphocytes are promoted by type 2 cell products *(15,17)*. The development and expression of allergic disease also are dependent upon selective activation of T-cell subpopulations. Allergic contact dermatitis, in common with other forms of DTH reaction, initially was viewed as a Th1-type response, mediated by CD4+ IFN-γ- producing effector cells *(22,23)*. It is becoming increasing apparent, however, that unlike DTH reactions to complex protein or cellular antigens, CD8+ (Tc1) effector cells may mediate or regulate the development and elicitation of allergic contact dermatitis to chemical allergens *(24,25)*. Conversely, the development of immediate type (IgE-mediated) hypersensitivity reactions, such as asthma/rhinitis induced by high molecular weight allergens, is promoted by type 2 cell activation. The type 2 cytokine IL-4 has been shown to be essential for the induction and maintenance of IgE antibody responses in both man and mouse *(26–28)*. In contrast, the type 1 cell product IFN-γ and inhibits IgE production *(29)*. Type 2 cytokines also favor other processes important to the development of immediate type hypersensitivity reactions, including the stimulation of mast cell and eosinophil growth and differentiation and the recruitment of eosinophils, cells that play important roles in the elicitation of the clinical manifestations of the allergic response *(11,30)*. Furthermore, it has been demonstrated that Th2 type cells predominate in allergic respiratory hypersensitivity reactions *(31–33)*.

However, in contrast to protein allergens, there has been considerable uncertainty regarding a universal association between IgE antibody and occupational chemical respiratory allergy *(12)*. It is true that for all known respiratory allergens, specific IgE antibody has been detected in at least some symptomatic patients. However, the frequencies of specific IgE detection can vary from 50% or more to as few as 5% in particular patient populations *(34–36)*. Despite the lack of evidence for an obligatory role for IgE antibody, there is an increasing consensus that chemical respiratory allergy is associated with the induction of a polarized type 2 immune response *(11,12)*. In common with the development of asthma to protein allergens *(31–33)*, eosinophils and T-lymphocytes are recruited into the site of inflammation *(37)*, and there is evidence that such lymphocytes selectively express type 2 cytokines *(38)*. The implication is that independent of the nature of the inducing allergen (high or low molecular weight) extrinsic asthma involves common immune and inflammatory processes. This evidence suggests, therefore, that even in the absence of a detectable IgE antibody response, chemical respiratory allergens will be characterized by their stimulation of preferential type 2 responses in experimental systems. In contrast, skin sensitization resulting in allergic contact dermatitis is apparently associated with the preferential activation of Th1 and Tc1 cells.

3. SELECTIVITY OF IMMUNE RESPONSES INDUCED BY CHEMICAL ALLERGENS IN RODENTS

Initial investigations focused on the ability of known human contact and chemical respiratory allergens to stimulate qualitatively divergent immune responses in mice. Comparisons were made between trimellitic anhydride (TMA), a known respiratory allergen *(39)*, and 2,4-dinitrochlorobenzene (DNCB), a potent contact allergen that is considered not to cause sensitization of the respiratory tract *(40)*. BALB/c-strain mice were exposed topically to concentrations of TMA or DNCB that induce equivalent levels of proliferative activity in lymph nodes draining the site of application. Under these conditions of exposure, both chemicals provoked anti-hapten IgG antibody responses of comparable vigor. Importantly, TMA, but not DNCB, caused a significant increase in the total serum concentration of IgE *(41)*. The divergent IgE antibody responses stimulated in mice by DNCB and TMA were not associated only with dermal exposure. Exposure to DNCB or to TMA via inhalation also resulted in the induction of IgE antibody production only following exposure to TMA against a background of similar IgG antibody responses *(42)*. These data demonstrate that the ability of chemical allergens to provoke divergent qualities of immune response in mice is not a function of the route through which primary sensitization is acquired.

In subsequent experiments, the ability of chemical allergens to provoke divergent cytokine secretion profiles was investigated. Under conditions of

exposure of equivalent immunogenicity with respect to lymphocyte proliferation and IgG antibody responses, the cytokine phenotypes induced by
prolonged topical exposure to DNCB and TMA were examined *(43–45)*.
Draining lymph node cells (LNCs) excised from DNCB-treated mice expressed
high levels of the type 1 cytokines IFN-γ and IL-12, but relatively low levels of
the type 2 cytokines IL-4, IL-5, IL-10, and IL-13. The converse type 2
cytokine secretion profile was provoked by topical exposure to TMA *(43–
45)*. The differentiated cytokine phenotypes induced by DNCB and TMA
were observed in the absence of restimulation in vitro with allergen or mitogen. The exception to this was IL-4; this cytokine was not detectable (using
an enzyme-linked immunosorbent assay [ELISA]) in supernatants harvested
from draining LNCs cultured in the absence of mitogen *(46)*. In the presence
of concanavalin A (con A; a potent T-cell mitogen), however, draining LNCs
from TMA- but not DNCB-sensitized mice secreted high levels of IL-4 *(43–
45)*. IL-4 production was not simply a result of mitogen activation because
cells derived from untreated control (naïve) mice failed to produce measurable levels of IL-4 (<100 pg/mL) even in the presence of con A *(45,46)*.
Furthermore, using a more sensitive detection system (based on a microbead
fluorescent end point) with limits of detection some 100-fold lower than the
ELISA, LNCs derived from TMA-treated mice expressed detectable, albeit
low, levels of this cytokine in the absence of mitogen stimulation. Indeed,
under these conditions higher levels of IL-4 were recorded for TMA-acti-
vated LNC compared with DNCB-stimulated cells *(47)*.

   The polarized cytokine phenotypes stimulated by topical exposure to
chemical contact or respiratory sensitizers take time to mature. After 3 d of
exposure, LNCs displayed a mixed Th0-like phenotype with type 1 and type
2 cytokines expressed after treatment with both chemical contact and respiratory allergens *(46,48,49)*. A more prolonged (13-d) exposure protocol was
required for the development of a differentiated cytokine phenotype
(43,46,48). These data are consistent with what is known of the development of differentiated type 1 and type 2 T-lymphocyte phenotypes during
the evolution of adaptive immune responses. Functional subpopulations of
Th (and Tc) cells apparently derive from common precursors that express an
unrestricted (non-selective) cytokine repertoire *(13)*. More differentiated
phenotypes of selective cytokine production develop with time as the
immune response matures *(13)*. Thus, early during immune responses to
chemical allergens, it is predicted that contact and respiratory sensitizers
would be associated with similar cytokine secretion patterns, whereas the
selective type 1 and type 2 cytokine phenotypes take time and/or repeated
exposure to develop.

4. RELATIVE CONTRIBUTION OF T-CELL
SUBSETS TO CYTOKINE PROFILES

As described previously, both Th (CD4$^+$) and Tc (CD8$^+$) cells display functional heterogeneity with respect to cytokine expression *(13–15)*. The relative contributions of these cell types to the cytokine secretion phenotypes of allergen-activated cells have been examined *(49,50)*. These experiments showed that early (3 d) after initiation of exposure to both DNCB and TMA, the mixed cytokine phenotype observed was a function of IFN-γ-secreting CD4$^+$ (Th1) and CD8$^+$ (Tc1) cells and IL-4-secreting CD4$^+$ (Th2) cells *(49)*. The preferential type 2 cytokine secretion profile observed after prolonged treatment with the respiratory sensitizer TMA was caused by the activation of CD4$^+$ (Th2) cells, with the relatively low levels of IFN-γ derived exclusively from CD8$^+$ (Tc1) cells. The selective type 1 cytokine pattern stimulated by more chronic exposure to DNCB was associated with both CD4$^+$ (Th1) and CD8$^+$ (Tc1) IFN-γ-expressing cells, with the low levels of type 2 cytokines being a result of CD4$^+$ (Th2) cell activation *(50)*. As the immune response to chemical allergen matures with time and becomes polarized, exposure to the contact allergen DNCB is associated with the selective development of Th1 and Tc1 cells, whereas treatment with the respiratory allergen TMA is associated primarily with Th2 cell development.

5. CONFIRMATION OF SELECTIVITY
OF IMMUNE RESPONSES
INDUCED BY ALLERGENS

Other investigators have confirmed the observation that topical exposure of rodents to different classes of chemical allergen stimulates divergent immune responses at the level of cytokine expression *(51–54)*. In addition, recent studies have revealed that intranasal exposure of mice to TMA, but not to DNCB, provokes increased levels of message for type 2 cytokines in both the nasal airways and the lung *(55)*. In the majority of the experiments described previously, BALB/c-strain mice have been used *(43–54)*. In theory, the use of this mouse strain that is predisposed to make Th2-type responses could have biased the observed cytokine phenotype toward type 2 responses. However, in the same experiments, topical treatment with the contact allergen DNCB under conditions of equivalent immunogenicity provokes selective type 1 cytokine responses *(43–50,52–54)*. Furthermore, recent investigations have demonstrated that in a mouse strain that is more predisposed to make type 1 responses (C57BL6), similar divergent patterns of cytokine expression are elicited by DNCB and TMA *(51)*. In some experiments, an alternative rodent strain has been used, the Brown Norway (BN) rat, which in common with the BALB/c strain mouse is regarded as having a predisposition to type 2 responses. Measurement of cytokine secretion profiles provoked in draining LNC by topical exposure of BN rats to DNCB or to TMA revealed the selective induction of type 1 and type 2 cytokine patterns, respectively *(45)*. These data suggest that chemical contact allergens and respiratory sensitizers exhibit an innate ability to stimulate type 1 and type 2 cytokine production profiles, respectively, which are independent of rodent strain or species.

There are conflicting reports that contact allergens, such as DNCB and oxazolone, can stimulate type 2 cytokine expression *(56)*. However, the divergent cytokine secretion profiles provoked by different classes of chemical allergen as described previously are selective, not absolute. It is the balance between Th1 and Th2 cell activation and cytokine products, not the absolute amounts of cytokine, which determines the nature of the developing immune response. Thus, the observation that contact allergens such as DNCB stimulated measurable IL-4 expression *(56)* is consistent with previous reports, inasmuch as DNCB did induce detectable type 2 cytokine expression, albeit at much lower levels than those stimulated by TMA *(43,47)*. Indeed, in the experiments conducted by Ulrich et al. *(56)*, it is not possible to compare directly the ability of DNCB and TMA to provoke selective type 1 and type 2 cytokine secretion patterns because the concentration of DNCB used was considerably less immunogenic than was the dose of TMA selected. Furthermore, in these experiments, in common with some other approaches to cytokine profiling *(52–54)*, the authors have chosen to measure cytokine expression following restimulation of LNC with mitogen in vitro. As described previously, although it is necessary to restimulate LNCs with the T-cell mitogen con A to detect expression of IL-4 protein, there is sufficient spontaneous production of the other cytokines (IFN-γ, IL-5, IL-10, IL-12, and IL-13) for measurement by ELISA in the absence of further restimulation *(44,45)*. Indeed, the addition of mitogens such as polyclonal anti-CD3 antibody as used by Ulrich et al. *(56)* may bias the cytokine secretion profile with respect to cytokines other than IL-4. Independent studies have shown that stimulation of naïve spleen cells with anti-CD3 antibody is sufficient to induce detectable cytokine production *(57)*. The experience of other investigators that use mitogen restimulation confirms that such cytokine secretion patterns are less polarized than those derived after the measurement of spontaneous cytokine production, particularly with respect to IFN-γ expression *(53,54)*. In contrast, without mitogen restimulation, resting control LNCs do not generally produce detectable levels of most cytokines (IFN-γ, IL-4, IL-5, IL-10, and IL-13), with the exception of constitutively expressed IL-12 *(45,46,48)*. Thus, allergen-induced changes in cytokine profiles may be measured against a background of low constitutive expression, providing a more robust assessment of sensitizing hazard.

6. CYTOKINE FINGERPRINTING

Taken together, these data provide compelling evidence that DNCB and TMA elicit qualitatively discrete immune responses in rodents that are consistent with what is known of their abilities to induce allergic disease in humans. On the basis of the accumulated data, it was proposed that measurement of induced cytokine secretion patterns in mice might provide a novel approach to the characterization of chemical allergens, a method that became known as cytokine fingerprinting *(44)*. To date, investigations using various skin sensitizers such as 2,4-dinitrofluorobenzene, isoeugenol, and hexyl cinnamic aldehyde (contact allergens that apparently lack respiratory sensitizing activity) have revealed the induction of a selective type 1 cytokine secretion pattern. Using the same exposure protocol as that used for the chemical contact allergens described above, researchers have confirmed that, for a wide range of additional chemical respiratory allergens, including isocyanates, platinum salts, glutaraldehyde, and various acid anhydrides, a preferential type 2 cytokine expression profile is a general property of respiratory sensitizing chemicals *(44)*.

7. PRACTICAL ISSUES
IN CYTOKINE FINGERPRINTING

These data demonstrate that, for a variety of chemical allergens, cytokine expression profiles can discriminate between chemical contact and respiratory sensitizers. In the commonest configuration of cytokine fingerprinting, LNCs have been pooled on an experimental group basis, single cell suspensions prepared and cultured for cytokine protein secretion in the absence of restimulation (with the exception of IL-4 as described above). Despite some interexperimental variation in absolute amounts of cytokine expressed, using this approach the phenotypes of DNCB- and TMA-stimulated LNCs were invariably type 1 and type 2, respectively *(44)*. These chemicals therefore provide negative and positive controls for respiratory sensitizing potential and it is recommended that cytokine secretion profiles induced by test chemicals are measured concurrently with those stimulated by treatment with DNCB and TMA. With respect to dose selection for test chemicals in cytokine fingerprinting, it is necessary that concentrations are used that are known to be immunogenic and to stimulate a cutaneous immune response of the vigor necessary for measurement of cytokine secretion. Currently, dose selection is based on prior conduct of an LLNA, in which skin sensitizing activity is measured as a function of LNC proliferative responses induced by topical exposure of mice to the test material *(5)*. In our experience those chemicals that are able to cause allergic sensitization of the respiratory tract also elicit positive responses in guinea pig and mouse predictive tests for skin sensitization (including the LLNA *[58]*). The reason for this is unclear, as many of those same chemical respiratory allergens only rarely, if ever, cause allergic contact dermatitis in humans *(58)*. In practice, this means that chemical respiratory allergens (and contact allergens) will be identified in tests for skin sensitization such as the LLNA and that chemicals that fail to elicit positive responses in the LLNA are very unlikely to have a significant potential to cause respiratory sensitization.

Another important issue is that of route of exposure. The dermal route of exposure has been chosen to facilitate analyses of cytokine responses induced in discrete draining lymph nodes and the exposure regimen designed to induce polarized cytokine secretion profiles *(44,47,48)*. Although inhalation is the most important and common route of exposure to chemical respiratory allergens, there is experimental evidence and some limited clinical evidence to suggest that skin contact may result in sensitization of the respiratory tract *(59)*. Furthermore it has been demonstrated that DNCB and TMA provoke type 1 and type 2 responses, respectively, with respect to antibody isotype and cytokine profiles, regardless of whether sensitization is via inhalation or the skin *(42,55)*. The accumulated experience confirms that the characterization of cytokine responses induced after topical application of different classes of chemical allergen is therefore an appropriate strategy for hazard identification.

One final consideration for the optimal design of cytokine profiling is whether cytokine responses are measured as a function of protein secretion, or at the level of messenger ribonucleic acid (mRNA) expression. Cytokine mRNA expression has been examined by ribonuclease protection assay (RPA *[48,52,60,61]*) or by reverse-transcription polymerase chain reaction (RT-PCR *[51,53,62]*). Measurement of cytokine transcripts by RT-PCR revealed that although increased IL-4 mRNA expression was associated with respiratory allergens, IFN-γ mRNA levels did not always discriminate between contact and respiratory allergens *(51,53,62)*. A similar pattern emerges when allergen-induced changes in cytokine gene expression are analyzed by RPA, a technique with a somewhat lower level of sensitivity than RT-PCR *(48,52,60,61)*. As shown in Fig. 1, topical exposure to TMA is associated with increased levels of transcripts for type 2 cytokines, including IL-4, IL-10, and

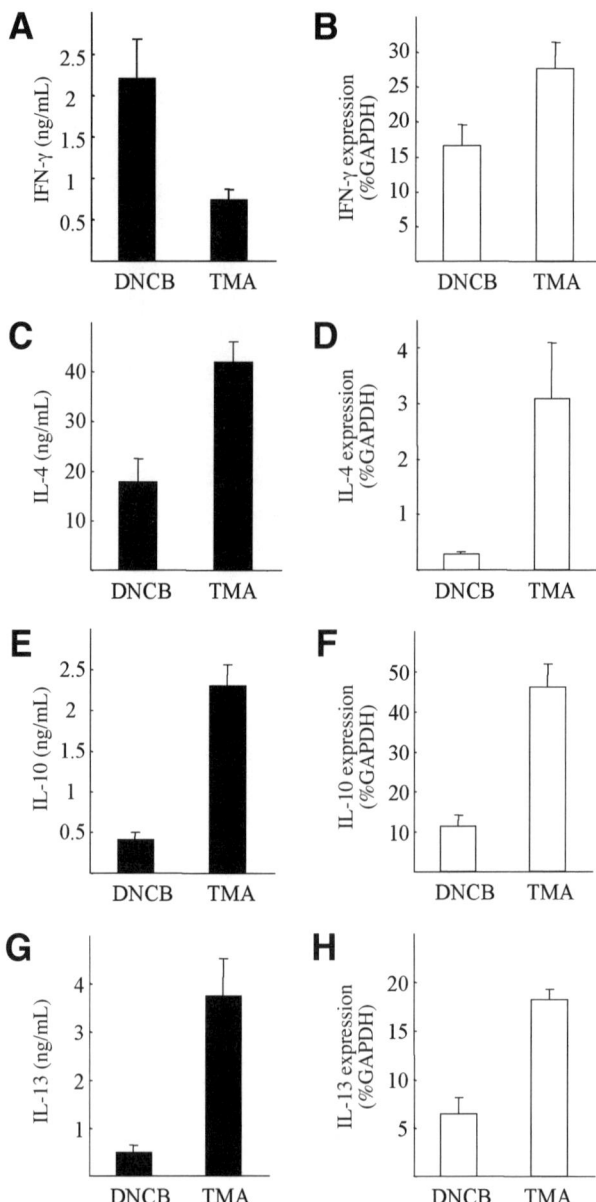

Fig. 1. Cytokine phenotypes induced by 2,4-dinitrochlorobenzene (DNCB) and trimellitic anhydride (TMA): comparison of message versus protein. Draining auricular lymph node cells were isolated 13 d after the initiation of topical exposure of BALB/c-strain mice to 1% DNCB or 10% TMA each dissolved in acetone: olive oil (4:1) vehicle. A single cell suspension was prepared and total RNA extracted from cell pellets using TRIZOL. Levels of cytokine and housekeeping (glyceraldehyde 3-

IL-13, compared with DNCB-activated LNCs. However, RNA isolated from DNCB-activated LNC did not express a selective type 1 cytokine phenotype, with lower levels of IFN-γ transcripts recorded than those found in TMA-stimulated tissue (Fig. 1). Under the same exposure conditions, the expected type 1 and type 2 phenotype of secreted cytokines was observed for DNCB- and TMA-stimulated LNC, respectively, which is consistent with previous reports, in which RNA isolated from DNCB- or 2,4-dinitrofluorobenzene-activated LNC tissue did not express a selective type 1 cytokine phenotype without restimulation in vitro *(52,60,61)*. The observation that IFN-γ mRNA expression is not increased despite robust secretion of this cytokine indicates that production of IFN-γ by draining LNC is controlled mainly at the level of secretion. Taken together these data suggest that cytokine profiling by RPA or by RT-PCR may identify those chemicals with respiratory sensitizing potential as a function of induced type 2 cytokine expression. However, neither method is appropriate to discriminate between respiratory and contact allergens.

8. CONCLUSIONS

There has been considerable interest in the application of cytokine profiling to the characterization of chemical allergy, with cytokine phenotypes analyzed in freshly isolated tissue or after culture in the presence or absence of mitogen at the level of protein secretion or mRNA expression. The most common configuration of cytokine fingerprinting (measurement of induced cytokine secretion patterns in the absence of restimulation) in particular shows considerable promise as an approach for the identification and characterization of chemical respiratory allergens.

Fig. 1. *(continued)* phosphate dehydrogenase [GAPDH] mRNA were measured by multiprobe ribonuclease protection assay and quantitated by densitometric analysis. Results are expressed as mean and SE relative levels of cytokine mRNA compared with GAPDH control transcripts from four independent experiments for IFN-γ (b), IL-4 (d), IL-10 (f), and IL-13 (h). In parallel, aliquots of cells were cultured at 10^7 cells/mL in the presence and absence of 2 μg/mL of the T-cell mitogen concanavalin A (Con A). Cytokine secretion was measured by enzyme-linked immunosorbent assay in supernatants prepared after culture for 24 h in the presence of Con A (IL-4; [c]) or for 120 h in the absence of Con A (IFN-γ [a]; IL-10 [e] and IL-13 [g]). Cytokine expression is shown in nanograms per milliliter as mean and SE of four independent experiments.

REFERENCES

1. Cronin E. Contact Dermatitis. London: Churchill Livingstone; 1980.
2. Chan-Yeung M, Malo JL. Compendium 1. Table of major inducers of occupational asthma. In: Bernstein IL, Chan-Yeung M, Malo JL, et al, eds. Asthma in the Workplace. New York: Marcel Dekker; 1993, pp. 595–623.
3. Magnusson B, Kligman AM. Allergic Contact Dermatitis in the Guinea Pig. Springfield, IL: Charles C. Thomas; 1970.
4. Buehler EV. (1965). Delayed contact hypersensitivity in the guinea pig. Arch Dermatol 1965;91:171–177.
5. Kimber I, Basketter DA. The murine local lymph node assay: a commentary on collaborative studies and new directions. Food Chem Toxicol 1992;30:165–169.
6. Kimber I, Bernstein IL, Karol MH, Robinson MK, Sarlo K, Selgrade MK. Identification of respiratory allergens. Fundam Appl Toxicol 1996;33:1–10.
7. Karol MH, Stadler J, Magreni C. Immunotoxicologic evaluation of the respiratory system: animal models for immediate and delayed-onset pulmonary hypersensitivity. Fundam Appl Toxicol 1985;5:459–472.
8. Pauluhn J. Predictive testing for respiratory sensitization. Toxicol Lett 1996;86:177–185.
9. Sarlo K, Ritz HL. Predictive assessment of respiratory sensitizing potential in guinea pigs. In: Kimber I, Dearman RJ, ed. Toxicology of Chemical Respiratory Hypersensitivity. London: Taylor and Francis; 1997, pp. 107–120.
10. Dearman RJ, Basketter DA, Coleman JW, Kimber I. The cellular and molecular basis for divergent allergic responses to chemicals. Chem-Biol Interact 1992;84:1–10.
11. Kimber I, Dearman RJ. Cell and molecular biology of chemical allergy. Clin Rev Allergy Immunol 1997;15:145–168.
12. Mapp C, Boschetto P, Miotto D, De Rosa E, Fabbri LM. Mechanisms of occupational asthma. Ann Allergy Asthma Immunol 1999; 83:645–664.
13. Bendelac A, Schwartz RH. Th0 cells in the thymus. The question of T-helper lineages. Immunol Rev 1991;123:169–188.
14. Sad S, Marcotte R, Mosmann TR. Cytokine-induced differentiation of precursor mouse CD8$^+$ T cells into cytotoxic CD8$^+$ T cells secreting Th1 or Th2 cytokines. Immunity 1995;2:271–279.
15. Mosmann TR, Sad S. The expanding universe of T cell subsets: Th1, Th2 and more. Immunol Today 1996;17:138–146.
16. Romagnani S. Human TH1 and TH2 subsets: doubt no more. Immunol Today 1991;12:256–257.
17. Abbas AK, Murphy KM, Sher A. Functional diversity of helper T lymphocytes. Nature 1996;383:787–793.
18. O'Garra A. Cytokines induce the development of functionally heterogenous T helper subsets. Immunity 1998;8:275–283.
19. Romagnani S. T-cell subsets (Th1 versus Th2). Ann Allergy Asthma Immunol 2000; 85:9–18.

20. Bucy RP, Karr L, Huang GQ, et al. Single cell analysis of cytokine gene co-expression during CD4⁺ T-cell phenotype development. Proc Natl Acad Sci USA 1995;92:7565–7569.

21. Kelso A. Th1 and Th2 subsets: paradigms lost? Immunol Today 1995;16:374–379.

22. Cher DJ, Mosmann TR. Two types of murine helper T cell clones. II. Delayed type hypersensitivity is mediated by Th1 clones. J Immunol 1987;138:3688–3694.

23. Diamantstein T, Eckert R, Volk HD, Kuper-Weglinski JW. Reversal by interferon-γ of inhibition of delayed-type hypersensitivity induction by anti-CD4 or anti-interleukin 2 receptor (CD25) monoclonal antibodies. Evidence for the physiological role of the CD4⁺Th1⁺ subset in mice. Eur J Immunol 1988; 181:2101–2103.

24. Gorbachev AV, Fairchild RL. Induction and regulation of T-cell priming for contact hypersensitivity. Crit Rev Immunol 2001;21:451–472.

25. Kimber I, Dearman RJ. Allergic contact dermatitis: the cellular effectors. Contact Dermatol 2002;46:1–5.

26. Finkelman FD, Katona IM, Urban JF, et al. IL-4 is required to generate and sustain in vivo IgE responses. J Immunol 1988;141:2335–2341.

27. Kuhn R, Rajewsky K, Muller W. Generation and analysis of interleukin-4-deficient mice. Science 1991;254:707–710.

28. Romagnani S, Del Prete G, Maggi E, et al. Role of interleukins in induction and regulation of human IgE synthesis. Clin Immunol Immunopathol 1989;50:S13–S23.

29. Finkelman FD, Katona IM, Mosmann TR, Coffman RL. IFN-γ regulates the isotypes of Ig secreted during in vivo humoral immune responses. J Immunol 1988;140:1022–1027.

30. Krishnan L, Mosmann TR. Functional subpopulations of CD4⁺ T lymphocytes. In: Kimber I, Selgrade MK, eds. T Lymphocyte Subpopulations in Immunotoxicology. Chichester: John Wiley & Sons, 1998, pp. 7–32.

31. Robinson DS, Hamid Q, Ying S, et al. Predominant Th2-like bronchoalveolar T-lymphocyte population in atopic asthma. N Eng J Med 1992;326:298–304.

32. Robinson DS, Hamid Q, Bentley A, Ying S, Kay AB, Durham SR. Activation of CD4⁺ T cells, increased Th2-type cytokine mRNA expression and eosinophil recruitment in bronchoalveolar lavage after allergen inhalation challenge in patients with atopic asthma. J Allergy Clin Immunol 1993;92:313–324.

33. Durham SR, Ying S, Varney VA, et al. Cytokine messenger RNA expression for IL-3, IL-4, IL-5 and granulocyte/macrophage colony- stimulating factor in the nasal mucosa after local allergen provocation: relationship to tissue eosinophilia. J Immunol 1992;148:2390–2394.

34. Topping MD, Venables KM, Luczynska CM, Howe W, Newman Taylor AJ. Specificity of the human IgE response to inhaled acid anhydrides. J Allergy Clin Immunol 1986;77:834–842.

35. Tarlo S M. Diisocyanate sensitization and antibody production. J Allergy Clin Immunol 1999;103:739–741.

36. Tee RD, Cullinan P, Welch J, Burge PS, Newman Taylor AJ. Specific IgE to isocyanates: a useful diagnostic role in occupational asthma. J Allergy Clin Immunol 1998;101:709–715.

37. Bentley AM, Maestrelli P, Saetta M, et al. Activated T-lymphocytes and eosinophils in the bronchial mucosa in isocyanate-induced asthma. J Allergy Clin Immunol 1992;89:821–829.

38. Del Prete GF, De Carli M, D'Elios MM, et al. Allergen exposure induces the activation of allergen-specific Th2 cells in the airway mucosa of patients with allergic respiratory disorders. Eur J Immunol 1993;23:1445–1449.

39. Zeiss CR, Patterson R, Pruzansky JJ, Miller M, Rosenberg M, Levitz D. Trimellitic anhydride-induced airway syndromes: clinical and immunologic studies. J Allergy Clin Immunol 1977;60:96–103.

40. Botham PA, Rattray NJ, Woodcock DR, Walsh ST, Hext PM. The induction of respiratory allergy in guinea pigs following intradermal injection of trimellitic anhydride: a comparison with the response to 2,4-dinitrochlorobenzene. Toxicol Lett 1989;47:25–39.

41. Dearman RJ, Kimber I. Differential stimulation of immune function by respiratory and contact chemical allergens. Immunology 1991;72:563–570.

42. Dearman RJ, Hegarty JM, Kimber I. Inhalation exposure of mice to trimellitic anhydride induces both IgG and IgE anti-hapten antibody. Int Arch Allergy Appl Immunol 1991;95:70–76.

43. Dearman RJ, Smith S, Basketter DA, Kimber I. Classification of chemical allergens according to cytokine secretion profiles in murine lymph node cells. J Appl Toxicol 1997;17:53–62.

44. Dearman RJ, Kimber I. Cytokine fingerprinting and hazard assessment of chemical respiratory allergy. J Appl Toxicol 2001;21:153–163.

45. Dearman RJ, Warbrick EV, Skinner R, Kimber I. Cytokine fingerprinting of chemical allergens : species comparisons and statistical analyses. Fd Chem Toxicol 2002;40:107–118.

46. Dearman RJ, Ramdin LSP, Basketter DA, Kimber I. Inducible interleukin-4-secreting cells provoked in mice during chemical sensitization. Immunology 1994;81:551–557.

47. Dearman RJ, Betts CJ, Humphreys N, et al. Chemical allergy: considerations for the practical application of cytokine profiling. Toxicol Sci 2003;71:137–145.

48. Betts CJ, Dearman RJ, Flanagan BF, Kimber I. Temporal changes in cytokine gene expression profiles induced in mice by trimellitic anhydride. Toxicol Lett 2002;136:121–132.

49. Moussavi A, Dearman RJ, Kimber I, Kemeny DM. Cytokine production by CD4$^+$ and CD8$^+$ T cells in mice following primary exposure to chemical allergens: evidence for functional differentiation of T lymphocytes in vivo. Int Arch Allergy Immunol 1998;116:116–123.

50. Dearman RJ, Moussavi A, Kemeny DM, Kimber I. Contribution of CD4$^+$ and CD8$^+$ T lymphocyte subsets to the cytokine secretion patterns induced in mice

during sensitization to contact and respiratory chemical allergens. Immunology 1996;89:502–510.

51. Hayashi M, Higashi K, Kato H, Kaneko H. Assessment of preferential Th1 or Th2 induction by low-molecular-weight compounds using a reverse transcription-polymerase chain reaction method: comparison of two mouse strains, C57BL/6 and BALB/c. Toxicol Appl Pharmacol 2001;177:38–45.

52. Manetz TC, Pettit DA, Meade BJ. The determination of draining lymph node cell cytokine mRNA levels in BALB/c mice following dermal sodium lauryl sulfate, dinitrofluorobenzene, and toluene diisocyanate exposure. Toxicol Appl Pharmacol 2001;171:174–183.

53. Vandebriel RJ, De Jong WH, Spiekstra SW, et al. Assessment of preferential T-helper 1 or T-helper 2 induction by low molecular weight compounds using the local lymph node assay in conjunction with RT-PCR and ELISA for interferon-γ and interleukin-4. Toxicol Appl Pharmacol 2000;162:77–85.

54. Van Och FMN, van Loveren H, de Jong WH, Vanderbriel R. Cytokine production induced by low molecular weight chemicals as a function of the stimulation index in a modified local lymph node assay : an approach to discriminate contact sensitizers from respiratory sensitizers. Toxicol Appl Pharmacol 2002;184:46–56.

55. Farraj AK, Harkema JR, Kaminski NE. Allergic rhinitis induced by intranasal sensitization and challenge with trimellitic anhydride but not with dinitrochlorobenzene or oxazolone in A/J mice. Toxicol Sci 2004;79:315–325.

56. Ulrich P, Grenet O, Bluemel J, et al. Cytokine expression profiles during murine contact allergy: T helper 2 cytokines are expressed irrespective of the type of contact allergen. Arch Toxicol 2001;75:470–479.

57. Wang H, Mohapatra SS, HayGlass KT. Evidence for the existence of IL-4 and IFN gamma secreting cells in the T cell repertoire of naive mice. Immunol Lett 1992;31:169–175.

58. Kimber I. Contact and respiratory sensitization by chemical allergens : uneasy relationships. Am J Contact Dermatol 1995;6:34–39.

59. Kimber I, Dearman RJ. Chemical respiratory allergy : role of IgE antibody and relevance of route of exposure. Toxicology 2002;181–182:311–315.

60. Plitnick LM, Loveless SL, Ladics GS, et al. Cytokine profiling for chemical sensitizers : application of the ribonuclease protection assay and effect of dose. Toxicol Appl Pharmacol 2002;179:145–154.

61. Plitnick LM, Loveless SE, Ladics GS, et al. Identifying airway sensitizers: cytokine mRNA profiles induced by various anhydrides. Toxicology 2003;193:191–201.

62. Warbrick EV, Dearman RJ, Basketter DA, Kimber I. Analysis of interleukin 12 protein production and mRNA expression in mice exposed topically to chemical allergens. Toxicology 1999;132:57–66.

6

Inflammatory Cytokines and Lung Toxicity

Debra L. Laskin, Vasanthi R. Sunil, Robert J. Laumbach, and Howard M. Kipen

SUMMARY

Exposure to airborne particles and gases including silica, asbestos, diesel exhaust, and ozone is associated with significant risk of pulmonary and cardiovascular morbidity and mortality. Increasing evidence suggests that macrophages and inflammatory mediators released, including cytokines, play a role in the pathogenic process. In response to lung injury, alveolar macrophages become activated and release increased quantities of cytokines such as TNF-α, IL-1, IL-6, and IL-10, as well as chemokines and growth factors such as TGF-β and PDGF. Although these mediators are released to protect the host and initiate wound repair, when generated in excessive amounts or at inappropriate times or places, they can damage host tissue and exacerbate or perpetuate injury. In this chapter, the role of inflammatory cytokines and growth factors released by macrophages in xenobiotic-induced pulmonary toxicity is reviewed. Potential mechanisms mediating expression of cytokine genes and the implications to human health are also discussed.

Key Words: Inflammatory cytokines; lung toxicity; xenobiotics; inflammatory mediators.

1. INTRODUCTION

The respiratory tract is particularly susceptible to injury induced by inhaled toxicants, largely because of its direct link to the external environment. The precise nature and site of initial injury, however, depends on the physical and chemical character, as well as the quantity of the inhaled toxicant (1). Host factors also often play a role. Depending on these parameters, injury may be acute and/or chronic and may result directly from the agent or from the host response. A number of agents have been identified that induce pulmonary injury, including pollutant gases such as ozone, nitrogen dioxide,

From: *Apoptosis Methods in Pharmacology and Toxicology: Cytokines in Human Health:
Immunotoxicology, Pathology, and Therapeutic Applications*
Edited by: R. V. House and J. Descotes © Humana Press Inc., Totowa, NJ

and sulfur dioxide, as well as various particulates, alone and in mixtures (e.g., asbestos, silica, cigarette smoke, and diesel exhaust). In general, whereas large particles (>5–10 μm in aerodynamic diameter) or highly reactive or water-soluble gases typically affect the upper airways and more often cause acute injury, smaller particles (<3–5 μm in aerodynamic diameter) or less reactive or less soluble gases are more toxic to lower more distal regions of the lung *(2)* and may act over the course of either an acute or chronic time frame.

The respiratory system possesses a number of important mechanisms to minimize injury induced by inhaled toxicants, including physical and chemical barriers, as well as cellular components of the immune system, in particular, alveolar macrophages. These cells constitute more than 95% of all the cells in the alveolar space and are known to play an essential role in nonspecific host defense and in the biological response to inhaled toxicants *(3,4)*. In addition to scavenging particles and debris, alveolar macrophages kill microorganisms, recruit and activate other inflammatory cells, function as accessory cells in immune responses, and maintain and repair the lung parenchyma. After injury to the lung, resident alveolar macrophages, along with inflammatory phagocytes derived from blood and bone marrow precursors that migrate into the lung, become activated resulting in increased functional responsiveness.

Secretory products released by macrophages mediate many of their functions. These include oxidants, proteases, bioactive lipids, and cytokines. Although released to protect the host and initiate wound repair, in excess amounts or when released at inappropriate times or places, these mediators can in fact damage host tissue and facilitate or perpetuate injury. Macrophages and the various inflammatory mediators that they release have been implicated in lung injury induced by a number of different pulmonary toxicants (reviewed in refs. *3–5*). The focus of this review is on inflammatory cytokines, many derived from macrophages. Several model toxicants are described to illustrate the role of inflammatory cytokines in pulmonary toxicity.

2. CYTOKINES AND LUNG INFLAMMATION

The success of the inflammatory response depends on intercellular communication for propagation, maintenance, and resolution. Although communication may occur through direct cell–cell contact via adhesion molecules, cells also communicate through the release of cytokines. These proteins act in an autocrine and paracrine manner to regulate cell behavior and functions, including proliferation, differentiation, recognition, and cellular recruitment. Alveolar macrophages are known to release a variety of cytokines and growth factors that influence the pathogenesis of lung injury. Whereas some of these promote the inflammatory response, for example, tumor necrosis factor-α (TNF-α), interleukin-1 (IL-1), IL-6, interferon-γ (IFN-γ), and chemokines, others such as IL-10, transforming growth factor-β (TGF-β), and platelet-derived growth factor (PDGF) exert anti-inflammatory activity, initiating wound repair or fibrosis. The overall outcome of the inflammatory response depends on the balance between levels of pro- and anti-inflammatory cytokines that are generated in the tissue.

Probably the best characterized of the proinflammatory cytokines involved in acute and chronic inflammation are TNF-α and IL-1. These low-molecular-weight multifunctional proteins induce a number of distinct and overlapping activities *(6–8)*. Considered early-response cytokines, they are produced in large part by resident macrophages and are thought to play a prominent role as initiators of the inflammatory response. Both TNF-α and IL-1 stimulate the production of chemotactic factors and upregulate expression of cell adhesion molecules on endothelial cells and leukocytes, thus promoting the margination of circulating phagocytes and emigration to sites of injury. IL-1 also exerts mitogenic effects on macrophages and endothelial cells and induces the release of prostaglandins, metalloproteinases, and colony-stimulating factor *(9)*. In the lung, TNF-α and IL-1 activate alveolar macrophages and stimulate the production of cytotoxic mediators, including reactive nitrogen intermediates and reactive oxygen intermediates *(9,10)*. They also induce the release of IL-6, colony-stimulating factor, platelet-activating factor, and eicosanoids from macrophages and epithelial cells *(6,11,12)*. TNF-α is unique among inflammatory cytokines in that it also has the capacity to directly induce cytotoxicity and apoptosis *(13,14)*. TNF-α is thought to mediate bronchial hyperresponsiveness after exposure to aerosolized endotoxin and, together with IL-1, may play a key role in the allergic reactions in human airways *(15–17)*. A number of studies have demonstrated an important role of TNF-α in wound healing. Thus, TNF-α stimulates the production of proteins that induce proliferation and are involved in matrix remodeling *(18,19)*. It appears that the biological effects of TNF-α are related to the quantity and timing of its release. Thus, high concentrations of TNF-α produced early in the inflammatory process cause damage to endothelium, microthrombosis, and tissue injury, whereas lower concentrations generated at later times exert homeostatic functions, such as initiation of tissue repair *(6–8)*.

Another cytokine that plays a role in initiation and propagation of the inflammatory process in the lung is IL-6. It is a multifunctional cytokine produced by a variety of lung cells, including macrophages and epithelial cells. In addition to promoting differentiation of monocytes toward macrophages, IL-6 stimulates monocytes to release macrophage chemotactic protein-1 (MCP-1), a C-C chemokine involved in mediating lung injury by activating monocytes. This leads to increased production of reactive oxygen intermediates and expression of cell adhesion molecules *(20,21)*. In some models, IL-6 also exhibits anti-inflammatory and protective activity, including inhibition of TNF-α, IL-1, macrophage inflammatory protein-2 (MIP-2), and intracellular superoxide anion production, as well as decreased neutrophil sequestration, and increased release of IL-1 receptor antagonist (IL-1RA) and soluble TNF receptor 2 (TNFR2 *[22–26]*). Thus, the ultimate effect of IL-6 in lung inflammation and injury is ambiguous. Whereas transgenic overexpression of IL-6 has been reported to decrease airway responsiveness and to protect against acute lung injury induced by chronic hyperoxia *(27–29)*, mice genetically deficient in IL-6 are reported to be protected from lung injury induced by oxidants *(23,30)*.

Proinflammatory cytokines that exhibit chemotactic activity have also received considerable attention as mediators of lung pathology. These so-called chemokines belong to a superfamily of low-molecular-weight proteins that play a key role in mediating the recruitment and activation of inflammatory cells to sites of injury. Chemotactic cytokines or chemokines are classified into four subfamilies according to the relative position of cysteine residues: C-X-C proteins (e.g., IL-8 or CINC), which are mainly neutrophil chemoattractants, C-C chemokines (e.g., MIP-1, MIP-2, MCP-1, MCP-2, MCP-3, RANTES), which induce migration and activation of macrophages/monocytes, basophils and eosinophils, C chemokines (e.g., lymphotactin), which stimulate lymphocyte migration, and CX3C chemokines (e.g., fractalkine), which activate lymphocytes and macrophages. Continuous local release of chemokines at sites of injury is thought to mediate the ongoing migration of effector cells into lesions during inflammatory responses.

Activated macrophages also release anti-inflammatory cytokines and growth factors including IL-10, TGF-β, and PDGF, which downregulate the inflammatory response and initiate wound healing. IL-10 is known to inhibit the production of TNF-α and IL-1 by macrophages and to decrease expression of inducible nitric oxide synthase and the production of reactive nitrogen intermediates *(21,31–33)*. IL-10 plays a role in a number of inflammatory conditions, including sepsis and chronic arthritis *(34–36)* and may be important in lung toxicity. This is supported by findings that intratracheal administration of IL-10 affords dose-dependent protection from immune complex-induced lung injury in rats *(37)*. TGF-β and PDGF contribute to normal pulmonary morphogenesis and function as well as in the pathogenesis of lung fibrosis *(38,39)*. Whereas PDGF stimulates replication, survival, and migration of myofibroblasts, leading to the development of fibrosis and ultimately fibrotic diseases, TGF-β increases the production of extracellular matrix proteins and decreases the degradation of connective tissue.

3. XENOBIOTICS, CYTOKINES, AND LUNG INJURY

3.1. Pollutant Gases: Ozone

Ozone is a ubiquitous urban air pollutant present in photochemical smog. It is generated by sunlight, largely from motor vehicle emissions of hydrocarbons and nitrogen oxides. Health effects associated with acute exposure to ozone include changes of obstructive lung function and associated airway hyperresponsiveness, as well as alveolar epithelial damage *(4)*. Epidemiologic and clinical studies have demonstrated that individuals with asthma and other preexisting respiratory disease are particularly sensitive to the effects of ozone *(30)*. Ozone-induced tissue injury is associated with airway inflammation characterized by an accumulation of neutrophils and macrophages in the lower lungs. A number of studies have demonstrated that these cells, as well as resident alveolar macrophages and epithelial cells, are activated to release cytotoxic mediators and inflammatory cytokines after ozone inhalation. That macrophages play a central role in ozone toxicity is most clearly evident from studies demonstrating that blocking the activity of these cells prevents ozone-induced tissue injury *(40)*.

The specific cytotoxic mediators responsible for the adverse effects of macrophages in ozone-induced lung injury include reactive oxygen and nitrogen intermediates and bioactive lipids (reviewed in ref. *4*). A number of studies have demonstrated that inflammatory cytokines, including IL-1, TNF-α, IL-6 and various chemokines, also contribute to ozone toxicity. After the exposure of humans or experimental animals to ozone, increased TNF-α and IL-1 levels, as well as soluble TNF receptor-1 (TNFR1, p55) and TNFR2 (p75) are found in the lung and/or in bronchoalveolar lavage (BAL) fluid *(30,41–50)*. This is evident prior to ozone-induced increases in permeability and lavageable inflammatory cells *(42)*. Macrophages isolated from ozone-treated animals also generate increased quantities of TNF-α and IL-1 *(41–43)*. These cytokines play an important role in inflammatory cell trafficking into sites of injury and also may contribute to the toxicity of ozone. This is supported by findings that pretreatment of animals with antibodies to TNF-α attenuates ozone-induced increases in macrophage adherence, tracheal permeability, neutrophil accumulation and IL-1 and IL-6 expression, and protects against tissue injury *(46,51,52)*. Similarly, ozone-induced airway hyperresponsiveness and neutrophilia, as well as IL-1β,

TNF-α, and MIP-2 expression are also prevented by pretreatment of mice with IL-1 receptor antagonist *(50)*. Abrogation of ozone-induced inflammation, macrophage activation and tissue injury, have been observed in TNF-α knockout mice *(53)* and in mice lacking TNFR1 and/or TNFR2 *(54)*. Deficiency in these receptors also afforded protection from ozone-induced airway hyperresponsiveness *(54,55)*. Moreover, exogenous administration of TNF-α causes changes similar to ozone such as airway hyperresponsiveness and lung injury, and neutrophil emigration into the lung of humans and rats *(56–58)*. Taken together, these data support the idea that TNF-α and IL-1β contribute to ozone-induced inflammation and toxicity.

Acute inhalation of ozone also has been reported to result in increased IL-6 in BAL fluid and IL-6 mRNA in the lung *(30,44,59–68)*. IL-6 deficiency is associated with attenuated pulmonary inflammation and tissue injury after ozone *(23)*. Protection has been observed in ozone-exposed mice treated with anti-IL-6 antibody and in IL-6 knockout mice, thus supporting a proinflammatory function of IL-6 in ozone-induced toxicity.

Several lines of evidence suggest that chemokines, including interferon-γ-inducible protein (IP-10), MCP-1, MCP-3, MIP-1α, KC, and MIP-2, are important in ozone-induced inflammation. First, the sequential increases in expression of these chemokines or their receptors in the lung after ozone inhalation are directly correlated with the time course of neutrophil and macrophage accumulation in the tissue *(44,47,50,60,65,69–76)*. Second, macrophages from ozone-treated mice generate MIP-2 and CINC *(42)*. Third, the anti-inflammatory corticosteroid dexamethasone suppresses ozone-induced expression of MIP-2 and CINC and neutrophilic accumulation in the lung *(77,78)*. Finally, administration of antibodies to diverse chemokines, including IP-10, KC, MCP-3, or CINC, abrogates ozone-induced inflammation *(74,79)*. Antibodies to the chemokine receptor, CXCR2 also prevent ozone-induced epithelial sloughing and hyperresponsiveness *(73)*. These findings suggest that the inflammatory response to ozone and resultant injury is mediated by chemokines generated in the lung.

Anti-inflammatory cytokine levels also change in the lung after ozone inhalation. Thus, whereas relatively high levels of IL-10 are detectable in lung sections from normal animals, these decreases significantly when animals are treated with ozone *(48,49,80)*. Moreover, when rats are instilled intratracheally with recombinant IL-10, ozone toxicity and inflammation is markedly attenuated *(80)*. These findings, together with the observation that ozone-induced toxicity is increased in transgenic mice lacking IL-10 (D. L. Laskin et al., unpublished observation), demonstrate the importance of this anti-inflammatory cytokine in the pathogenesis of ozone-induced toxicity.

3.2. Inorganic Particles: Silica and Asbestos

Asbestos and silica are complex naturally occurring minerals that are chemically and physically distinct. However, both induce pulmonary fibrosis after chronic exposure. Silica or silicon dioxide, mostly in its quartz form, is commonly found throughout the earth's crust and is particularly hazardous in underground mining or operations using sand. Asbestos is a general term for a variety of naturally occurring fibrous silicates that have the common property of high tensile strength, high heat resistance and relatively high chemical resistance *(81)*. Whereas the histological presentations of asbestosis and silicosis are distinct, the events leading to pulmonary fibrosis are similar. Both minerals stimulate the production of oxidants, chemokines, and cytokines in the lung *(82–84)*. These factors act in concert to induce chemotaxis, cell injury, proliferation, and collagen synthesis. Although alveolar macrophages are considered the primary source of these mediators, increasing evidence suggests that other lung cells, including type II epithelial cells, contribute to cytokine generation in the lung *(85–87)*. With both silica and asbestos, the latency period for development of pulmonary fibrosis is inversely proportional to exposure levels and generally ranges from years to decades.

The persistence of inflammation in the deep lung appears to be a key factor in the development of both silicosis and asbestosis *(81,88)*. Inflammatory cells are recruited to the lung by chemotactic factors released from silica or asbestos damaged tissue and/or particle-activated alveolar macrophages and epithelial cells. Chemokines reported to be upregulated in lung cells, whole lung, or BAL fluid after exposure of humans or rodents to silica or asbestos include IL-8, MIP-2, MIP-1α, MIP-1β, IP-10, MCP-1, and CINC *(89–103)*. The demonstration that anti-MIP-2 antiserum attenuates neutrophilia associated with exposure to silica supports the concept that chemokines are intrinsic to inflammation *(104)*.

Accumulating evidence suggests that TNF-α also plays an important role in silica and asbestos-induced inflammation. Expression of TNF-α, as well as TNFR1, increases in the lung after silica or asbestos exposure *(93,103,105)*. TNF-α production by isolated alveolar macrophages is also augmented. Moreover, increases in TNF-α production and TNFR expression correlate with the degree of inflammation in both humans and in animal models *(96,97,106–115)*. Clear evidence for a role of TNF-α in silica- and asbestos-induced inflammation and injury comes from studies demonstrating that these responses are markedly attenuated in mice pretreated with anti-TNF-α antibody *(105,116,117)* and in TNFR1 knockout mice *(118)*. Silica and asbestos-induced increases in lung chemokine expression are also reduced by TNF-α antibody treatment and in TNFR knockout mice *(93,119)*.

TNF-α also appears to be an important factor in the fibrotic response to silica and asbestos. Using a murine model of silicosis, Piguet et al. *(120,121)* demonstrated that passive immunization against TNF-α prevented silica-induced fibrosis. A similar abrogation of fibrosis was observed in TNFR1/ TNFR2 double knockout mice treated with silica or asbestos *(108,122)*. Mice overexpressing TNF-α in alveolar type II cells have also been reported to spontaneously develop pulmonary fibrosis that is similar to asbestosis *(123)*. In TNFR1/TNFR2 knockout mice treated with silica or asbestos, expression of TGF-β and PDGF, growth factors important in fibrogenesis are reduced suggesting a potential regulatory mechanism mediating the actions of TNF-α *(118,122)*.

IL-1 and IL-6 also have been implicated in the pathogenesis of pulmonary fibrosis induced by asbestos and silica. Elevated levels of these cytokines have been identified in BAL fluid from patients diagnosed with lung fibrosis and having histories of long-term asbestos exposure *(103,124,125)*. Similar increases in IL-1 and IL-6 have been described in BAL fluid and lung tissue from rodents and humans exposed to silica *(96,100,114,115,126)*. Acute intratracheal instillation of asbestos stimulates IL-1 and IL-6 release from alveolar macrophages *(113,127)*. This is associated with inflammatory cell recruitment into the lung. Moreover, IL-1 knockout mice exhibit reduced inflammation, apoptosis, and fewer silicotic lesions *(14)*, and an IL-1 receptor antagonist blocked the upregulatory effects of IL-1 on PDGF receptor expression during asbestos-induced fibrosis. Alveolar macrophage-derived IL-1β released following particle exposure is thought to trigger proliferation by upregulating PDGF receptors on fibroblasts *(128)*.

Recent studies suggest a role of IL-10 in silica-induced inflammation and injury. IL-10 levels are reported to be increased in BAL fluid and cells isolated from silica exposed animals *(116,129)*. Protein levels and numbers of inflammatory cells in BAL fluid are also attenuated in mice deficient in IL-10 *(129,130)*. Driscoll et al. *(131)* reported that treatment of animals with recombinant IL-10 attenuated quartz-induced inflammation and injury. Moreover, administration of anti-IL-10 antibody enhanced the responses to quartz. The mechanisms may be due, in part, to attenuation of MIP-2 expression. In contrast to these findings, Barbarin et al. *(130)* reported that pulmonary overexpression of IL-10 augments lung fibrosis induced by silica. These data suggest that IL-10 synthesis induced after silica exposure can limit the amplitude of the inflammatory response but may promote the fibrotic responses. This idea is supported by findings that silica-induced production of TGF-β is reduced in IL-10 knockout mice *(130)*. These data indicate that the profibrotic activity of IL-10 may be mediated by its ability to upregulate expression of TGF-β.

It is clear that one mechanism mediating the actions of cytokines involves interaction with growth factors such as TGF-α, TGF-β, and PDGF, which induce fibroblast proliferation and extracellular matrix protein production. TGF-α and TGF-β expression is upregulated in the lungs of developing fibrotic lesions induced by asbestos as well as silica *(107,118,132–134)*. Increased TGF-α immunoreactivity also is temporally related to increased proliferating cell nuclear antigen expression and bromodeoxyuridine uptake, which are markers of cellular proliferation *(107,134)*. These data, together with the findings that overexpression of TGF-α in the lungs of mice leads to the development of spontaneous pulmonary fibrosis *(135)*, demonstrate the importance of these growth factors in mineral-induced lung pathology.

To date, this mechanistic understanding has not yielded effective or reliable therapies for asbestosis or silicosis. Standard immunosuppressive therapies such as corticosteroids are not clinically effective for either condition, although dexamethasone has been shown to block activation of the transcription factor NF-κB in silica-exposed BAL cells *(136)*. Some specific anti-cytokine therapies have been tried in experimental models *(137,138)*. In humans, the Chinese traditional anti-inflammatory compound tetrandrine is reported to be effective in the treatment of radiographically evident silicosis; however, these reports have not yet been documented in rigorous clinical trials, nor have potential side-effects been systemically examined. Tetrandrine has been reported to, among other activities, inhibit IL-1, TNF-α, IL-6, and IL-8 production by monocytes *(139–141)*.

3.3. Diesel Exhaust: A Model for Particulate Matter Air Pollution

Epidemiological studies have linked ambient particulate matter (PM) air pollution with a number of health effects, including asthma, allergy, chronic obstructive pulmonary disease, lung cancer, and cardiopulmonary morbidity and mortality *(142,143)*. The "fine" PM fraction, consisting of particles with aerodynamic diameter less than 2.5 μm ($PM_{2.5}$), is readily deposited in the airways and alveoli and appears to be more toxic than larger particles *(144)*. $PM_{2.5}$ is a complex mixture of particles from a variety of sources, each with distinct physical and chemical characteristics. $PM_{2.5}$ in industrialized countries comes mainly from combustion of fossil fuels. Diesel exhaust particles (DEPs) are a major component of urban PM in the United States *(145)*. DEPs consist of a carbon core onto which more than several hundred identified compounds, including toxic hydrocarbons and small amounts of sulfate, nitrate, and trace metals, are adsorbed *(145)*. The characteristics of DEP vary with engine type, operating conditions, and fuel composition. Investigators have studied either collected DEP or diesel exhaust (DE), which includes both particle phase and gas phase pollutants, such as carbon monoxide and nitrogen oxides. Although there is clearly no "typical" PM, DEPs are a commonly used model of PM that has been studied extensively. Experimental studies of mechanisms of action of DE and DEP have yielded inconsistent and at times seemingly contradictory findings, which may, at least in part, be explained by the fact that DEP can exert various effects depending on the site of action in the lung and the concentration and duration of exposure. Variations in DEP characteristics and experimental methods also may explain differing results *(146,147)*. Increases in chemoattractant cytokines, such as IL-8 and RANTES, are the most consistent response in in vivo and in vitro studies of acute exposure to DEP *(148–153)*. In both human volunteers and animal models, increased IL-8 protein in lung fluid and messenger RNA in bronchial epithelium have been observed after exposure to DE and DEP *(146,154,155)*. Increases in IL-6 also are detected in lung fluid after DE exposure in human subjects *(154)*. These changes are accompanied by neutrophil influx into the airways and alveoli of human volunteers *(154,155)*, and in animal models *(146,156)*, suggesting that these cytokines may play a role in initiating and maintaining inflammation, which is a key feature of asthma and chronic obstructive pulmonary disease. Although mechanisms by which DEP and other PM may play a role in cardiovascular morbidity and mortality remain unclear, pulmonary inflammation is an initiating event in some hypothetical models *(157)*. In addition to chemokines, aspiration of DEP increases the concentration of the proinflammatory cytokines TNF-α, IL-1, and IL-6 in lung fluid in mice *(146,151–153)*, and human bronchial epithelial cells have been reported to release TNF-α, IL-1, IL-6, GM-CSF, and eotaxin in response to acute exposure to DEP *(158,159)*. In contrast, in a study of chronic exposure of mice to DE at concentrations at the upper end of human occupational exposure and far higher than ambient concentrations (up to 1000 μg/m^3 6 h per day for 6 mo), the only significant change among the lung fluid cytokines measured (TNF-α, MIP-2, IL-1β) was a decrease in TNF-α *(160)*. Differences between acute and chronic exposure effects may reflect adaptive homeostatic responses.

In addition to bronchial epithelial cells, alveolar macrophages are likely to be important targets for DEP. However, in vitro studies of macrophage responses to DEP have also produced inconsistent results. Some studies have shown that DEP enhance secretion of IL-8 by human and murine lung macrophages *(148)*, whereas others report suppression of IL-8 and other proinflammatory cytokines, such as IL-1, TNF-α, and IL-6 *(161,162)*. Suppression of macrophage function and production of cytokines that activate lung phagocytes in response to pathogens may help to explain the increased risk of respiratory infection, particularly pneumonia, which appears to contribute substantially to PM-associated increases in morbidity and mortality among elderly individuals *(163)*. In vitro exposure of rat alveolar macrophages to DEP decreased lipopolysaccharide (LPS)-induced production of IL-1 and TNF-α *(164)*. DEP also suppressed LPS-stimulated production of TNF-α and IL-6 by human alveolar macrophages, and inhibited IL-8 production from murine alveolar macrophages *(161)*. These findings are consistent with studies showing that intratracheal instillation of DEP suppressed macrophage function and pulmonary clearance of *L. monocytogenes* in rats *(165)*. Inhalation of DE also increased inflammation caused by respiratory syncytial virus, and DEP synergistically enhanced LPS-induced lung injury in mice *(166,167)*. The significance of these findings remains to be determined.

Exposure to DE and DEP may underlie epidemiological associations between increased incidence and exacerbation of asthma and living near roadways with heavy traffic (reviewed in ref. *164*). DEP has an adjuvant effect on responses to nasal allergen exposure in human volunteers with pre-existing allergy, in addition to potentiating responses to a neo-antigen *(169,170)*. In rodent sensitization models, DEP enhances airway inflammation and bronchial hyperreactivity when co-administered with the sensitizing agent *(171–174)*. DEP may exert these effects by shifting cytokine responses away from T helper-1 and toward T helper-2 responses. In human and animal models, DEP suppresses IFN-γ and increases TH2-type cytokines, including IL-4, IL-5, IL-10, and IL-13 *(169,175,176)*.

4. MECHANISMS REGULATING THE PRODUCTION OF INFLAMMATORY MEDIATORS IN THE LUNG AFTER XENOBIOTIC EXPOSURE

Recent studies have focused on analyzing potential biochemical mechanisms controlling excessive and/or dysregulated production of inflammatory cytokines in the lung after exposure to inhaled toxicants. Various stress activated signaling pathways, including phosphoinositide 3-kinase (PI 3-K) and its target, protein kinase B (PKB), as well as mitogen-activated protein (MAP) kinases, such as extracellular signal-regulated kinase (ERK), c-jun N-terminal kinase (JNK) and p38 MAPK and transcription factors like NF-κB, NF-IL-6, C/EBP, and activator protein-1 (AP-1), which are known to regulate the activity of inflammatory genes have been investigated. Results from these studies suggest that diverse signaling pathways and their crosstalk control inflammatory gene expression in the lung after toxicant exposure.

NF-κB is composed of a family of transcription factors known to control the expression of numerous cellular genes crucial for immunity, inflammation, and stress responses *(177,178)*. In mammalian cells, the NF-κB family is divided into Class I proteins (p50/p105 and p52/p100) and Class II proteins (RelA or p65, c-Rel, and RelB). Through a conserved Rel homology domain, family members form homodimers and heterodimers and bind to κB consensus sequences in promoter and enhancer regions of responsive genes *(178)*. The activity of NF-κB is controlled by inhibitory IκB proteins, which retain NF-κB in the cytoplasm in an inactive complex. After the binding of stimuli such as TNF-α or IL-1β to receptors on responsive cell types, or exposure to oxidants, NF-κB translocates into the nucleus. This process is dependent on stimulus-induced phosphorylation of IκB followed by its ubiquitination and degradation *(178)*. Increases in NF-κB nuclear binding activity have been observed in lung tissue and macrophages after exposure of animals to ozone, silica, asbestos, and DEP *(71,78,84,94,110,111,159, 167,177,179–182)*. Whereas early NF-κB activation in vivo is correlated with expression of chemokines such as CINC, IL-8, MCP-1, and MIP-2, secondary increases are associated with expression of inducible nitric oxide synthase (NOS II) and cyclooxygenase-2 (COX-2), which regulate production of inflammatory mediators like nitric oxide and eicosanoids, respectively. Ozone, silica, asbestos, and DEP have also been reported to activate NF-κB in cultured macrophages, epidermal cells, and/or bronchial epithelial cells *(124,182–192)*. Moreover, inflammatory gene expression induced by these toxicants in cultured cells is dependent on NF-κB activation. Thus, asbestos-induced TNF-α, IL-1, IL-6, and IL-8 expression in lung epithelial cells, and macrophages is prevented by antioxidants that inhibit NF-κB *(85,124,190,193,194)*. Similarly, DEP-induced IL-8, IL-18, and TGF-β production is abrogated by blocking NF-κB *(159,182,194–196)*. These findings, together with the observations that macrophages from transgenic mice lacking NF-κB p50 do not generate reactive nitrogen intermediates or TNF-α, and that these mice are protected from lung injury induced by ozone *(49)*, demonstrate the importance of the NF-κB signaling pathway in inflammatory mediator production and toxicity.

AP-1 is a family of transcription factors comprised of homodimers and heterodimers of the Jun and Fos early response protooncogenes. It is a redox-sensitive transcription factor classically associated with cell proliferation and tumor promotion *(197)*. AP-1 proteins are known to be phosphorylated and activated by MAP kinase signaling cascades *(198)*. Exposure of rodents to asbestos or silica leads to increases in AP-1 nuclear binding activity in the lung *(84,179,199–202)*. Increased AP-1 binding activity has also been observed in cultured alveolar macrophages and epithelial cells after exposure to asbestos, silica, or ozone *(183,191,203)*. In contrast, DEP had no effect on AP-1 nuclear binding in lung cells suggesting that other transcription factors regulate inflammatory genes activated by this toxicant *(182,186,204)*.

Studies have also demonstrated upregulation of the IFN-γ-inducible transcription factor, signal transducer and activator of transcription (STAT)-1 in alveolar macrophages following ozone inhalation *(180)*. This is associated with increased nuclear expression of p91, one of the major STAT-1 subunits. IFN-γ is known to synergize with TNF-α and IL-1β in inducing nitric oxide production by macrophages *(205)*. This appears to be due to the fact that the promoter region of the NOS II gene contains binding sites for both NF-κB and STAT-1, which act cooperatively to activate the gene *(205)*. In addition to enhancing expression of specific genes such as NOS II, synergistic interactions between transcription factors may contribute to sustained or prolonged expression of inflammatory mediators in the lung.

A question arises about the biochemical pathways regulating transcription factor activity in the lung after xenobiotic exposure. Although oxidative stress induced directly by pulmonary toxicants or indirectly by reactive oxygen and nitrogen species released from inflammatory cells can activate redox sensitive transcription factors like NF-κB and AP-1, the early response cytokines, TNF-α and IL-1β may also be involved in inducing their activity. PI 3-K/PKB, as well as p38 MAP kinase, ERK1/2 or p44/42, and JNK are upstream signaling molecules implicated in the regulation of transcription factor activity. These proteins are readily activated by cytokines and oxidants and they may play a role in regulating inflammatory gene expression in macrophages and lung cells after exposure to inhaled toxicants. Following ozone inhalation, PI 3-K and PKB expression are increased in alveolar macrophages *(206)*. PI 3-K/PKB is also increased in cultured macrophages exposed to silica *(207)*, and in epidermal cells treated with DEP *(186)*. The observation that blocking PI 3-K inhibits ozone-induced production of nitric oxide and NF-κB activation demonstrates the importance of these enzymes in macrophage production of inflammatory mediators in this model of toxicity.

ERK, JNK, and p38 MAP kinase also have been reported to be activated in lung macrophages, epithelial cells, and/or mesothelial cells after exposure of animals to asbestos, ozone and silica *(208–213)*. DEP also induces phosphorylation of ERK, JNK, and p38 in lung epithelial cells *(146,148,208)*, whereas silica and asbestos activate ERK and p38 MAP kinase in cultured macrophages and epithelial cells *(148,150,192,194)*. Blocking the activity of p38 MAP kinase has been reported to abrogate DEP-induced IL-8, RANTES, ICAM, and GM-CSF expression *(182,195,215–217)*, as well as TGF-β and procollagen *(190)*. p38 MAP kinase inhibitors also prevent silica-induced IL-8 release from epithelial cells *(196)*. Similarly, asbestos-induced PDGF and TGF-β gene expression in tracheal explants is ERK dependent *(218)*. These findings suggest that protein kinases may be early signaling molecules in the pathway leading to the generation of inflammatory mediators in the lung following ozone exposure. A schematic representation of potential signaling pathways activated following exposure to pulmonary toxicants is shown in Figure 1. Further in vivo investigation of these signaling molecules is necessary in order to elucidate their precise role in macrophage activation, cytokine production and lung injury.

5. CONCLUSIONS AND IMPLICATIONS FOR HUMAN HEALTH

Despite our understanding of the mechanisms leading to pulmonary disease after exposure to air pollutants, and the role of inflammatory cytokines in this process, at present, the only reliable way to intervene in these diseases remains to limit exposure. Nevertheless, elucidation of early biological responses to exposures may facilitate the development of strategies for early recognition and prevention of disease. For example, increased serum TNF-α and IL-8 have been reported in workers occupationally exposed to polypropylene (nylon) flock used to make microfiber garments *(220)*. Although the precise role of these cytokines in toxicity induced by nylon flock is unknown, preventive actions have already been initiated. For environmental exposures, the development and implementation of effective engineering and administrative protective measures is generally more complicated and difficult. Thus, translation of mechanisms of disease into prevention or therapeutics may be especially valuable, as for example, in the case of DEP exposure and asthma. New therapeutic approaches will have to both be proven efficacious and to exhibit acceptable levels of toxicity, a considerable challenge given the pleiotropic actions of many of the cytokines. Prevention will always be the most desirable approach, and as our mechanistic understanding improves so can the specificity of our preventive efforts.

ACKNOWLEDGMENTS

This work was supported by USPHS National Institutes of Health grants GMO34310, ESO04738, ESO05022, ESO13520, EPA grant R832144, and grants from the ALA-NJ and the Department of Defense (DAMD17–03–1–0537).

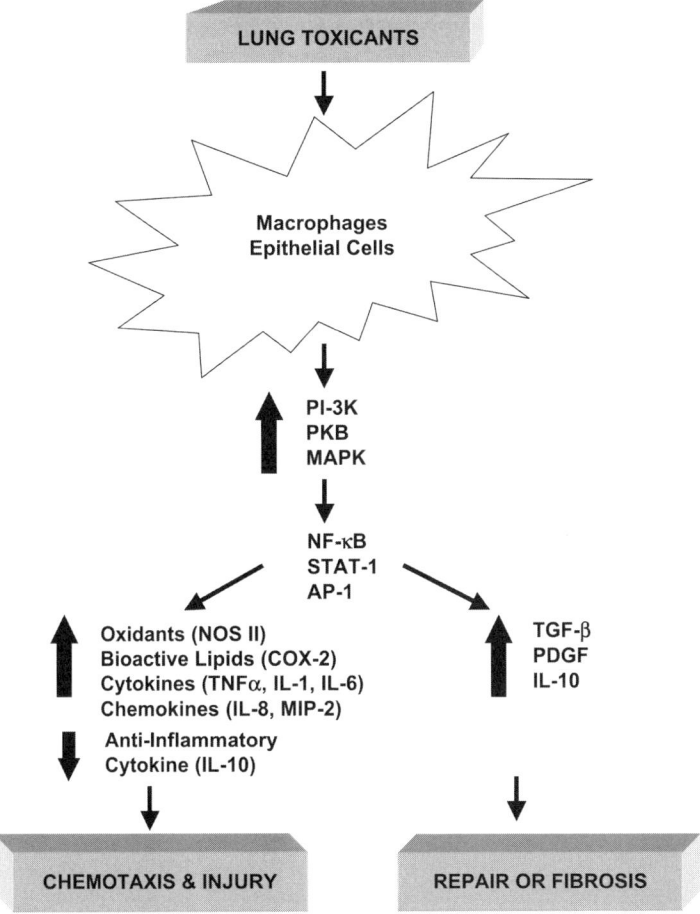

Fig. 1. Cellular responses to inhaled toxicants and plausible signaling pathways leading to injury, repair, or fibrosis in the lung.

REFERENCES

1. Cross CE, Halliwell B. Biological consequences of general environmental contaminants. In: Crystal RG,West JB, eds. The Lung: Scientific foundations, Ed. New York: 1991, pp. 1961–1973.
2. Lippman M, Yeates DB, Albert RE. Deposition, retention and clearance of inhaled particles. Br J Ind Med 1980;37:337–362.
3. Laskin DL, Laskin JD. Macrophages, Inflammatory Mediators, and Lung Injury. Methods 1996;10:61–70.
4. Laskin DL, Gardner CR, Gerecke DR, Laskin JD. Ozone-induced lung injury: role of macrophages and inflammatory mediators. In: Vallyathan V, Shi

X,Castranova V, eds. Reactive Oxygen/Nitrogen Species: Lung Injury and Disease, Ed. New York: 2004, pp. 289–316.

5. Laskin DL, Pendino KJ. Macrophages and inflammatory mediators in tissue injury. Annu Rev Pharmacol Toxicol 1995;35:655–677.

6. Larrick JW, Kunkel SL. The role of tumor necrosis factor and interleukin 1 in the immunoinflammatory response. Pharm Res 1988;5:129–139.

7. Whicher JT, Evans SW. Cytokines in disease. Clin Chem 1990;36:1269–1281.

8. Cerami A. Inflammatory cytokines. Clin Immunol Immunopathol 1992;62: S3–S10.

9. Zhang P, Summer WR, Bagby GJ, Nelson S. Innate immunity and pulmonary host defense. Immunol Rev 2000;173:39–51.

10. Pathania V, Syal N, Pathak CM, Khanduja KL. Vitamin E suppresses the induction of reactive oxygen species release by lipopolysaccharide, interleukin-1beta and tumor necrosis factor-alpha in rat alveolar macrophages. J Nutr Sci Vitaminol (Tokyo) 1999;45:675–686.

11. Lavnikova N, Drapier JC, Laskin DL. A single exogenous stimulus activates resident rat macrophages for nitric oxide production and tumor cytotoxicity. J Leukoc Biol 1993;54:322–328.

12. Larrick JW, Wright SC. Cytotoxic mechanism of tumor necrosis factor-alpha. FASEB J 1990;4:3215–3223.

13. Hirano S. Nitric oxide-mediated cytotoxic effects of alveolar macrophages on transformed lung epithelial cells are independent of the beta 2 integrin-mediated intercellular adhesion. Immunology 1998;93:102–108.

14. Srivastava KD, Rom WN, Jagirdar J, Yie TA, Gordon T, Tchou-Wong KM. Crucial role of interleukin-1beta and nitric oxide synthase in silica-induced inflammation and apoptosis in mice. Am J Respir Crit Care Med 2002;165: 527–533.

15. Kips JC, Tavernier J, Pauwels RA. Tumor necrosis factor causes bronchial hyperresponsiveness in rats. Am Rev Respir Dis 1992;145:332–336.

16. Ohkawara Y, Yamauchi K, Tanno Y, et al. Human lung mast cells and pulmonary macrophages produce tumor necrosis factor-alpha in sensitized lung tissue after IgE receptor triggering. Am J Respir Cell Mol Biol 1992;7:385–392.

17. Schmitz N, Kurrer M, Kopf M. The IL-1 receptor 1 is critical for Th2 cell type airway immune responses in a mild but not in a more severe asthma model. Eur J Immunol 2003;33:991–1000.

18. Dubaybo BA. Role of tumor necrosis factor-alpha in regulating fibrotic lung repair. Res Commun Mol Pathol Pharmacol 1998;101:69-83.

19. Sasaki M, Kashima M, Ito T, et al. Differential regulation of metalloproteinase production, proliferation and chemotaxis of human lung fibroblasts by PDGF, interleukin-1beta and TNF-alpha. Mediators Inflamm 2000;9:155–160.

20. Ward PA. Oxygen radicals, cytokines, adhesion molecules, and lung injury. Environ Health Perspect 1994;102 Suppl 10:13–26.

21. Ward PA. Role of complement, chemokines, and regulatory cytokines in acute lung injury. Ann N Y Acad Sci 1996;796:104–112.

22. Tilg H, Trehu E, Atkins MB, Dinarello CA, Mier JW. Interleukin-6 (IL-6) as an anti-inflammatory cytokine: induction of circulating IL-1 receptor antagonist and soluble tumor necrosis factor receptor p55. Blood 1994;83:113–118.
23. Yu M, Zheng X, Witschi H, Pinkerton KE. The role of interleukin-6 in pulmonary inflammation and injury induced by exposure to environmental air pollutants. Toxicol Sci 2002;68:488–497.
24. Rollwagen FM, Yu ZY, Li YY, Pacheco ND. IL-6 rescues enterocytes from hemorrhage induced apoptosis in vivo and in vitro by a bcl-2 mediated mechanism. Clin Immunol Immunopathol 1998;89:205–213.
25. Shingu M, Isayama T, Yasutake C et al. Role of oxygen radicals and IL-6 in IL-1-dependent cartilage matrix degradation. Inflammation 1994;18:613–623.
26. Ulich TR, Yin S, Guo K, Yi ES, Remick D, del Castillo J. Intratracheal injection of endotoxin and cytokines. II. Interleukin-6 and transforming growth factor beta inhibit acute inflammation. Am J Pathol 1991;138:1097–1101.
27. Ward NS, Waxman AB, Homer RJ et al. Interleukin-6-induced protection in hyperoxic acute lung injury. Am J Respir Cell Mol Biol 2000;22:535–542.
28. DiCosmo BF, Geba GP, Picarella D et al. Airway epithelial cell expression of interleukin-6 in transgenic mice. Uncoupling of airway inflammation and bronchial hyperreactivity. J Clin Invest 1994;94:2028–2035.
29. Kuhn C, 3rd, Homer RJ, Zhu Z et al. Airway hyperresponsiveness and airway obstruction in transgenic mice. Morphologic correlates in mice overexpressing interleukin (IL)-11 and IL-6 in the lung. Am J Respir Cell Mol Biol 2000;22: 289–295.
30. Johnston RA, Schwartzman IN, Flynt L, Shore SA. Role of interleukin-6 in murine airway responses to ozone. Am J Physiol Lung Cell Mol Physiol 2005; 288:L390–L397.
31. Fiorentino DF, Zlotnik A, Mosmann TR, Howard M, O'Garra A. IL-10 inhibits cytokine production by activated macrophages. J Immunol 1991;147:3815–3822.
32. Wang P, Ba ZF, Chaudry IH. Nitric oxide. To block or enhance its production during sepsis? Arch Surg 1994;129:1137–1142.
33. Wang P, Wu P, Siegel MI, Egan RW, Billah MM. Interleukin (IL)-10 inhibits nuclear factor kappaB (NF kB) activation in human monocytes. IL-10 and IL-4 suppress cytokine synthesis by different mechanisms. J Biol Chem 1995;270:9558–9563.
34. Oberholzer A, Oberholzer C, Moldawer LL. Interleukin-10: a complex role in the pathogenesis of sepsis syndromes and its potential as an anti-inflammatory drug. Crit Care Med 2002;30:S58–S63.
35. Kumar A, Creery WD. The therapeutic potential of interleukin 10 in infection and inflammation. Arch Immunol Ther Exp (Warsz) 2000;48:529–538.
36. Opal SM, Huber CE. The role of interleukin-10 in critical illness. Curr Opin Infect Dis 2000;13:221–226.
37. Mulligan MS, Jones ML, Vaporciyan AA, Howard MC, Ward PA. Protective effects of IL-4 and IL-10 against immune complex-induced lung injury. J Immunol 1993;151:5666–5674.

38. Bartram U, Speer CP. The role of transforming growth factor beta in lung development and disease. Chest 2004;125:754–765.
39. Bonner JC. Regulation of PDGF and its receptors in fibrotic diseases. Cytokine Growth Factor Rev 2004;15:255–273.
40. Pendino KJ, Meidhof TM, Heck DE, Laskin JD, Laskin DL. Inhibition of macrophages with gadolinium chloride abrogates ozone-induced pulmonary injury and inflammatory mediator production. Am J Respir Cell Mol Biol 1995;13: 125–132.
41. Pendino KJ, Shuler RL, Laskin JD, Laskin DL. Enhanced production of interleukin-1, tumor necrosis factor-alpha, and fibronectin by rat lung phagocytes following inhalation of a pulmonary irritant. Am J Respir Cell Mol Biol 1994;11:279–286.
42. Ishii Y, Yang H, Sakamoto T et al. Rat alveolar macrophage cytokine production and regulation of neutrophil recruitment following acute ozone exposure. Toxicol Appl Pharmacol 1997;147:214–223.
43. Weller BL, Witschi H, Pinkerton KE. Quantitation and localization of pulmonary manganese superoxide dismutase and tumor necrosis factor alpha following exposure to ozone and nitrogen dioxide. Toxicol Sci 2000;54:452–461.
44. Johnston CJ, Oberdorster G, Gelein R, Finkelstein JN. Newborn mice differ from adult mice in chemokine and cytokine expression to ozone, but not to endotoxin. Inhal Toxicol 2000;12:205–224.
45. Cohen MD, Sisco M, Li Y, Zelikoff JT, Schlesinger RB. Ozone-induced modulation of cell-mediated immune responses in the lungs. Toxicol Appl Pharmacol 2001;171:71–84.
46. Bhalla DK, Reinhart PG, Bai C, Gupta SK. Amelioration of ozone-induced lung injury by anti-tumor necrosis factor alpha. Toxicol Sci 2002;69:400–408.
47. Shore SA, Rivera-Sanchez YM, Schwartzman IN, Johnston RA. Responses to ozone are increased in obese mice. J Appl Physiol 2003;95:938–945.
48. Fakhrzadeh L, Laskin JD, Gardner CR, Laskin DL. Superoxide dismutase-overexpressing mice are resistant to ozone-induced tissue injury and increases in nitric oxide and tumor necrosis factor-alpha. Am J Respir Cell Mol Biol 2004;30:280–287.
49. Fakhrzadeh L, Laskin JD, Laskin DL. Ozone-induced production of nitric oxide and TNF-alpha and tissue injury are dependent on NF-kappaB p50. Am J Physiol Lung Cell Mol Physiol 2004;287:L279–L285.
50. Park JW, Taube C, Swasey C et al. Interleukin-1 receptor antagonist attenuates airway hyperresponsiveness following exposure to ozone. Am J Respir Cell Mol Biol 2004;30:830–836.
51. Young C, Bhalla DK. Effects of ozone on the epithelial and inflammatory responses in the airways: role of tumor necrosis factor. J Toxicol Environ Health 1995;46:329–342.
52. Pearson AC, Bhalla DK. Effects of ozone on macrophage adhesion in vitro and epithelial and inflammatory responses in vivo: the role of cytokines. J Toxicol Environ Health 1997;50:143–157.

53. Fakhrzadeh L, Laskin JD, Laskin DL. Deficiency in inducible nitric oxide synthase protects mice from ozone-induced lung inflammation and tissue injury. Am J Respir Cell Mol Biol 2002;26:413–419.
54. Cho HY, Zhang LY, Kleeberger SR. Ozone-induced lung inflammation and hyperreactivity are mediated via tumor necrosis factor-alpha receptors. Am J Physiol Lung Cell Mol Physiol 2001;280:L537–L546.
55. Shore SA, Schwartzman IN, Le Blanc B, Murthy GG, Doerschuk CM. Tumor necrosis factor receptor 2 contributes to ozone-induced airway hyperresponsiveness in mice. Am J Respir Crit Care Med 2001;164:602–607.
56. Thomas PS, Yates DH, Barnes PJ. Tumor necrosis factor-alpha increases airway responsiveness and sputum neutrophilia in normal human subjects. Am J Respir Crit Care Med 1995;152:76–80.
57. Wesselius LJ, Smirnov IM, O'Brien-Ladner AR, Nelson ME. Synergism of intratracheally administered tumor necrosis factor with interleukin-1 in the induction of lung edema in rats. J Lab Clin Med 1995;125:618–625.
58. Koh Y, Hybertson BM, Jepson EK, Repine JE. Tumor necrosis factor induced acute lung leak in rats: less than with interleukin-1. Inflammation 1996;20:461–469.
59. Dye JA, Madden MC, Richards JH, Lehmann JR, Devlin RB, Costa DL. Ozone effects on airway responsiveness, lung injury, and inflammation. Comparative rat strain and in vivo/in vitro investigations. Inhal Toxicol 1999;11:1015–1040.
60. Johnston CJ, Stripp BR, Reynolds SD, Avissar NE, Reed CK, Finkelstein JN. Inflammatory and antioxidant gene expression in C57BL/6J mice after lethal and sublethal ozone exposures. Exp Lung Res 1999;25:81–97.
61. Johnston CJ, Reed CK, Avissar NE, Gelein R, Finkelstein JN. Antioxidant and inflammatory response after acute nitrogen dioxide and ozone exposures in C57Bl/6 mice. Inhal Toxicol 2000;12:187–203.
62. Jonsson LM, Edlund T, Marklund SL, Sandstrom T. Increased ozone-induced airway neutrophilic inflammation in extracellular-superoxide dismutase null mice. Respir Med 2002;96:209–214.
63. Jorres RA, Holz O, Zachgo W et al. The effect of repeated ozone exposures on inflammatory markers in bronchoalveolar lavage fluid and mucosal biopsies. Am J Respir Crit Care Med 2000;161:1855–1861.
64. Bornholdt J, Dybdahl M, Vogel U, Hansen M, Loft S, Wallin H. Inhalation of ozone induces DNA strand breaks and inflammation in mice. Mutat Res 2002;520:63–71.
65. Shore SA, Johnston RA, Schwartzman IN, Chism D, Krishna Murthy GG. Ozone-induced airway hyperresponsiveness is reduced in immature mice. J Appl Physiol 2002;92:1019–1028.
66. Samet JM, Hatch GE, Horstman D et al. Effect of antioxidant supplementation on ozone-induced lung injury in human subjects. Am J Respir Crit Care Med 2001;164:819–825.
67. Vincent R, Vu D, Hatch G et al. Sensitivity of lungs of aging Fischer 344 rats to ozone: assessment by bronchoalveolar lavage. Am J Physiol Lung Cell Mol Physiol 1996;271:L555–L565.

68. Johnston C, Holm B, Finkelstein JN. Differential proinflammatory cytokine responses of the lung to ozone and lipopolysaccharide exposure during postnatal development. Exp Lung Res 2004;30:599–614.

69. Bhalla DK, Gupta SK. Lung injury, inflammation, and inflammatory stimuli in rats exposed to ozone. J Toxicol Environ Health 2000;59:211–228.

70. Driscoll KE, Simpson L, Carter J, Hassenbein D, Leikauf GD. Ozone inhalation stimulates expression of a neutrophil chemotactic protein, macrophage inflammatory protein 2. Toxicol Appl Pharmacol 1993;119:306–309.

71. Zhao Q, Simpson LG, Driscoll KE, Leikauf GD. Chemokine regulation of ozone-induced neutrophil and monocyte inflammation. Am J Physiol Lung Cell Mol Physiol 1998;274:L39–L46.

72. Johnston CJ, Oberdorster G, Finkelstein JN. Recovery from oxidant-mediated lung injury: response of metallothionein, MIP-2, and MCP-1 to nitrogen dioxide, oxygen, and ozone exposures. Inhal Toxicol 2001;13:689–702.

73. Johnston RA, Mizgerd JP, Shore SA. CXCR2 is essential for maximal neutrophil recruitment and methacholine responsiveness after ozone exposure. Am J Physiol Lung Cell Mol Physiol 2005;288:L61–L67.

74. Michalec L, Choudhury BK, Postlethwait E et al. CCL7 and CXCL10 orchestrate oxidative stress-induced neutrophilic lung inflammation. J Immunol 2002;168:846–852.

75. Kenyon NJ, van der Vliet A, Schock BC, Okamoto T, McGrew GM, Last JA. Susceptibility to ozone-induced acute lung injury in iNOS-deficient mice. Am J Physiol Lung Cell Mol Physiol 2002;282:L540–L545.

76. Park JW, Taube C, Joetham A et al. Complement activation is critical to airway hyperresponsiveness after acute ozone exposure. Am J Respir Crit Care Med 2004;169:726–732.

77. Haddad EB, Salmon M, Sun J et al. Dexamethasone inhibits ozone-induced gene expression of macrophage inflammatory protein-2 in rat lung. FEBS Lett 1995;363:285–288.

78. Haddad EB, Salmon M, Koto H, Barnes PJ, Adcock I, Chung KF. Ozone induction of cytokine-induced neutrophil chemoattractant (CINC) and nuclear factor-kappaB in rat lung: inhibition by corticosteroids. FEBS Lett 1996;379:265–268.

79. Koto H, Salmon M, Haddad el B, Huang TJ, Zagorski J, Chung KF. Role of cytokine-induced neutrophil chemoattractant (CINC) in ozone-induced airway inflammation and hyperresponsiveness. Am J Respir Crit Care Med 1997;156:234–239.

80. Reinhart PG, Gupta SK, Bhalla DK. Attenuation of ozone-induced lung injury by interleukin-10. Toxicol Lett 1999;110:35–42.

81. Osinubi OY, Gochfeld M, Kipen HM. Health effects of asbestos and nonasbestos fibers. Environ Health Perspect 2000;108:665–674.

82. Mossman BT, Churg A. Mechanisms in the pathogenesis of asbestosis and silicosis. Am J Respir Crit Care Med 1998;157:1666–1680.

83. Castranova V, Vallyathan V. Silicosis and coal workers' pneumoconiosis. Environ Health Perspect 2000;108 Suppl 4:675–684.

84. Castranova V. Signaling pathways controlling the production of inflammatory mediators in response to crystalline silica exposure: role of reactive oxygen/ nitrogen species. Free Radic Biol Med 2004;37:916–925.

85. Simeonova PP, Luster MI. Asbestos induction of nuclear transcription factors and interleukin 8 gene regulation. Am J Respir Cell Mol Biol 1996;15:787–795.

86. Driscoll KE, Carter JM, Hassenbein DG, Howard B. Cytokines and particle-induced inflammatory cell recruitment. Environ Health Perspect 1997;105 Suppl 5:1159–1164.

87. Kamp DW, Weitzman SA. The molecular basis of asbestos induced lung injury. Thorax 1999;54:638–652.

88. Shaw RJ. The role of lung macrophages at the interface between chronic inflammation and fibrosis. Respir Med 1991;85:267–273.

89. Broser M, Zhang Y, Aston C, Harkin T, Rom WN. Elevated interleukin-8 in the alveolitis of individuals with asbestos exposure. Int Arch Occup Environ Health 1996;68:109–114.

90. Johnston CJ, Driscoll KE, Finkelstein JN et al. Pulmonary chemokine and mutagenic responses in rats after subchronic inhalation of amorphous and crystalline silica. Toxicol Sci 2000;56:405–413.

91. Driscoll KE, Guthrie GD. Crystalline silica and silicosis. In: Roth RA, eds. Comprehensive Toxicology, First Ed. New York: Toxicology of the respiratory system, 1997, pp. 373–391.

92. Yucesoy B, Vallyathan V, Landsittel DP, Simeonova P, Luster MI. Cytokine polymorphisms in silicosis and other pneumoconioses. Mol Cell Biochem 2002;234-235:219–224.

93. Pryhuber GS, Huyck HL, Baggs R, Oberdorster G, Finkelstein JN. Induction of chemokines by low-dose intratracheal silica is reduced in TNFR I (p55) null mice. Toxicol Sci 2003;72:150–157.

94. Hubbard AK, Timblin CR, Shukla A, Rincon M, Mossman BT. Activation of NF-kappaB-dependent gene expression by silica in lungs of luciferase reporter mice. Am J Physiol Lung Cell Mol Physiol 2002;282:L968–L975.

95. Desaki M, Sugawara I, Iwakura Y, Yamamoto K, Takizawa H. Role of interferon-gamma in the development of murine bronchus-associated lymphoid tissues induced by silica in vivo. Toxicol Appl Pharmacol 2002;185:1–7.

96. Davis GS, Pfeiffer LM, Hemenway DR. Persistent overexpression of interleukin-1 beta and tumor necrosis factor-alpha in murine silicosis. J Environ Pathol Toxicol Oncol 1998;17:99–114.

97. Zeidler P, Hubbs A, Battelli L, Castranova V. Role of inducible nitric oxide synthase-derived nitric oxide in silica-induced pulmonary inflammation and fibrosis. J Toxicol Environ Health A 2004;67:1001–1026.

98. Hata J, Aoki K, Mitsuhashi H, Uno H. Change in location of cytokine-induced neutrophil chemoattractants (CINCs) in pulmonary silicosis. Exp Mol Pathol 2003;75:68–73.

99. Rao KM, Porter DW, Meighan T, Castranova V. The sources of inflammatory mediators in the lung after silica exposure. Environ Health Perspect 2004;112:1679–1686.

100. Yuen IS, Hartsky MA, Snajdr SI, Warheit DB. Time course of chemotactic factor generation and neutrophil recruitment in the lungs of dust-exposed rats. Am J Respir Cell Mol Biol 1996;15:268–274.

101. Hill GD, Mangum JB, Moss OR, Everitt JI. Soluble ICAM-1, MCP-1, and MIP-2 protein secretion by rat pleural mesothelial cells following exposure to amosite asbestos. Exp Lung Res 2003;29:277–290.

102. Mascagni P, Corsini E, Pettazzoni M et al. Determination of interleukin-8 in induced sputum of workers exposed to low concentrations of asbestos. G Ital Med Lav Ergon 2003;25 Suppl:137.

103. Vanhee D, Gosset P, Boitelle A, Wallaert B, Tonnel AB. Cytokines and cytokine network in silicosis and coal workers' pneumoconiosis. Eur Respir J 1995;8:834–842.

104. Driscoll KE, Howard BW, Carter JM et al. Alpha-quartz-induced chemokine expression by rat lung epithelial cells: effects of in vivo and in vitro particle exposure. Am J Pathol 1996;149:1627–1637.

105. Gossart S, Cambon C, Orfila C et al. Reactive oxygen intermediates as regulators of TNF-alpha production in rat lung inflammation induced by silica. J Immunol 1996;156:1540–1548.

106. Choe N, Tanaka S, Xia W, Hemenway DR, Roggli VL, Kagan E. Pleural macrophage recruitment and activation in asbestos-induced pleural injury. Environ Health Perspect 1997;105 Suppl 5:1257–1260.

107. Brass DM, Hoyle GW, Poovey HG, Liu JY, Brody AR. Reduced tumor necrosis factor alpha and transforming growth factor beta 1 expression in the lungs of inbred mice that fail to develop fibroproliferative lesions consequent to asbestos exposure. Am J Pathol 1999;154:853–862.

108. Ortiz LA, Lasky J, Lungarella G et al. Upregulation of the p75 but not the p55 TNF-alpha receptor mRNA after silica and bleomycin exposure and protection from lung injury in double receptor knockout mice. Am J Respir Cell Mol Biol 1999;20:825–833.

109. Corsini E, Giani A, Peano S, Marinovich M, Galli CL. Resistance to silica-induced lung fibrosis in senescent rats: role of alveolar macrophages and tumor necrosis factor-alpha (TNF). Mech Ageing Dev 2004;125:145–146.

110. Porter DW, Ye J, Ma J et al. Time course of pulmonary .response of rats to inhalation of crystalline silica: NF-kappa B activation, inflammation, cytokine production, and damage. Inhal Toxicol 2002;14:349–367.

111. Castranova V, Porter D, Millecchia L, Ma JY, Hubbs AF, Teass A. Effect of inhaled crystalline silica in a rat model: time course of pulmonary reactions. Mol Cell Biochem 2002;234-235:177–184.

112. Driscoll KE, Maurer JK, Lindenschmidt RC, Romberger D, Rennard SI, Crosby L. Respiratory tract responses to dust: relationships between dust burden, lung injury, alveolar macrophage fibronectin release, and the development of pulmonary fibrosis. Toxicol Appl Pharmacol 1990;106:88–101.

113. Driscoll KE, Maurer JK, Higgins J, Poynter J. Alveolar macrophage cytokine and growth factor production in a rat model of crocidolite-induced pulmonary inflammation and fibrosis. J Toxicol Environ Health 1995;46:155–169.

114. Orfila C, Lepert JC, Gossart S, Frisach MF, Cambon C, Pipy B. Immunocytochemical characterization of lung macrophage surface phenotypes and expression of cytokines in acute experimental silicosis in mice. Histochem J 1998;30:857–867.

115. Zhang Y, Lee TC, Guillemin B, Yu MC, Rom WN. Enhanced IL-1 beta and tumor necrosis factor-alpha release and messenger RNA expression in macrophages from idiopathic pulmonary fibrosis or after asbestos exposure. J Immunol 1993;150:4188–4196.

116. Huaux F, Arras M, Vink A, Renauld JC, Lison D. Soluble tumor necrosis factor (TNF) receptors p55 and p75 and interleukin-10 downregulate TNF-alpha activity during the lung response to silica particles in NMRI mice. Am J Respir Cell Mol Biol 1999;21:137–145.

117. Miller MD, Krangel MS. Biology and biochemistry of the chemokines: a family of chemotactic and inflammatory cytokines. Crit Rev Immunol 1992;12:17–46.

118. Liu JY, Brody AR. Increased TGF-beta1 in the lungs of asbestos-exposed rats and mice: reduced expression in TNF-alpha receptor knockout mice. J Environ Pathol Toxicol Oncol 2001;20:97–108.

119. Driscoll KE, Hassenbein DG, Carter JM, Kunkel SL, Quinlan TR, Mossman BT. TNF alpha and increased chemokine expression in rat lung after particle exposure. Toxicol Lett 1995;82-83:483–489.

120. Piguet PF. Is "tumor necrosis factor" the major effector of pulmonary fibrosis? Eur Cytokine Netw 1990;1:257–258.

121. Piguet PF, Collart MA, Grau GE, Sappino AP, Vassalli P. Requirement of tumour necrosis factor for development of silica-induced pulmonary fibrosis. Nature 1990;344:245–247.

122. Liu JY, Brass DM, Hoyle GW, Brody AR. TNF-alpha receptor knockout mice are protected from the fibroproliferative effects of inhaled asbestos fibers. Am J Pathol 1998;153:1839–1847.

123. Miyazaki Y, Araki K, Vesin C et al. Expression of a tumor necrosis factor-alpha transgene in murine lung causes lymphocytic and fibrosing alveolitis. A mouse model of progressive pulmonary fibrosis. J Clin Invest 1995;96:250–259.

124. Simeonova PP, Toriumi W, Kommineni C et al. Molecular regulation of IL-6 activation by asbestos in lung epithelial cells: role of reactive oxygen species. J Immunol 1997;159:3921–3928.

125. Kline JN, Schwartz DA, Monick MM, Floerchinger CS, Hunninghake GW. Relative release of interleukin-1 beta and interleukin-1 receptor antagonist by alveolar macrophages. A study in asbestos-induced lung disease, sarcoidosis, and idiopathic pulmonary fibrosis. Chest 1993;104:47–53.

126. Li XY, Lamb D, Donaldson K. The production of TNF-alpha and IL-1-like activity by bronchoalveolar leucocytes after intratracheal instillation of crocidolite asbestos. Int J Exp Pathol 1993;74:403–410.

127. Lemaire I, Ouellet S. Distinctive profile of alveolar macrophage-derived cytokine release induced by fibrogenic and nonfibrogenic mineral dusts. J Toxicol Environ Health 1996;47:465–478.
128. Lindroos PM, Coin PG, Badgett A, Morgan DL, Bonner JC. Alveolar macrophages stimulated with titanium dioxide, chrysotile asbestos, and residual oil fly ash upregulate the PDGF receptor-alpha on lung fibroblasts through an IL-1beta-dependent mechanism. Am J Respir Cell Mol Biol 1997;16:283–292.
129. Huaux F, Louahed J, Hudspith B et al. Role of interleukin-10 in the lung response to silica in mice. Am J Respir Cell Mol Biol 1998;18:51–59.
130. Barbarin V, Arras M, Misson P et al. Characterization of the effect of interleukin-10 on silica-induced lung fibrosis in mice. Am J Respir Cell Mol Biol 2004;31:78–85.
131. Driscoll KE, Carter JM, Howard BW et al. Interleukin-10 regulates quartz-induced pulmonary inflammation in rats. Am J Physiol Lung Cell Mol Physiol 1998;275:L887–L894.
132. Brody AR, Liu JY, Brass D, Corti M. Analyzing the genes and peptide growth factors expressed in lung cells in vivo consequent to asbestos exposure and in vitro. Environ Health Perspect 1997;105 Suppl 5:1165–1171.
133. Perdue TD, Brody AR. Distribution of transforming growth factor-beta1, fibronectin, and smooth muscle actin in asbestos-induced pulmonary fibrosis in rats. J Histochem Cytochem 1994;42:1061–1070.
134. Liu JY, Morris GF, Lei WH, Corti M, Brody AR. Up-regulated expression of transforming growth factor-alpha in the bronchiolar-alveolar duct regions of asbestos-exposed rats. Am J Pathol 1996;149:205–217.
135. Korfhagen TR, Swantz RJ, Wert SE et al. Respiratory epithelial cell expression of human transforming growth factor-alpha induces lung fibrosis in transgenic mice. J Clin Invest 1994;93:1691–1699.
136. Sacks M, Gordon J, Bylander J et al. Silica-induced pulmonary inflammation in rats: activation of NF-kappaB and its suppression by dexamethasone. Biochem Biophys Res Commun 1998;253:181–184.
137. Piguet PF, Rosen H, Vesin C, Grau GE. Effective treatment of the pulmonary fibrosis elicited in mice by bleomycin or silica with anti-CD-11 antibodies. Am Rev Respir Dis 1993;147:435–441.
138. Piguet PF, Vesin C, Grau GE, Thompson RC. Interleukin 1 receptor antagonist (IL-1ra) prevents or cures pulmonary fibrosis elicited in mice by bleomycin or silica. Cytokine 1993;5:57–61.
139. Xie QM, Tang HF, Chen JQ, Bian RL. Pharmacological actions of tetrandrine in inflammatory pulmonary diseases. Acta Pharmacol Sin 2002;23:1107–1113.
140. Seow WK, Ferrante A, Li SY, Thong YH. Suppression of human monocyte interleukin 1 production by the plant alkaloid tetrandrine. Clin Exp Immunol 1989;75:47–51.
141. Chang DM, Chang WY, Kuo SY, Chang ML. The effects of traditional anti-rheumatic herbal medicines on immune response cells. J Rheumatol 1997;24:436–441.

142. Donaldson K, Gilmour MI, MacNee W. Asthma and PM10. Respiratory Research 2000;1:12–15.

143. Pope CA, III, Burnett RT, Thun MJ et al. Lung cancer, cardiopulmonary mortality, and long-term exposure to fine particulate air pollution. JAMA 2002;287:1132–1141.

144. Air quality criteria for particulate matter. U. S. Environmental Protection Agency Research Triangle Park, NC: 2004.

145. Health assessment document for diesel engine exhaust. U. S. Envirinmental Protection Agency Washington DC: 2002.

146. Singh P, DeMarini DM, Dick CA et al. Sample characterization of automobile and forklift diesel exhaust particles and comparative pulmonary toxicity in mice. Environ Health Perspect 2004;112:820–825.

147. Kongerud J, Madden MC, Hazucha M, Peden D. Nasal responses in asthmatic and nonasthmatic subjects following exposure to diesel exhaust particles. Inhal Toxicol 2006;18:589–594.

148. Li N, Wang M, Oberley TD, Sempf JM, Nel AE. Comparison of the pro-oxidative and proinflammatory effects of organic diesel exhaust particle chemicals in bronchial epithelial cells and macrophages. J Immunol 2002;169: 4531–4541.

149. Abe S, Takizawa H, Sugawara I et al. Diesel exhaust (DE)-induced cytokine expression in human bronchial epithelial cells: a study with a new cell exposure system to freshly generated DE in vitro. Am J Resp Cell Mol Biol 2000; 22:296–303.

150. Hashimoto S, Gon Y, Takeshita I et al. Diesel exhaust particles activate p38 MAP kinase to produce interleukin 8 and RANTES by human bronchial epithelial cells and N-acetylcysteine attenuates p38 MAP kinase activation. Am J Respir Crit Care Med 2000;161:280–285.

151. Saber AT, Jacobsen NR, Bornholdt J et al. Cytokine expression in mice exposed to diesel exhaust particles by inhalation. Role of tumor necrosis factor. Part Fibre Toxicol 2006;3:4–11.

152. Inoue K, Takano H, Yanagisawa R et al. The role of toll-like receptor 4 in airway inflammation induced by diesel exhaust particles. Arch Toxicol 2006; 80:275–279.

153. Rao KM, Ma JY, Meighan T, Barger MW, Pack D, Vallyathan V. Time course of gene expression of inflammatory mediators in rat lung after diesel exhaust particle exposure. Environ Health Perspect 2005;113:612–617.

154. Nordenhall C, Pourazar J, Blomberg A, Levin JO, Sandstrom T, Adelroth E. Airway inflammation following exposure to diesel exhaust: a study of time kinetics using induced sputum. Eur Respir J 2000;15:1046–1051.

155. Stenfors N, Nordenhall C, Salvi SS et al. Different airway inflammatory responses in asthmatic and healthy humans exposed to diesel. Eur Respir J 2004;23:82–86.

156. Ichinose T, Furuyama A, Sagai M. Biological effects of diesel exhaust particles (DEP). II. Acute toxicity of DEP introduced into lung by intratracheal instillation. Toxicology 1995;99:153–167.

157. Donaldson K, Stone V, Seaton A, MacNee W. Ambient particle inhalation and the cardiovascular system: potential mechanisms. Environ Health Perspect 2001;109 Suppl 4:523–527.

158. Boland S, Bonvallot V, Fournier T et al. Mechanisms of GM-CSF increase by diesel exhaust particles in human airway epithelial cells. Am J of Physiol Lung Cell Mol Physiol 2000;278:L25–L32.

159. Takizawa H, Abe S, Okazaki H et al. Diesel exhaust particles upregulate eotaxin gene expression in human bronchial epithelial cells via nuclear factor-kappaB-dependent pathway. Am J Physiol Lung Cell Mol Physiol 2003; 284:L1055–L1062.

160. Reed MD, Gigliotti AP, McDonald JD et al. Health effects of subchronic exposure to environmental levels of diesel exhaust. Inhal Toxicol 2004;16:177–193.

161. Amakawa K, Terashima T, Matsuzaki T, Matsumaru A, Sagai M, Yamaguchi K. Suppressive effects of diesel exhaust particles on cytokine release from human and murine alveolar macrophages. Exp Lung Res 2003;29:149–164.

162. Becker S, Mundandhara S, Devlin RB, Madden M. Regulation of cytokine production in human alveolar macrophages and airway epithelial cells in response to ambient air pollution particles: Further mechanistic studies. Toxicol Appl Pharmacol 2005;207:269–275.

163. Pope CA, III. Epidemiology of fine particulate air pollution and human health: biologic mechanisms and who's at risk?. Environ Health Perspect 2000;108 Suppl 4:713–723.

164. Castranova V, Ma JY, Yang HM et al. Effect of exposure to diesel exhaust particles on the susceptibility of the lung to infection. Environ Health Perspect 2001;109 Suppl 4:609–612.

165. Yang HM, Antonini JM, Barger MW et al. Diesel exhaust particles suppress macrophage function and slow the pulmonary clearance of Listeria monocytogenes in rats. Environ Health Perspect 2001;109:515–521.

166. Harrod KS, Jaramillo RJ, Rosenberger CL et al. Increased susceptibility to RSV infection by exposure to inhaled diesel engine emissions. Am J Resp Cell Mol Biol 2003;28:451–463.

167. Takano H, Yanagisawa R, Ichinose T et al. Diesel exhaust particles enhance lung injury related to bacterial endotoxin through expression of proinflammatory cytokines, chemokines, and intercellular adhesion molecule-1. Am J Respir Crit Care Med 2002;165:1329–1335.

168. Delfino RJ. Epidemiologic evidence for asthma and exposure to air toxics: linkages between occupational, indoor, and community air pollution research. Environ Health Perspect 2002;110 Suppl 4:573–589.

169. Diaz-Sanchez D, Tsien A, Fleming J, Saxon A. Combined diesel exhaust particulate and ragweed allergen challenge markedly enhances human in vivo nasal ragweed-specific IgE and skews cytokine production to a T helper cell 2-type pattern. J Immunol 1997;158:2406–2413.

170. Diaz-Sanchez D, Garcia MP, Wang M, Jyrala M, Saxon A. Nasal challenge with diesel exhaust particles can induce sensitization to a neoallergen in the human mucosa. J Allergy Clin Immunol 1999;104:1183–1188.

171. Takano H, Yoshikawa T, Ichinose T, Miyabara Y, Imaoka K, Sagai M. Diesel exhaust particles enhance antigen-induced airway inflammation and local cytokine expression in mice. Am J Resp Crit Care Med 1997;156:36–42.

172. Takano H, Ichinose T, Miyabara Y, Yoshikawa T, Sagai M. Diesel exhaust particles enhance airway responsiveness following allergen exposure in mice. Immunopharmacol Immunotoxicol 1998;20:329–336.

173. Ichinose T, Takano H, Sadakane K et al. Mouse strain differences in eosinophilic airway inflammation caused by intratracheal instillation of mite allergen and diesel exhaust particles. J Appl Toxicol 2004;24:69–76.

174. Sadakane K, Ichinose T, Takano H et al. Murine strain differences in airway inflammation induced by diesel exhaust particles and house dust mite allergen. Int Arch Allergy Immunol 2002;128:220–228.

175. Finkelman FD, Yang M, Orekhova T et al. Diesel exhaust particles suppress in vivo IFN-gamma production by inhibiting cytokine effects on NK and NKT cells. J Immunol 2004;172:3808–3813.

176. Wang M, Saxon A, Diaz-Sanchez D. Early IL-4 production driving Th2 differentiation in a human in vivo allergic model is mast cell derived. Clin Immunol 1999;90:47–54.

177. Krishna MT, Chauhan AJ, Frew AJ, Holgate ST. Toxicological mechanisms underlying oxidant pollutant-induced airway injury. Rev Environ Health 1998;13:59–71.

178. May MJ, Ghosh S. Rel/NF-kappaB and I kappaB proteins: an overview. Semin Cancer Biol 1997;8:63–73.

179. Ortiz LA, Lasky J, Gozal E et al. Tumor necrosis factor receptor deficiency alters matrix metalloproteinase 13/tissue inhibitor of metalloproteinase 1 expression in murine silicosis. Am J Respir Crit Care Med 2001;163:244–252.

180. Laskin DL, Sunil V, Guo Y, Heck DE, Laskin JD. Increased nitric oxide synthase in the lung after ozone inhalation is associated with activation of NF-kappaB. Environ Health Perspect 1998;106 Suppl 5:1175–1178.

181. Hisada T, Adcock IM, Nasuhara Y et al. Inhibition of ozone-induced lung neutrophilia and nuclear factor-kappaB binding activity by vitamin A in rat. Eur J Pharmacol 1999;377:63–68.

182. Takizawa H, Ohtoshi T, Kawasaki S et al. Diesel exhaust particles induce NF-kappaB activation in human bronchial epithelial cells in vitro: importance in cytokine transcription. J Immunol 1999;162:4705–4711.

183. Jaspers I, Flescher E, Chen LC. Ozone-induced IL-8 expression and transcription factor binding in respiratory epithelial cells. Am J Physiol Lung Cell Mol Physiol 1997;272:L504–L511.

184. Nichols BG, Woods JS, Luchtel DL, Corral J, Koenig JQ. Effects of ozone exposure on nuclear factor-kappaB activation and tumor necrosis factor-alpha expression in human nasal epithelial cells. Toxicol Sci 2001;60:356–362.

185. Yun YP, Joo JD, Lee JY et al. Induction of nuclear factor-kappaB activation through TAK1 and NIK by diesel exhaust particles in L2 cell lines. Toxicol Lett 2005;155:337–342.

186. Ma C, Wang J, Luo J. Activation of nuclear factor kappaB by diesel exhaust particles in mouse epidermal cells through phosphatidylinositol 3-kinase/Akt signaling pathway. Biochem Pharmacol 2004;67:1975–1983.

187. Takizawa H. Diesel exhaust particles and their effect on induced cytokine expression in human bronchial epithelial cells. Curr Opin Allergy Clin Immunol 2004;4:355–359.

188. Kang JL, Go YH, Hur KC, Castranova V. Silica-induced nuclear factor-kappaB activation: involvement of reactive oxygen species and protein tyrosine kinase activation. J Toxicol Environ Health A 2000;60:27–46.

189. Rojanasakul Y, Ye J, Chen F et al. Dependence of NF-kappaB activation and free radical generation on silica-induced TNF-alpha production in macrophages. Mol Cell Biochem 1999;200:119–125.

190. Chen F, Sun SC, Kuh DC, Gaydos LJ, Demers LM. Essential role of NF-kappaB activation in silica-induced inflammatory mediator production in macrophages. Biochem Biophys Res Commun 1995;214:985–992.

191. Gambelli F, Di P, Niu X et al. Phosphorylation of tumor necrosis factor receptor 1 (p55) protects macrophages from silica-induced apoptosis. J Biol Chem 2004;279:2020–2029.

192. Pourazar J, Mudway IS, Samet JM et al. Diesel exhaust activates redox-sensitive transcription factors and kinases in human airways. Am J Physiol Lung Cell Mol Physiol 2005;289:L724–L730.

193. Brown DM, Beswick PH, Donaldson K. Induction of nuclear translocation of NF-kappaB in epithelial cells by respirable mineral fibres. J Pathol 1999;189:258–264.

194. Cheng N, Shi X, Ye J et al. Role of transcription factor NF-kappaB in asbestos-induced TNF alpha response from macrophages. Exp Mol Pathol 1999; 66:201–210.

195. Kawasaki S, Takizawa H, Takami K et al. Benzene-extracted components are important for the major activity of diesel exhaust particles: effect on interleukin-8 gene expression in human bronchial epithelial cells. Am J Respir Cell Mol Biol 2001;24:419–426.

196. Dai J, Xie C, Vincent R, Churg A. Air pollution particles produce airway wall remodeling in rat tracheal explants. Am J Respir Cell Mol Biol 2003; 29:352–358.

197. Albrecht C, Borm PJ, Unfried K. Signal transduction pathways relevant for neoplastic effects of fibrous and non-fibrous particles. Mutat Res 2004;553:23–35.

198. Rahman I. Oxidative stress and gene transcription in asthma and chronic obstructive pulmonary disease: antioxidant therapeutic targets. Curr Drug Targets Inflamm Allergy 2002;1:291–315.

199. Ding M, Dong Z, Chen F et al. Asbestos induces activator protein-1 transactivation in transgenic mice. Cancer Res 1999;59:1884–1889.

200. Quinlan TR, Marsh JP, Janssen YM et al. Dose-responsive increases in pulmonary fibrosis after inhalation of asbestos. Am J Respir Crit Care Med 1994;150:200–206.

201. Quinlan TR, BeruBe KA, Marsh JP et al. Patterns of inflammation, cell proliferation, and related gene expression in lung after inhalation of chrysotile asbestos. Am J Pathol 1995;147:728–739.

202. Mossman BT. Signal transduction by oxidants: look who's talking. Free Radic Biol Med 2000;28:1315–1316.

203. Flaherty DM, Monick MM, Carter AB, Peterson MW, Hunninghake GW. Oxidant-mediated increases in redox factor-1 nuclear protein and activator protein-1 DNA binding in asbestos-treated macrophages. J Immunol 2002; 168:5675–5681.

204. Li N, Venkatesan MI, Miguel A et al. Induction of heme oxygenase-1 expression in macrophages by diesel exhaust particle chemicals and quinones via the antioxidant-responsive element. J. Immunol 2000;165:3393–3401.

205. Ganster RW, Taylor BS, Shao L, Geller DA. Complex regulation of human inducible nitric oxide synthase gene transcription by Stat 1 and NF-kappaB. Proc Natl Acad Sci U S A 2001;98:8638–8643.

206. Laskin DL, Fakhrzadeh L, Heck DE, Gerecke D, Laskin JD. Upregulation of phosphoinositide 3-kinase and protein kinase B in alveolar macrophages following ozone inhalation. Role of NF-kappaB and STAT-1 in ozone-induced nitric oxide production and toxicity. Mol Cell Biochem 2002;234–235:91–98.

207. Kang JL, Lee HS, Pack IS, Hur KC, Castranova V. Phosphoinositide 3-kinase activity leads to silica-induced NF-kappaB activation through interacting with tyrosine-phosphorylated IkappaB-alpha and contributing to tyrosine phosphorylation of p65 NF-kappaB. Mol Cell Biochem 2003;248:17–24.

208. Cummins AB, Palmer C, Mossman BT, Taatjes DJ. Persistent localization of activated extracellular signal-regulated kinases (ERK1/2) is epithelial cell-specific in an inhalation model of asbestosis. Am J Pathol 2003;162:713–720.

209. Fubini B, Hubbard A. Reactive oxygen species (ROS) and reactive nitrogen species (RNS) generation by silica in inflammation and fibrosis. Free Radic Biol Med 2003;34:1507–1516.

210. Robledo R, Mossman B. Cellular and molecular mechanisms of asbestos-induced fibrosis. J Cell Physiol 1999;180:158–166.

211. Yuan Z, Taatjes DJ, Mossman BT, Heintz NH. The duration of nuclear extracellular signal-regulated kinase 1 and 2 signaling during cell cycle reentry distinguishes proliferation from apoptosis in response to asbestos. Cancer Res 2004;64:6530–6536.

212. Geist LJ, Powers LS, Monick MM, Hunninghake GW. Asbestos stimulation triggers differential cytokine release from human monocytes and alveolar macrophages. Exp Lung Res 2000;26:41–56.

213. Robledo RF, Buder-Hoffmann SA, Cummins AB, Walsh ES, Taatjes DJ, Mossman BT. Increased phosphorylated extracellular signal-regulated kinase immunoreactivity associated with proliferative and morphologic lung alterations after chrysotile asbestos inhalation in mice. Am J Pathol 2000;156:1307–1316.

214. Zhang Q, Kleeberger SR, Reddy SP. DEP-induced fra-1 expression corre-lates with a distinct activation of AP-1-dependent gene transcription in the lung. Am J Physiol Lung Cell Mol Physiol 2004;286:L427–L436.

215. Hashimoto S, Matsumoto K, Gon Y et al. p38 MAP kinase regulates TNF alpha-, IL-1 alpha- and PAF-induced RANTES and GM-CSF production by human bronchial epithelial cells. Clin Exp Allergy 2000;30:48–55.

216. Bonvallot V, Baeza-Squiban A, Baulig A et al. Organic compounds from die-sel exhaust particles elicit a proinflammatory response in human airway epi-thelial cells and induce cytochrome p450 1A1 expression. Am J Respir Cell Mol Biol 2001;25:515–521.

217. Ohtoshi T, Takizawa H, Okazaki H et al. Diesel exhaust particles stimulate human airway epithelial cells to produce cytokines relevant to airway inflam-mation in vitro. J Allergy Clin Immunol 1998;101:778–785.

218. Ovrevik J, Lag M, Schwarze P, Refsnes M. p38 and Src-ERK1/2 pathways regulate crystalline silica-induced chemokine release in pulmonary epithelial cells. Toxicol Sci 2004;81:480–490.

219. Dai J, Churg A. Relationship of fiber surface iron and active oxygen species to expression of procollagen, PDGF-A, and TGF-beta1 in tracheal explants exposed to amosite asbestos. Am J Respir Cell Mol Biol 2001;24:427–435.

220. Atis S, Tutluoglu B, Levant E et al. The respiratory effects of occupational polypropylene flock exposure. Eur Respir J. 2005;25:110–117.

7

Cytokine Polymorphisms and Relationship to Disease

Berran Yucesoy, Victor J. Johnson, Michael L. Kashon, and Michael I. Luster

SUMMARY

Cytokines are polypeptide mediators produced by a variety of cell types that play crucial roles in immune and inflammatory responses. Genes that code for cytokines are highly polymorphic, and some of these polymorphisms directly or indirectly influence cytokine expression. The most frequent types of mutations are characterized by a change in a single nucleotide base pair and are called single nucleotide polymorphisms (SNPs). Many SNPs that affect cytokine expression represent disease modifiers and influence the severity or progression of immune-mediated and chronic inflammatory diseases. SNPs in cytokine genes have been associated with common diseases, including cardiovascular diseases, cancer, neurodegenerative diseases, allergy, and asthma, and data are now accumulating on their role in occupational/environmental diseases. All these diseases are multigenic and multifactorial in nature and involve interactions between genetic, physiological, and environmental factors. Currently, there exist inconsistencies in association studies examining relationships between cytokine SNPs and disease because of known limitations in population-based studies. Recent advances in genotyping platforms for large-scale genetic studies, more robust study designs, and haplotype analysis should help reduce the amount of spurious and inconsistent associations in the literature and allow for incorporating genetic information into the risk assessment process although challenges still remain.

Key Words: Cytokine; polymorphism; SNP; common diseases; autoimmune diseases; occupational diseases; epidemiology.

From: *Methods in Pharmacology and Toxicology: Cytokines in Human Health:*
Immunotoxicology, Pathology, and Therapeutic Applications
Edited by: R. V. House and J. Descotes © Humana Press Inc., Totowa, NJ

1. INTRODUCTION

Although many rare genetic variants exist in the human genome, most of the heterozygosity can be attributed to common alleles. The rare mutations usually are the cause of rare monogenetic diseases, such as Huntington's disease. These diseases are of recent origin and highly penetrant. In contrast, the common allelic variants are present in high frequencies (>1%) in the general population. Among these, the most represented type of mutations is single nucleotide substitutions, referred to as single nucleotide polymorphisms (SNPs). Other polymorphisms, such as repeat sequences or insertion/deletions, are found less frequently in the genome. The widespread availability of human DNA sequence data has suggested that literally millions of SNPs exist but that the vast majority do not alter gene structure or function and therefore are unlikely to be associated with phenotypic changes *(1)*. Those that do affect phenotype can be referred to as "functional" polymorphisms or variants. Many of these functional variants are believed to contribute to the risk of common diseases that are polygenic in nature, and this has led to the common disease–common variant hypothesis. Several examples of important associations between common variants and common diseases include APOE*E4 and Alzheimer's disease *(2)*, CCR5Δ32 and resistance to HIV infection *(3)*, and α1-antitrypsin deficiency and chronic obstructive pulmonary disease *(4,5)*.

The potential applications for SNP studies include gene discovery and mapping, association-based candidate polymorphism testing, pharmacogenetics, diagnostics and risk profiling, homogeneity testing, and the prediction of response to environmental stimuli *(6)*. In toxicology, the focus of SNP studies has been in their role in chemical detoxification/drug metabolism, including pharmacogenetics and, to a lesser extent, receptor binding or expression of biological mediators *(7,8)*. Efforts to incorporate SNP studies into environmental epidemiology investigations have been undertaken sparingly and, when examined, have focused primarily on examining hypothesis-driven associations between environmental/occupational diseases and specific polymorphisms such as silicosis and tumor necrosis factor (TNF)-α-238, -308 *(9)*, chronic beryllium disease and HLA-DPB1 Glu69 *(10)*, and pesticide-related cancers and CYP1A1 mutations *(11)*. Although most of the major diseases of interest are polygenic, there have been limited efforts to examine the effect of multiple polymorphisms on disease modification (i.e., gene–gene–environmental interactions; Fig. 1). Furthermore, there has been little effort in incorporating genetic information into the risk-assessment process, although the advantage of such data in improving accuracy has been discussed *(12,13)*. Foremost among these would be an opportunity to pro-

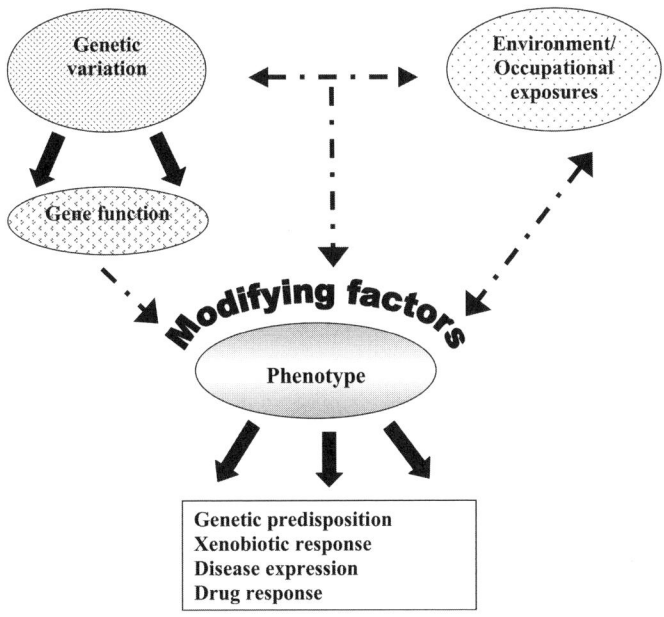

Fig. 1. A model showing genetic variants–environment interactions.

vide more accurate quantitative information on the interindividual variability likely to occur in the population.

Cytokines are not only important mediators of the immune response, but also of inflammatory responses and, thus, play a central role in the pathogenesis of chronic inflammatory diseases. The genes that control cytokine expression are highly polymorphic, and their role as modifiers of common diseases is receiving considerable attention. In this respect, epidemiological studies have identified associations between specific cytokine polymorphisms and cardiovascular diseases, cancer, neurodegenerative diseases, periodontal disease, and immune-mediated diseases such as allergic asthma and autoimmunity. Although the majority of studies published so far have focused on polymorphisms in the interleukin (IL)-1 and TNF-α gene families, most cytokines and chemokines have been examined to some extent. Population studies demonstrate, for the most part, odds ratio (OR) values for any single variant only approx 2. The relatively low ORs are probably the result of the polygenic nature of the disease under study. Accordingly, well-designed and relatively large population studies are necessary to obtain meaningful data. With respect to environmental or occupational diseases, several factors need to be considered. First, if the disease is a result of a chemical agent, consideration must be given to polymorphisms associated with chemical metabolism in addition to those associated with disease. Second, exposure characteristics (dose, duration, etc.) are of primary consideration and, in some cases, may overwhelm differences associated with the polymorphism. Third, since genetic polymorphisms act primarily as disease modifiers, it would be important to consider disease severity in the analyses.

2. CYTOKINE POLYMORPHISMS AND THEIR ASSOCIATION WITH DISEASE

Genes that code for cytokines have been identified as candidates for many common polygenic diseases. Perturbations of the balance among cytokines, with both diverse and overlapping functions, have been implicated in the clinical course of many common disorders. One of the striking observations that has been recognized since the human genome has been sequenced is that cytokine genotypes often are associated with the course of immune or inflammatory diseases. However, with only a few exceptions, such as the association of TNF-α receptor polymorphisms with periodic fevers and the IL-2-type cytokine-receptor γ-chain family variants with severe combined immunodeficiency diseases *(14–17)*, cytokine or cytokine receptor polymorphisms have not been directly linked to disease causation. Rather, genetic polymorphisms in cytokine genes act as disease modifiers by influencing disease condition, such as severity or response to specific treatment regimens. With this in mind, examples presented here represent cytokine polymorphisms which modify inflammatory, allergic, autoimmune or immunodeficiency diseases with specific attention to those that affect environmental/occupational diseases.

3. COMMON INFLAMMATORY AND AUTOIMMUNE DISEASES

Although sometimes conflicting results have been reported, numerous investigations have suggested that polymorphisms in cytokine genes may predispose individuals to chronic inflammatory or immune-mediated diseases and affect their clinical outcomes (Table 1 *[18–48]*). For example, allergic asthma is representative of complex disorders resulting from the interaction of genetic and environmental factors in which cytokine polymorphisms influence disease susceptibility and severity. Numerous loci and candidate genes, including cytokines, have been reported to be associated with asthma, atopy, increased immunoglobulin E (IgE) levels, and bronchial hyperresponsiveness. For asthma, the studies have focused primarily on chromosomes 5, 6, 12, and 13 *(49)*. Although major histocompatability

Table 1
Examples of Associations Between Cytokine
Polymorphisms and Common Complex Diseases

Disease	Cytokine SNPs	References
Asthma	TNF-α -308	*18*
	IL-10 -627	*19*
	IL-13 -1055	*20*
	TGF-β -509	*21*
Allergic and irritant	TNF-α -308	*22–24*
contact dermatitis	IL-16 -295	
Alzheimer's disease	TGF-β -509	*25*
	IL-1-α -889	*26*
	IL-6 -174	*27*
Cancer	IL-1RN VNTR	*28*
	TNF-α -308	*29*
	IL-6 -174	*30*
Coronary artery disease	IL-1RN VNTR	*31*
	IL-6 -174	*32*
	TNF-α -308	*33*
Diabetes	IL-1β +3953	*34*
	IL-6 -174	*35*
	IL-18 -137	*36*
Inflammatory bowel disease	IL-1β -511	*37,38*
Periodontitis	IL-1β +3953, -511	*39*
	IL-6 -174	*40*
	IL-10 -819, -592	*41*
Rheumatoid arthritis	IL-6 -174	*42*
	TNF-α -308	*43*
	IL-4 VNTR	*44*
	TGF-β1 codon 10	*45*
Systemic lupus erythematosus	IL-10 -1082, -592, -819	*46*
	TNF-α -308	*47*
Ulcerative colitis	IL-1RN VNTR	*48*

(MHC) variants are strongly associated with asthma, other inflammatory and allergic components also are involved. For example, TNF-α plays a role in the initiation of allergic asthmatic airway inflammation and in the generation of airway hyperreactivity *(50)*. A number of reports indicated an association between the TNF-α-308 variant and asthma *(18,51,52)*. Regulatory cytokines also are determining/modifying factors in immunological responses related to asthma. In this respect, IL-4 -589, -33, IL-13 -1055, IL-10 -627, and transforming growth factor (TGF)-β1-509 polymorphisms are associated with asthma phenotypes and severity *(19–21,53–58)*. There is also evidence that susceptibility to irritant and allergic contact dermatitis may be influenced by cytokine gene polymorphisms. An association between TNF-α -308 variant and irritant contact dermatitis was reported *(22)*. Recently, associations between IL-16 -295 and TNF-α -308 polymorphisms and allergic contact dermatitis were found *(23,24)*.

Association studies have shown that the IL-1α -889, IL-1β +3953, and IL-1β-511 variants are differentially associated with an increased risk of developing progressive neurodegenerative diseases such as Alzheimer's disease (AD) and Parkinson's disease (PD), and are associated with the age of disease onset *(26,59–63)*. Other cytokine SNPs, including TNF-α -308, IL-10 -1082, TGF-β1 -509, and IL-6 -174, also have been reported to be additive and independent risk factors for AD *(25,27,64,65)*.

Although contradictory findings have been reported, severe chronic periodontitis appears to be associated with the presence of composite IL-1β +3953 and IL-1α -889 genotypes *(39,66,67)*. Likewise, TNF-α -1031, -863 and -857, IL-2 -330 and IL-10 -1082 SNPs were reported to be disease modifiers in inflammatory periodontal diseases *(68–70)*.

Cytokines that regulate either adaptive or innate immunity influence autoimmune disease processes at multiple levels. In addition to MHC alleles, SNPs within genes that regulate proinflammatory cytokines have been extensively studied. For example, TNF-α -308, -857 and micro-satellite a6 alleles have been associated with rheumatoid arthritis *(43,71)*, systemic lupus erythematosus (SLE *[47]*), and type 2 diabetes *(72,73)*. IL-1 gene family (i.e., IL-1β and IL-receptor agonist [RN] 1) polymorphisms also have been associated with autoimmune diseases *(74–76)*. Increased levels of IL-1β are reported to play a role in the maintenance of autoimmune damage of pancreatic β cells, and the high producer IL-1β +3953 genotype is associated with increased risk of type-1 diabetes *(77)*. IL-1RN VNTR variations have been associated with a variety of autoimmune diseases, including diabetes and SLE *(78–80)*. High levels of IL-6 are observed in several autoimmune diseases, and the IL-6 –174 polymorphism has been implicated in systemic juvenile chronic arthritis, SLE and diabetes *(42,81–83)*. Table 1 provides further examples of associations between cytokine polymorphisms and chronic inflammatory and immune-mediated diseases.

4. OCCUPATIONAL AND ENVIRONMENTAL DISEASES

Occupational or environmental diseases also are multifactorial in nature, being under the influence of polygenic, physiological, and environmental factors. Assessing interactions between genes and exposure variables, such

Table 2
Examples of Associations Between Cytokine Polymorphisms
and Occupational/Environmental Diseases

Disease	Cytokine SNPs	References
Alcohol and chemical-induced Hepatitis	TNF-α -308, -238	*84*
	IL-1β +3953, -511	*85*
Chemical-induced neurotoxicity	IL-1α -889	*86*
	TNF-α -308	*87*
Chronic beryllium disease	TNF-α -308	*88*
Chronic obstructive pulmonary disease	TNF-α -308, +489	*89,90*
	TGF-β codon 10	*91*
Coal workers' pneumoconiosis	TNF-α -308	*92*
Farmer's lung disease	TNF-α -308	*93*
Sarcoidosis	TNF-α -308	*94*
Silicosis	IL-1RN +2018	*95*
	TNF-α -238, -308	*9*

as duration and dose, are particularly important for understanding these diseases. Table 2 *(9,84–95)* provides examples of the associations reported between cytokine polymorphisms and various diseases with occupational/environmental origin. Representative examples are discussed herein.

Allergic occupational asthma, which is a result of workplace exposure to high- (flour, grain dust) and low-molecular weight substances (diisocyanates, metals, dyes), requires a latency period for acquiring sensitization and manifests pathologies similar to allergic asthma *(96)*. Isocyanates are common causes of occupational asthma, and HLA class II genes have been found to be involved in the risk of developing disease *(97)*. TNF-α also plays a major role in the immune and inflammatory responses of isocyanate-induced asthma. High levels of TNF-α were observed in the peripheral blood mononuclear cells of subjects exposed to isocyanates *(98)*. However, in contrast to the reports indicating an association between TNF-α -308 SNP and allergic asthma *(52,99)*, no association was found in the case of toluene diisocyanate-induced asthma *(100)*.

Silicosis, an interstitial lung disease resulting from inhalation of crystalline silica, is characterized by chronic inflammation leading to severe pulmonary fibrotic changes that are prevalent among miners, sand blasters and quarry workers. Proinflammatory cytokines, such as TNF-α and IL-1, have been implicated in the formation of these lesions. A strong association was found between disease severity and the TNF-α -238 variant *(9,95)*. Irrespective of disease severity, the TNF-α (-308) and IL-1RN(+2018) variants conferred an increased risk for the presence of disease. In studies of South African miners, TNF-α promoter polymorphisms (-308, -238, -376) were found to be associated with severe silicosis helping to confirm these associations *(101)*.

Chronic beryllium disease (CBD) is caused by hypersensitivity to beryllium in a variety of industrial processes, including computer, automotive, and ceramics. It has pathological and clinical features similar to sarcoidosis *(102)*. Recent studies investigating the contribution of HLA alleles to disease processes revealed an association between HLA-DPB1 (Glu69) variation and CBD *(103–105)*. The TNF-α -308 allele was also reported to be associated with a high level of beryllium-stimulated TNF-α, and appears to be linked to disease severity *(106)*.

5. IMMUNOMODULATION

Cytokines, like growth factors, play a central role in vascular repair and allograft survival. Cytokines, including interferon (IFN)-γ and TNF-α, are intimately involved in allograft rejection, whereas the Th2 type cytokines, IL-4, -5, and -10 are involved in allograft tolerance *(107)*. Cytokine polymorphisms have been shown to affect the survival of kidney, heart, and lung in transplant patients. TNF-α variants have been studied most extensively, and TNF-α -308 SNP is associated with acute heart, kidney, and liver rejection *(108)*. The IL-10 -1082, -819, and -592 haplotype GCC, which confers increased IL-10 levels, is protective against early acute heart graft rejection *(109)*. The IFN-γ +874 polymorphism and CA repeat in the first intron are associated with acute kidney rejection, the occurrence of kidney infections after transplant, and the development of fibrosis after lung transplantation *(110–112)*. A strong association has also been found between the TGF-β codon 25 variant and tissue rejection. Development of fibrosis after lung transplantation is associated with the TGF-β1 homozygous high producer genotype *(113)*. Cytokine polymorphisms also play a role in vaccination efficacy and are thought to be a factor responsible from inadequate responses. In this respect, the immune responses after hepatitis B vaccination are influenced by the IL-1β +3953 polymorphism *(114)*. UVB-induced immunomodulation also involves cytokines in regulatory capacity, and the IL-1β +3953 polymorphism was found to suppress antibody responses to hepatitis B in individuals exposed to UVB radiation *(115)*.

6. EPIDEMIOLOGICAL
AND STATISTICAL ISSUES

The ability to accurately detect associations between genetic variants in cytokine genes and common disease in current scientific practice is faced with multiple challenges that probably account for a large number of nonreplicable associations in the published literature *(13,116)*. Although several experimental designs are available for assessing potential associations between cytokine SNPs and disease, the case-control association study, because of its relative simplicity, is the most widely used design for detecting common disease alleles with modest risk. Depending upon factors such as the true effect size, the extent of linkage disequilibrium, the frequency of the disease in question, and the frequency of markers associated with the disease, the sample size required to detect true associations often becomes intractable *(117)*. When interactions between genes and the environment, as well as between multiple genes, are included in the research question, the challenges become even greater. There are statistical challenges as well with the problem of multiple testing being foremost. As high-throughput genetic screening of multiple SNPs becomes commonplace, the multiple testing of various SNPs on a single population often requires that *p* values for significance be adjusted so low that only the largest effect sizes are detected. Conversely, failing to adjust the significance level reduces the confidence that significant associations, when found, are actually real and meaningful. Recent reviews have provided excellent discussions of the issues and challenges involved in association studies and provide relevant guidelines for carrying out these types of studies *(118–120)*.

Recently, meta-analysis techniques have been implemented to combine genetic association data from multiple populations in both case–control *(13)* and linkage studies *(121)*. Although this technique has some limitations imposed by the heterogeneous populations examined, publication bias, and sampling biases, it may prove to be a useful analytic technique particularly when sufficient raw data are available for examining cytokine polymorphisms. Other statistical techniques are becoming available for enhancing case–control genetic association studies, including sample pooling *(122)*, genetic control *(123)*, the use of haplotypes *(124)*, and whole-genome association studies *(120)*. An overview of the various testing methods for genetic association studies, including family-based methods is presented by Schulze and McMahon *(125)*. In all situations, it is important that sound epidemiological and statistical principles be incorporated in the design and sampling of individuals from the population to maximize the power of the study and reduce the number of spurious associations.

7. GENOTYPING TECHNIQUES

Traditionally, genotyping technologies used for allelic discrimination included direct sequencing analysis and polymerase chain reaction-restriction fragment length polymorphism (PCR-RFLP). These techniques are robust and sensitive and represent the mainstay technologies for genotyping cytokine SNPs. However, because of the laborious and low-throughput nature of these technologies, they are only amenable to analyses of small numbers of SNPs in relatively small populations. The completion of the human genome sequencing project has been accompanied by an exponential increase in genotyping demands. A comprehensive review of all genotyping platforms is beyond the scope of this discussion and has been presented elsewhere *(126,127)*. Instead, selected non-gel-based technologies that display potential for increased throughput currently being used in our laboratory will be discussed. Homogeneous solution hybridization using fluorescence resonance energy transfer is one of the most used and validated methodologies and is typified by the Taqman allelic discrimination assay *(128)*. Being homogeneous, all aspects, including amplification, detection, and identification of genotypes, are performed in a single tube. We have recently validated this assay for several important functional cytokine SNPs, and the results showed high concordance with PCR-RFLP and reduced error rates *(129)*.

Allele-specific fluorescent nucleotide incorporation techniques also have to be used to determine genotypes. Single-base extension is the most common technique used and is based on the extension of a primer by a single dideoxy (terminator) nucleotide (ddNTP) with a distinct label, fluorescence being the most common. The primer anneals immediately adjacent to the SNP site, making the incorporated ddNTP specific for one allele. Fluorescence polarization can then be used to determine genotypes *(130,131)*.

Recently, solid-phase detection platforms have been identified that are amenable to high-throughput genotyping. Two such platforms are microarray and matrix-assisted laser desorption ionization (MALDI)-mass spectrometry. Microarray detection involves the fixation of allele specific probes to a solid support and then applying samples amplified by PCR for the arrayed SNPs *(132)*. Several commercial platforms are currently available that can be used to genotype a genome-wide panel of SNPs for a single individual, some having more than 500,00 features. Therefore, this technique is applicable to projects involving genotyping of a limited number of individuals for thousands of SNPs. Recently, a microarray-based approach has been developed that is capable of genotyping very large populations for a small number of SNPs *(133)*. Genotyping using mass spectrometry has the advantage of accurately determining molecular mass. As such, the end products of allele-specific single base extension reactions can be discriminated by mass using MALDI-MS *(134)*.

8. CONCLUSION

In recent years, there has been increased attention on the effects of polymorphisms in cytokine genes and their role as modifiers of common diseases. These alleles have been implicated in influencing the clinical course of common multifactorial diseases, including cardiovascular and neurodegenerative diseases, cancer, asthma, and immune-mediated diseases. Despite some contradictory findings in SNP-disease association studies, the use of high-throughput genotyping technology and statistically robust association study designs will lead to a better understanding of disease mechanisms and identify novel therapeutic strategies as well as provide opportunities to improve the risk assessment process.

ACKNOWLEDGMENTS

This work was supported in part by interagency grant from NIEHS, Y1-ES–0001.

REFERENCES

1. Collins A, Lonjou C, Morton NE. Genetic epidemiology of single-nucleotide polymorphisms. Proc Natl Acad Sci USA 1999;96:15,173–15,177.
2. Saunders AM, Strittmatter WJ, Schmechel D, et al. Association of apolipoprotein E allele epsilon 4 with late-onset familial and sporadic Alzheimer's disease. Neurology 1993;43:1467–1472.
3. Dean M, Carrington M, Winkler C, et al. Genetic restriction of HIV-1 infection and progression to AIDS by a deletion allele of the CKR5 structural gene. Hemophilia Growth and Development Study, Multicenter AIDS Cohort Study, Multicenter Hemophilia Cohort Study, San Francisco City Cohort, ALIVE Study. Science 1996;273:1856–1862.
4. Poller W, Meisen C, Olek K. DNA polymorphisms of the alpha 1-antitrypsin gene region in patients with chronic obstructive pulmonary disease. Eur J Clin Invest 1990;20:1–7.
5. DeMeo DL, Silverman EK. Alpha1-antitrypsin deficiency. 2: genetic aspects of alpha(1)-antitrypsin deficiency: phenotypes and genetic modifiers of emphysema risk. Thorax 2004;59:259–264.
6. Palmer LJ, Cookson WO. Using single nucleotide polymorphisms as a means to understanding the pathophysiology of asthma. Respir Res 2001;2:102–112.
7. Kelada SN, Eaton DL, Wang SS, Rothman NR, Khoury MJ. The role of genetic polymorphisms in environmental health. Environ Health Perspect 2003;111:1055–1064.

8. Knudsen LE, Loft SH, Autrup H. Risk assessment: the importance of genetic polymorphisms in man. Mutat Res 2001;482:83–88.
9. Yucesoy B, Vallyathan V, Landsittel DP, et al. Association of tumor necrosis factor-alpha and interleukin-1 gene polymorphisms with silicosis. Toxicol Appl Pharmacol 2001;172:75–82.
10. Lombardi G, Germain C, Uren J, et al. HLA-DP allele-specific T cell responses to beryllium account for DP-associated susceptibility to chronic beryllium disease. J Immunol 2001;166:3549–3555.
11. Infante-Rivard C, Labuda D, Krajinovic M, Sinnett D. Risk of childhood leukemia associated with exposure to pesticides and with gene polymorphisms. Epidemiology 1999;10:481–487.
12. Zhu Y, Spitz MR, Amos CI, Lin J, Schabath MB, Wu X. An evolutionary perspective on single-nucleotide polymorphism screening in molecular cancer epidemiology. Cancer Res 2004;64:2251–2257.
13. Lohmueller KE, Pearce CL, Pike M, Lander ES, Hirschhorn JN. Meta-analysis of genetic association studies supports a contribution of common variants to susceptibility to common disease. Nat Genet 2003;33:177–182.
14. Leonard WJ. Cytokines and immunodeficiency diseases. Nat Rev Immunol 2001;1:200–208.
15. Gilmour KC, Fujii H, Cranston T, Davies EG, Kinnon C, Gaspar HB. Defective expression of the interleukin-2/interleukin-15 receptor beta subunit leads to a natural killer cell-deficient form of severe combined immunodeficiency. Blood 2001;98:877–879.
16. Aganna E, Aksentijevich I, Hitman GA, et al. Tumor necrosis factor receptor-associated periodic syndrome (TRAPS) in a Dutch family: evidence for a TNFRSF1A mutation with reduced penetrance. Eur J Hum Genet 2001;9:63–66.
17. Aksentijevich I, Galon J, Soares M, et al. The tumor-necrosis-factor receptor-associated periodic syndrome: new mutations in TNFRSF1A, ancestral origins, genotype-phenotype studies, and evidence for further genetic heterogeneity of periodic fevers. Am J Hum Genet 2001;69:301–314.
18. Moffatt MF, Cookson WO. Tumour necrosis factor haplotypes and asthma. Hum Mol Genet 1997;6:551–554.
19. Hang LW, Hsia TC, Chen WC, Chen HY, Tsai JJ, Tsai FJ. Interleukin-10 gene -627 allele variants, not interleukin-1 beta gene and receptor antagonist gene polymorphisms, are associated with atopic bronchial asthma. J Clin Lab Anal 2003;17:168–173.
20. van der Pouw Kraan TC, van Veen A, Boeije LC, et al. An IL-13 promoter polymorphism associated with increased risk of allergic asthma. Genes Immun 1999;1:61–65.
21. Pulleyn LJ, Newton R, Adcock IM, Barnes PJ. TGFbeta1 allele association with asthma severity. Hum Genet 2001;109:623–627.
22. Allen MH, Wakelin SH, Holloway D, et al. Association of TNFA gene polymorphism at position -308 with susceptibility to irritant contact dermatitis. Immunogenetics 2000;51:201–205.

23. Reich K, Westphal G, Konig IR, et al. Association of allergic contact dermatitis with a promoter polymorphism in the IL16 gene. J Allergy Clin Immunol 2003;112:1191–1194.

24. Westphal GA, Schnuch A, Moessner R, et al. Cytokine gene polymorphisms in allergic contact dermatitis. Contact Dermatitis 2003;48:93–98.

25. Luedecking EK, DeKosky ST, Mehdi H, Ganguli M, Kamboh MI. Analysis of genetic polymorphisms in the transforming growth factor-beta1 gene and the risk of Alzheimer's disease. Hum Genet 2000;106:565–569.

26. Griffin WS, Mrak RE. Interleukin-1 in the genesis and progression of and risk for development of neuronal degeneration in Alzheimer's disease. J Leukoc Biol 2002;72:233–238.

27. Faltraco F, Burger K, Zill P, et al. Interleukin-6–174 G/C promoter gene polymorphism C allele reduces Alzheimer's disease risk. J Am Geriatr Soc 2003;51:578–579.

28. Hulkkonen J, Vilpo J, Vilpo L, Koski T, Hurme M. Interleukin-1 beta, interleukin-1 receptor antagonist and interleukin-6 plasma levels and cytokine gene polymorphisms in chronic lymphocytic leukemia: correlation with prognostic parameters. Haematologica 2000;85:600–606.

29. Landi S, Moreno V, Gioia-Patricola L, et al. Association of common polymorphisms in inflammatory genes interleukin (IL)6, IL8, tumor necrosis factor alpha, NFKB1, and peroxisome proliferator-activated receptor gamma with colorectal cancer. Cancer Res 2003;63:3560–3566.

30. DeMichele A, Martin AM, Mick R, et al. Interleukin-6 -174G—>C polymorphism is associated with improved outcome in high-risk breast cancer. Cancer Res 2003;63:8051–8056.

31. Francis SE, Camp NJ, Dewberry RM, et al. Interleukin-1 receptor antagonist gene polymorphism and coronary artery disease. Circulation 1999;99:861–866.

32. Humphries SE, Luong LA, Ogg MS, Hawe E, Miller GJ. The interleukin-6 -174 G/C promoter polymorphism is associated with risk of coronary heart disease and systolic blood pressure in healthy men. Eur Heart J 2001;22:2243–2252.

33. Szalai C, Fust G, Duba J, et al. Association of polymorphisms and allelic combinations in the tumour necrosis factor-alpha-complement MHC region with coronary artery disease. J Med Genet 2002;39:46–51.

34. Pociot F, Molvig J, Wogensen L, Worsaae H, Nerup J. A TaqI polymorphism in the human interleukin-1 beta (IL-1 beta) gene correlates with IL-1 beta secretion in vitro. Eur J Clin Invest 1992;22:396–402.

35. Jahromi MM, Millward BA, Demaine AG. A polymorphism in the promoter region of the gene for interleukin-6 is associated with susceptibility to type 1 diabetes mellitus. J Interferon Cytokine Res 2000;20:885–888.

36. Kretowski A, Mironczuk K, Karpinska A, et al. Interleukin-18 promoter polymorphisms in type 1 diabetes. Diabetes 2002;51:3347–3349.

37. Heresbach D, Ababou A, Bourienne A, et al. Polymorphism of the microsatellites and tumor necrosis factor genes in chronic inflammatory bowel diseases. Gastroenterol Clin Biol 1997;21:555–561.

38. Nemetz A, Kope A, Molnar T, et al. Significant differences in the interleukin-1beta and interleukin-1 receptor antagonist gene polymorphisms in a Hungarian population with inflammatory bowel disease. Scand J Gastroenterol 1999;34:175–179.

39. Kornman KS, Crane A, Wang HY, et al. The interleukin-1 genotype as a severity factor in adult periodontal disease. J Clin Periodontol 1997;24:72–77.

40. Trevilatto PC, Scarel-Caminaga RM, de Brito RB, Jr., de Souza AP, Line SR. Polymorphism at position -174 of IL-6 gene is associated with susceptibility to chronic periodontitis in a Caucasian Brazilian population. J Clin Periodontol 2003;30:438–442.

41. Scarel-Caminaga RM, Trevilatto PC, Souza AP, Brito RB, Camargo LE, Line SR. Interleukin 10 gene promoter polymorphisms are associated with chronic periodontitis. J Clin Periodontol 2004;31:443–448.

42. Fishman D, Faulds G, Jeffery R, et al. The effect of novel polymorphisms in the interleukin-6 (IL-6) gene on IL-6 transcription and plasma IL-6 levels, and an association with systemic-onset juvenile chronic arthritis. J Clin Invest 1998;102:1369–1376.

43. Waldron-Lynch F, Adams C, Amos C, et al. Tumour necrosis factor 5' promoter single nucleotide polymorphisms influence susceptibility to rheumatoid arthritis (RA) in immunogenetically defined multiplex RA families. Genes Immun 2001;2:82–87.

44. Buchs N, Silvestri T, di Giovine FS, et al. IL-4 VNTR gene polymorphism in chronic polyarthritis. The rare allele is associated with protection against destruction. Rheumatology (Oxford) 2000;39:1126–1131.

45. Sugiura Y, Niimi T, Sato S, et al. Transforming growth factor beta1 gene polymorphism in rheumatoid arthritis. Ann Rheum Dis 2002;61:826–828.

46. D'Alfonso S, Rampi M, Bocchio D, Colombo G, Scorza-Smeraldi R, Momigliano-Richardi P. Systemic lupus erythematosus candidate genes in the Italian population: evidence for a significant association with interleukin-10. Arthritis Rheum 2000;43:120–128.

47. Rood MJ, van Krugten MV, Zanelli E, et al. TNF-308A and HLA-DR3 alleles contribute independently to susceptibility to systemic lupus erythematosus. Arthritis Rheum 2000;43:129–134.

48. Mansfield JC, Holden H, Tarlow JK, et al. Novel genetic association between ulcerative colitis and the anti-inflammatory cytokine interleukin-1 receptor antagonist. Gastroenterology 1994;106:637–642.

49. Cookson WO. Asthma genetics. Chest 2002;121:7S–13S.

50. Thomas PS. Tumour necrosis factor-alpha: the role of this multifunctional cytokine in asthma. Immunol Cell Biol 2001;79:132–140.

51. Moffatt MF, Cookson WO. Linkage and candidate gene studies in asthma. Am J Respir Crit Care Med 1997;156:S110–S112.

52. Albuquerque RV, Hayden CM, Palmer LJ, et al. Association of polymorphisms within the tumour necrosis factor (TNF) genes and childhood asthma. Clin Exp Allergy 1998;28:578–584.

53. Chouchane L, Sfar I, Bousaffara R, El Kamel A, Sfar MT, Ismail A. A repeat polymorphism in interleukin-4 gene is highly associated with specific clinical phenotypes of asthma. Int Arch Allergy Immunol 1999;120:50–55.

54. Noguchi E, Nukaga-Nishio Y, Jian Z, et al. Haplotypes of the 5' region of the IL-4 gene and SNPs in the intergene sequence between the IL-4 and IL-13 genes are associated with atopic asthma. Hum Immunol 2001;62:1251–1257.

55. Kabesch M, Tzotcheva I, Carr D, et al. A complete screening of the IL4 gene: novel polymorphisms and their association with asthma and IgE in childhood. J Allergy Clin Immunol 2003;112:893–898.

56. Graves PE, Kabesch M, Halonen M, et al. A cluster of seven tightly linked polymorphisms in the IL-13 gene is associated with total serum IgE levels in three populations of white children. J Allergy Clin Immunol 2000;105:506–513.

57. Hobbs K, Negri J, Klinnert M, Rosenwasser LJ, Borish L. Interleukin-10 and transforming growth factor-beta promoter polymorphisms in allergies and asthma. Am J Respir Crit Care Med 1998;158:1958–1962.

58. Lim S, Crawley E, Woo P, Barnes PJ. Haplotype associated with low interleukin-10 production in patients with severe asthma. Lancet 1998;352:113.

59. Combarros O, Sanchez-Guerra M, Infante J, Llorca J, Berciano J. Gene dose-dependent association of interleukin-1A [-889] allele 2 polymorphism with Alzheimer's disease. J Neurol 2002;249:1242–1245.

60. Grimaldi LM, Casadei VM, Ferri C, et al. Association of early-onset Alzheimer's disease with an interleukin-1alpha gene polymorphism. Ann Neurol 2000;47:361–365.

61. Sciacca FL, Ferri C, Licastro F, et al. Interleukin-1B polymorphism is associated with age at onset of Alzheimer's disease. Neurobiol Aging 2003;24:927–931.

62. Mattila KM, Rinne JO, Lehtimaki T, Roytta M, Ahonen JP, Hurme M. Association of an interleukin 1B gene polymorphism (-511) with Parkinson's disease in Finnish patients. J Med Genet 2002;39:400–402.

63. McGeer PL, Yasojima K, McGeer EG. Association of interleukin-1 beta polymorphisms with idiopathic Parkinson's disease. Neurosci Lett 2002;326: 67–69.

64. Alvarez V, Mata IF, Gonzalez P, et al. Association between the TNFalpha-308 A/G polymorphism and the onset-age of Alzheimer disease. Am J Med Genet 2002;114:574–577.

65. Lio D, Scola L, Crivello A, et al. Inflammation, genetics, and longevity: further studies on the protective effects in men of IL-10 -1082 promoter SNP and its interaction with TNF-alpha -308 promoter SNP. J Med Genet 2003;40: 296–299.

66. Gore EA, Sanders JJ, Pandey JP, Palesch Y, Galbraith GM. Interleukin-1beta+3953 allele 2: association with disease status in adult periodontitis. J Clin Periodontol 1998;25:781–785.

67. Parkhill JM, Hennig BJ, Chapple IL, Heasman PA, Taylor JJ. Association of interleukin-1 gene polymorphisms with early-onset periodontitis. J Clin Periodontol 2000;27:682–689.

68. Soga Y, Nishimura F, Ohyama H, Maeda H, Takashiba S, Murayama Y. Tumor necrosis factor-alpha gene (TNF-alpha) -1031/-863, -857 single-nucleotide polymorphisms (SNPs) are associated with severe adult periodontitis in Japanese. J Clin Periodontol 2003;30:524–531.

69. Berglundh T, Donati M, Hahn-Zoric M, Hanson LA, Padyukov L. Association of the -1087 IL 10 gene polymorphism with severe chronic periodontitis in Swedish Caucasians. J Clin Periodontol 2003;30:249–254.

70. Scarel-Caminaga RM, Trevilatto PC, Souza AP, Brito RB, Line SR. Investigation of an IL-2 polymorphism in patients with different levels of chronic periodontitis. J Clin Periodontol 2002;29:587–591.

71. Ozen S, Alikasifoglu M, Bakkaloglu A, et al. Tumour necrosis factor alpha G—>A -238 and G—>A -308 polymorphisms in juvenile idiopathic arthritis. Rheumatology (Oxford) 2002;41:223–227.

72. Kubaszek A, Pihlajamaki J, Komarovski V, et al. Promoter polymorphisms of the TNF-alpha (G-308A) and IL-6 (C-174G) genes predict the conversion from impaired glucose tolerance to type 2 diabetes: the Finnish Diabetes Prevention Study. Diabetes 2003;52:1872–1876.

73. Heijmans BT, Westendorp RG, Droog S, Kluft C, Knook DL, Slagboom PE. Association of the tumour necrosis factor alpha -308G/A polymorphism with the risk of diabetes in an elderly population-based cohort. Genes Immun 2002;3:225–228.

74. McDowell TL, Symons JA, Ploski R, Forre O, Duff GW. A genetic association between juvenile rheumatoid arthritis and a novel interleukin-1 alpha polymorphism. Arthritis Rheum 1995;38:221–228.

75. Ploski R, McDowell TL, Symons JA, et al. Interaction between HLA-DR and HLA-DP, and between HLA and interleukin 1 alpha in juvenile rheumatoid arthritis indicates heterogeneity of pathogenic mechanisms of the disease. Hum Immunol 1995;42:343–347.

76. Jouvenne P, Chaudhary A, Buchs N, Giovine FS, Duff GW, Miossec P. Possible genetic association between interleukin-1alpha gene polymorphism and the severity of chronic polyarthritis. Eur Cytokine Netw 1999;10:33–36.

77. Krikovsky D, Vasarhelyi B, Treszl A, et al. Genetic polymorphism of interleukin-1beta is associated with risk of type 1 diabetes mellitus in children. Eur J Pediatr 2002;161:507–508.

78. Tjernstrom F, Hellmer G, Nived O, Truedsson L, Sturfelt G. Synergetic effect between interleukin-1 receptor antagonist allele (IL1RN*2) and MHC class II (DR17,DQ2) in determining susceptibility to systemic lupus erythematosus. Lupus 1999;8:103–108.

79. Blakemore AI, Tarlow JK, Cork MJ, Gordon C, Emery P, Duff GW. Interleukin-1 receptor antagonist gene polymorphism as a disease severity factor in systemic lupus erythematosus. Arthritis Rheum 1994;37:1380–1385.

80. Blakemore AI, Cox A, Gonzalez AM, et al. Interleukin-1 receptor antagonist allele (IL1RN*2) associated with nephropathy in diabetes mellitus. Hum Genet 1996;97:369–374.

81. Ogilvie EM, Fife MS, Thompson SD, et al. The -174G allele of the interleukin-6 gene confers susceptibility to systemic arthritis in children: a multicenter study using simplex and multiplex juvenile idiopathic arthritis families. Arthritis Rheum 2003;48:3202–3206.

82. Linker-Israeli M, Wallace DJ, Prehn J, et al. Association of IL-6 gene alleles with systemic lupus erythematosus (SLE) and with elevated IL-6 expression. Genes Immun 1999;1:45–52.

83. Kristiansen OP, Nolsoe RL, Larsen L, et al. Association of a functional 17beta-estradiol sensitive IL6–174G/C promoter polymorphism with early-onset type 1 diabetes in females. Hum Mol Genet 2003;12:1101–1110.

84. Yee LJ, Tang J, Herrera J, Kaslow RA, van Leeuwen DJ. Tumor necrosis factor gene polymorphisms in patients with cirrhosis from chronic hepatitis C virus infection. Genes Immun 2000;1:386–390.

85. Takamatsu M, Yamauchi M, Maezawa Y, Saito S, Maeyama S, Uchikoshi T. Genetic polymorphisms of interleukin-1beta in association with the development of alcoholic liver disease in Japanese patients. Am J Gastroenterol 2000;95:1305–1311.

86. Rebeck GW. Confirmation of the genetic association of interleukin-1A with early onset sporadic Alzheimer's disease. Neurosci Lett 2000;293:75–77.

87. Boin F, Zanardini R, Pioli R, Altamura CA, Maes M, Gennarelli M. Association between -G308A tumor necrosis factor alpha gene polymorphism and schizophrenia. Mol Psychiatry 2001;6:79–82.

88. Maier LA, Sawyer RT, Bauer RA, et al. High beryllium-stimulated TNF-alpha is associated with the -308 TNF-alpha promoter polymorphism and with clinical severity in chronic beryllium disease. Am J Respir Crit Care Med 2001; 164:1192–1199.

89. Sakao S, Tatsumi K, Igari H, Shino Y, Shirasawa H, Kuriyama T. Association of tumor necrosis factor alpha gene promoter polymorphism with the presence of chronic obstructive pulmonary disease. Am J Respir Crit Care Med 2001;163:420–422.

90. Kucukaycan M, Van Krugten M, Pennings HJ, et al. Tumor Necrosis Factor-alpha +489G/A gene polymorphism is associated with chronic obstructive pulmonary disease. Respir Res 2002;3:29.

91. Wu L, Chau J, Young RP, et al. Transforming growth factor-beta1 genotype and susceptibility to chronic obstructive pulmonary disease. Thorax 2004;59: 126–129.

92. Zhai R, Jetten M, Schins RP, Franssen H, Borm PJ. Polymorphisms in the promoter of the tumor necrosis factor-alpha gene in coal miners. Am J Ind Med 1998;34:318–324.

93. Schaaf BM, Seitzer U, Pravica V, Aries SP, Zabel P. Tumor necrosis factor-alpha -308 promoter gene polymorphism and increased tumor necrosis factor serum bioactivity in farmer's lung patients. Am J Respir Crit Care Med 2001; 163:379–382.

94. Seitzer U, Swider C, Stuber F, et al. Tumour necrosis factor alpha promoter gene polymorphism in sarcoidosis. Cytokine 1997;9:787–90.

95. Yucesoy B, Vallyathan V, Landsittel DP, et al. Polymorphisms of the IL-1 gene complex in coal miners with silicosis. Am J Ind Med 2001;39:286–291.

96. Youakim S. Work-related asthma. Am Fam Physician 2001;64:1839–1848.

97. Mapp CE, Beghe B, Balboni A, et al. Association between HLA genes and susceptibility to toluene diisocyanate-induced asthma. Clin Exp Allergy 2000;30:651–656.

98. Lummus ZL, Alam R, Bernstein JA, Bernstein DI. Diisocyanate antigen-enhanced production of monocyte chemoattractant protein-1, IL-8, and tumor necrosis factor-alpha by peripheral mononuclear cells of workers with occupational asthma. J Allergy Clin Immunol 1998;102:265–274.

99. Witte JS, Palmer LJ, O'Connor RD, Hopkins PJ, Hall JM. Relation between tumour necrosis factor polymorphism TNFalpha-308 and risk of asthma. Eur J Hum Genet 2002;10:82–85.

100. Beghe B, Padoan M, Moss CT, et al. Lack of association of HLA class I genes and TNF alpha-308 polymorphism in toluene diisocyanate-induced asthma. Allergy 2004;59:61–64.

101. Corbett EL, Mozzato-Chamay N, Butterworth AE, et al. Polymorphisms in the tumor necrosis factor-alpha gene promoter may predispose to severe silicosis in black South African miners. Am J Respir Crit Care Med 2002;165:690–693.

102. Infante PF, Newman LS. Beryllium exposure and chronic beryllium disease. Lancet 2004;363:415–416.

103. McCanlies EC, Ensey JS, Schuler CR, Kreiss K, Weston A. The association between HLA-DPB1(Glu69) and chronic beryllium disease and beryllium sensitization. Am J Ind Med 2004;46:95–103.

104. Sawyer RT, Parsons CE, Fontenot AP, et al. Beryllium-induced tumor necrosis factor-alpha production by CD4+ T cells is mediated by HLA-DP. Am J Respir Cell Mol Biol 2004;31:122–130.

105. Maier LA, McGrath DS, Sato H, et al. Influence of MHC class II in susceptibility to beryllium sensitization and chronic beryllium disease. J Immunol 2003;171:6910–6918.

106. Maier LA, Sawyer RT, Bauer RA, et al. High beryllium-stimulated TNF-alpha is associated with the -308 TNF-alpha promoter polymorphism and with clinical severity in chronic beryllium disease. Am J Respir Crit Care Med 2001;164:1192–1199.

107. McDaniel DO, Barber WH, Nguyan C, et al. Combined analysis of cytokine genotype polymorphism and the level of expression with allograft function in African-American renal transplant patients. Transpl Immunol 2003;11:107–119.

108. Hutchinson IV, Turner D, Sankaran D, Awad M, Pravica V, Sinnott P. Cytokine genotypes in allograft rejection: guidelines for immunosuppression. Transplant Proc 1998;30:3991–3992.

109. Densem CG, Hutchinson IV, Yonan N, Brooks NH. Influence of interleukin-10 polymorphism on the development of coronary vasculopathy following cardiac transplantation. Transpl Immunol 2003;11:223–228.

110. Sankaran D, Asderakis A, Ashraf S, et al. Cytokine gene polymorphisms predict acute graft rejection following renal transplantation. Kidney Int 1999;56:281–288.

111. Pelletier R, Pravica V, Perrey C, et al. Evidence for a genetic predisposition towards acute rejection after kidney and simultaneous kidney-pancreas transplantation. Transplantation 2000;70:674–680.

112. Awad M, Pravica V, Perrey C, et al. CA repeat allele polymorphism in the first intron of the human interferon-gamma gene is associated with lung allograft fibrosis. Hum Immunol 1999;60:343–346.

113. El-Gamel A, Awad MR, Hasleton PS, et al. Transforming growth factor-beta (TGF-beta1) genotype and lung allograft fibrosis. J Heart Lung Transplant 1999;18:517–523.

114. Yucesoy B, Sleijffers A, Kashon M, et al. IL-1beta gene polymorphisms influence hepatitis B vaccination. Vaccine 2002;20:3193–3196.

115. Sleijffers A, Yucesoy B, Kashon M, et al. Cytokine polymorphisms play a role in susceptibility to ultraviolet B-induced modulation of immune responses after hepatitis B vaccination. J Immunol 2003;170:3423–3428.

116. Ioannidis JP, Ntzani EE, Trikalinos TA, Contopoulos-Ioannidis DG. Replication validity of genetic association studies. Nat Genet 2001;29:306–309.

117. Zondervan KT, Cardon LR. The complex interplay among factors that influence allelic association. Nat Rev Genet 2004;5:89–100.

118. Cardon LR, Bell JI. Association study designs for complex diseases. Nat Rev Genet 2001;2:91–99.

119. Risch N. Searching for genes in complex diseases: lessons from systemic lupus erythematosus. J Clin Invest 2000;105:1503–1506.

120. Carlson CS, Eberle MA, Kruglyak L, Nickerson DA. Mapping complex disease loci in whole-genome association studies. Nature 2004;429:446–452.

121. Dempfle A, Loesgen S. Meta-analysis of linkage studies for complex diseases: an overview of methods and a simulation study. Ann Hum Genet 2004; 68:69–83.

122. Risch N, Teng J. The relative power of family-based and case-control designs for linkage disequilibrium studies of complex human diseases I. DNA pooling. Genome Res 1998;8:1273–1288.

123. Bacanu SA, Devlin B, Roeder K. The power of genomic control. Am J Hum Genet 2000;66:1933–1944.

124. Johnson GC, Esposito L, Barratt BJ, et al. Haplotype tagging for the identification of common disease genes. Nat Genet 2001;29:233–237.

125. Schulze TG, McMahon FJ. Genetic association mapping at the crossroads: which test and why? Overview and practical guidelines. Am J Med Genet 2002;114:1–11.

126. Chen X, Sullivan PF. Single nucleotide polymorphism genotyping: biochemistry, protocol, cost and throughput. Pharmacogenomics J 2003;3:77–96.

127. Shi MM. Enabling large-scale pharmacogenetic studies by high-throughput mutation detection and genotyping technologies. Clin Chem 2001;47:164–172.

128. Livak KJ. Allelic discrimination using fluorogenic probes and the 5' nuclease assay. Genet Anal 1999;14:143–149.

129. Johnson VJ, Yucesoy B, Luster MI. Genotyping of single nucleotide polymorphisms in cytokine genes using real-time PCR allelic discrimination technology. Cytokine 2004;27:135–141.

130. Akula N, Chen YS, Hennessy K, Schulze TG, Singh G, McMahon FJ. Utility and accuracy of template-directed dye-terminator incorporation with fluorescence-polarization detection for genotyping single nucleotide polymorphisms. Biotechniques 2002;32:1072–1076, 1078.

131. Chen X, Levine L, Kwok PY. Fluorescence polarization in homogeneous nucleic acid analysis. Genome Res 1999;9:492–498.

132. Hacia JG. Resequencing and mutational analysis using oligonucleotide microarrays. Nat Genet 1999;21:42–47.

133. Ji M, Hou P, Li S, He N, Lu Z. Microarray-based method for genotyping of functional single nucleotide polymorphisms using dual-color fluorescence hybridization. Mutat Res 2004;548:97–105.

134. Gut IG. DNA analysis by MALDI-TOF mass spectrometry. Hum Mutat 2004;23:437–441.

8

Effects of Drugs of Abuse on Cytokine Responses

Stephen B. Pruett

SUMMARY

In the United States, the association between drugs of abuse and increased risk of infection has been suspected since colonial times. More recent studies have confirmed this relationship for most commonly abused drugs. Studies in animal models and in human subjects indicate that this increased susceptibility to infection is associated with decreased effectiveness of important innate and acquired immune defense mechanisms. All of the major drugs of abuse have been reported to affect cytokine and chemokine responses, which are critical in resistance to most infections. In general, proinflammatory, proimmune cytokine responses are decreased and anti-inflammatory, immunosuppressive cytokine responses are increased. The mechanisms by which this occurs have not been fully delineated, but receptor mediated effects, which probably act by interfering with immune stimulus-mediated signaling and indirect effects mediated by stress hormones, clearly are involved in the action of most drugs of abuse.

Key Words: Cytokine; chemokine; ethanol; cannabinoid; marijuana; nicotine; tobacco; cocaine; opioid.

1. INTRODUCTION

Cytokines are critical components of innate and acquired immune defense mechanisms. The term cytokine refers to an agent produced by a cell that acts on a target cell. A particular cytokine may act on the same cell or the same cell type that produced it (autocrine), it may act locally (paracrine), or it may act systemically (endocrine). Most cytokines are produced by and act on more than one cell type, and some can act in an autocrine, paracrine, or

From: *Methods in Pharmacology and Toxicology: Cytokines in Human Health:*
Immunotoxicology, Pathology, and Therapeutic Applications
Edited by: R. V. House and J. Descotes © Humana Press Inc., Totowa, NJ

endocrine manner, depending on the circumstances. The effects of cytokines on target cells are mediated by specific receptor molecules that transduce signals into the target cell. Most cytokines induce more than one functional change in their target cells, and they may produce one response in one cell type (e.g., apoptosis) and a different or even opposite response (e.g., preventing apoptosis) in another cell type. One of the functions induced by several cytokines is the production and release of other cytokines. Thus, it is often stated that a "cytokine network" exists. This concept is indeed a useful one, but it would be a mistake to assume that this network operates independently of other systems. For example, it has been well documented that a number of cytokines can act on the central nervous system to promote sleep or to activate the hypothalamic–pituitary–adrenal axis, leading to a classical stress response. Some of the products of this response (e.g., glucocorticoids) can in turn decrease the expression of the genes coding for a number of cytokines. This action constitutes an important feedback loop that should be considered when evaluating the effects of drugs on cytokine responses.

A number of different classification schemes have been devised for cytokines and their receptors and are based on the physical structure or amino acid sequence of the cytokine molecule, its major functional characteristics, the major cell type that produces it, or the major cell type that responds to it, or the key signaling mediators activated by it. For example, one classification scheme refers to type I and type II cytokines based on a protein structure in the type I cytokines that includes a bundle of four alpha helical regions and based on structural similarities in another family of cytokines that includes the interferons (IFNs; type II; *see* Table 1; background information on cytokines and information in Table 1 was derived from Leonard *[1]*). Most of the cytokines are designated by the prefix "IL-" followed by a number. The "IL" designates interleukin, and the term was derived and began to be used before it was fully appreciated that a wide variety of cells in addition to leukocytes can produce many of the interleukins. The number indicates the time sequence of discovery and characterization. Other important cytokine families that do not fall within the type I or type II classification include the tumor necrosis factor (TNF) superfamily (which includes at least 19 distinct cytokines) and the interleukin 1 family of ligands. Both families also include receptor antagonists or soluble receptors that act to inhibit the action of ligands in these families.

The functions of cytokines are remarkably diverse. In fact, the list in Table 1 indicates that one protein is labeled "hormone," because it was first identified as a small protein that is produced at one location and acts primarily at other locations (endocrine) to produce growth of the body. However, it was determined later that his molecule has structural homology with the type I

Table 1
An Overview of Cytokine Classification

Cytokines classified by protein structure

Type I cytokines (four α helix family)		Type II cytokines (Interferon family)
IL-2	GM-CSF	IFN-α
IL-3	SCF	IFN-β
IL-4	M-CSF	IFN-ω
IL-5	LIF	IFN-τ
IL-6	EPO	IFN-γ
IL-7	Prolactin	IL-10
IL-9	Growth Hormone	IL-19
IL-11	Leptin	IL-20
IL-13		IL-22
IL-15		IL-24
IL-21		IL-26
		IL-28
		IL-29

Cytokines classified by cellular source/function

Th1 cytokines (produced by T-helper 1 cells)	Th2 cytokines (produced by T-helper 2 cells)
IL-2	IL-4
IFN-γ	IL-5
LT	IL-6
	IL-9
	IL-10
	IL-13

cytokines. At one time, there were vigorous discussions about fine distinctions between hormones and cytokines, but the terms clearly now overlap. Cytokines exist that promote apoptosis of cancer cells (some members of the TNF superfamily), activate macrophages to increase their microbicidal capabilities (IFN-γ), permit B-lymphocyte proliferation or antibody class switching (IL-4, IL-5, IL-6), activate the acute phase response (IL-6), facilitate T-cell proliferation or activation (IL-2), activate natural killer cells to more effectively kill tumor cells or virus infected cells (IL-12, IFN-α, IFN-β), directly inhibit virus replication in a variety of host cells (all IFNs), induce fever (IL-1), enhance or initiate inflammatory responses (TNF-α and many others), induce the production of other cytokines or chemokines (IL-12 and others), or inhibit immune or inflammatory responses (IL-10 and transforming growth factor [TGF]-β). This small sample of examples serves to illustrate the importance and complexity of the cytokine network, which also is illustrated by the observation that transgenic knockout mice that lack one of them or the corresponding receptor typically display markedly decreased resistance to at least one type of infectious agent or an increased propensity to develop autoimmune disease or other immunopathological conditions.

Another important family of small proteins is also critically important in innate and acquired immunity: the chemokines (background information on chemokines was derived primarily from Murphy *[2]*). This term is derived from "chemotactic" and "cytokine." It refers to molecules produced by cells that attract other cells to the region where they are secreted. Fortunately, only one classifications scheme exists for chemokines, and it is based on unique protein sequence motifs that include cysteine residues. For example, the CXC chemokines share a cysteine an adjacent amino acid that is not cysteine (X) followed at a later position by another cysteine. On this basis, there are six classes of chemokines and a total of at least 48 distinct members. Although many chemokines were first named on the basis of a particular function (e.g., eotaxin-3, which attracts eosinophils), a uniform nomenclature has been implemented in which the broad family (for example CXC or CC) is followed by "L" (ligand) and a number, which indicates individual members within that family. Thus, CXCL1 is an example of a CXC chemokine. The issue is confused somewhat by a several older names that are still used. For example, the protein first designated IL-8 was later found to fit the structural and functional criteria of a chemokine and it is now also referred to as CXCL8. An additional complication arises from species differences. For example, humans have CXCL8, but mice do not have a structurally comparable protein.

2. GENERAL ISSUES

The species of interest in this volume is the human. However, there are serious ethical limitations in the use of human subjects in research on drugs of abuse. Risk of toxicity and of addiction generally limits the use of nonaddicted or nonexperienced human subjects to acute exposure to low doses of the drug. More realistic results could theoretically be obtained from drug-dependent persons, but these persons often abuse more than one drug and are subject to a number of confounding issues, such as poor nutrition, altered sleep patterns, or mental illness. Thus, animal models have been invaluable in understanding the effects and the mechanisms of action of drugs of abuse on the immune system. Therefore, this chapter will discuss animal studies before relating them to pertinent findings in human subjects and synthesizing both to consider the mechanisms of immunomodulation. Most of the drugs discussed in this chapter produce profound physical dependence. When drug-dependent persons withdraw from the drug a stress response, as well as numerous other physiological adaptations, occur, and ample evidence exists that some of these effects can alter immune function. A few examples are given, but this is not the focus of the chapter, so this issue is not considered in a systematic way.

As already mentioned, many drug abusers are actually polydrug abusers. There is at least one report indicating that two drugs of abuse may interact to produce greater immunomodulation than either alone *(3)*. However, the effects of drug mixtures been the subject of a relatively small number of studies and will not be considered systematically in this chapter.

3. OPIOIDS

3.1. Background and Effects of Opioids In Vitro

Endogenous, plant-derived, and synthetic opioids affect a wide range of immune functions, including cytokine production, in vitro. In most cases, these effects would suggest generally immunosuppressive or anti-inflammatory actions of opioids on the production or response to cytokines or chemokines. Some of these effects are only evident at concentrations higher than attainable in vivo, but significant effects also have been reported at relevant opioid concentrations *(4–6)*, which are up to 400 ng/mL (approx 0.6 μM) for morphine in humans and probably somewhat greater in rodents. Interpretation of in vitro studies in immunology also can be difficult because agents may work in vitro by unexpected mechanisms, such as restoration of appropriate nutritional status or restoration of other parameters to values that are more similar to those observed in vivo *(7)*. Thus, immunological effects and mechanisms observed only in vitro are of uncertain relevance with regard to effects and mechanisms of action in vivo. However, there have been a number of studies of opioid effects involving cytokine expression in vitro that also have been documented in vivo. For example, opioids were found to induce macrophage apoptosis in vitro by inducing TGF-β and nitric oxide, and similar results were noted in vivo *(8)*. Others have reported similar findings and have implicated upregulation of heme oxygenase 1 (HO-1) in opioid-induced apoptosis of macrophages *(9)*. In vitro studies are particularly useful in distinguishing direct effects of opioids on immune system cells from indirect effects mediated through other cell types. For example, morphine can directly inhibit the activation of some lymphocytes by mitogens in vitro, but other types of lymphocytes are only inhibited in vivo, probably by glucocorticoids induced by the opioid used in the study *(10)*.

Suppression of both T-cell and macrophage-derived cytokine production by fentanyl in vitro has been reported, and these effects could not be reversed by naltrexone, suggesting that classical opioid receptors were not involved in this effect *(11)*. In addition, it has been noted that not only morphine but some of its metabolites decrease the production of IL-2, IL-4, and IL-6 by lymphocytes or macrophages *(12)*.

One of the most interesting effects of opioids in vitro is increasing the infectivity and replication of HIV-1 and SIV-1 in appropriate immune cell cultures. In the case of SIV, increased expression of the chemokine receptor CCR5, a coreceptor for SIV and HIV, may be one of the mechanisms by which opioids act *(13)*. However, evidence also has been presented in support of other mechanisms, such as decreased expression of anti-viral type I and type II interferons *(14)*. In addition, in vitro studies have suggested that morphine may prevent the anti-inflammatory actions of retinoids and thus enhance the chronic inflammatory response associated with enhanced HIV progression *(15)*. Although some of these conclusions are consistent with results from limited in vivo studies in monkey models for HIV disease, epidemiological studies do not clearly indicate that consistent opioid abuse (addiction) enhances the progression of the virus *(16)*. However, decreased resistance to HIV or to common coinfecting viruses is indicated by in vitro studies *(17)*, and limited epidemiological data have been reported that support this idea *(18)*. In contrast to the generally suppressive effects of opioids on the induction of proinflammatory cytokines in vitro, opioids increase the production of several chemokines by human monocytes *(5)*. The effects are somewhat different in monocytes infected with HIV, suggesting a possible role for these alterations in the pathogenesis of HIV. Considering the complexity of the pathogenesis of HIV disease, it is possible that enhancement of chemokine production, which would tend to promote the chronic inflammatory state reported in HIV-infected persons, may exacerbate rather than enhance resistance to the disease.

3.2. In Vivo Effects of Opioids in Animal Models

In vivo experimental results strongly support a connection between opioid administration and altered cytokine responses. In most cases, the effects reported would be expected to suppress specific and innate immune responses. Recent reviews summarize a number of studies and presents evidence in support of a central role for suppressed cytokine responses in decreased innate resistance to infection, decreased antibody responses, and decreased NK cell function *(19,20)*.

Specific examples of suppressed cytokine responses have been reported in animal (mostly rodent) models. Restoration of adherent cell (macrophage)-derived IL-6 restores morphine-mediated suppression of antibody responses to a T-dependent antigen in mice *(21)*, suggesting that suppression of IL-6 production is one mechanism by which opioids suppress antibody responses. Chronic morphine exposure using a timed-release pellet decreases production of IFN-γ, IL-1β, and TNF-α in mice *(22,23)*. Although evidence has presented suggesting that type I interferons (IFN-α and -β) can bind to and activate opioid receptors in the central nervous system *(24,25)*, the effect of opioids on the expression of IFN-α and -β in response to immunological stimuli has not been examined. Suppression of both IL-10 and IL-12 production by lipopolysaccharide (LPS)-stimulated macrophages in vitro after morphine treatment in vivo has been reported, but this could not be prevented by naltrexone, suggesting an effect not mediated by classical opioid receptors *(26)*. In contrast, others have reported an increase in IL-12 production and no effect on IL-10 production by LPS-stimulated macrophages ex vivo after treatment of mice by implantation of a 75-mg timed-release morphine pellet *(27)*. The difference in results may reflect administration of a single bolus dose of morphine in the former study in contrast with continuous dosing in the latter. Other conflicting results have also been reported *(28,29)*, but a possible explanation is proved by a study of the time course of effects of a single dose of morphine or heroin in vivo on subsequent cytokine production ex vivo. In this study, a biphasic effect was noted in which production of IL-1β, IL-2, TNF-α, and IFN-γ produced by splenocytes were significantly increased when the spleen was obtained up to 40 min after dosing. However, these cytokines were substantially suppressed and IL-10 and TGF-β (generally anti-inflammatory cytokines) were increased in splenocytes obtained 24 to 48 h after dosing with morphine or heroin *(30)*. This suggests the possibility that some conflicting results may be explained by differences in the time after dosing at which cells were obtained for analysis. Both μ and κ-opioid receptor agonists have been consistently reported to have anti-inflammatory properties, with potential therapeutic efficacy in inflammatory conditions such as arthritis or inflammatory bowel disease *(31,32)*.

A key issue with regard to the effects of opioids on cytokine responses is the role of these changes in resistance to infection or tumor development. There is a broad consensus that opioid abuse is associated with decreased innate immunity and inflammation and an increased risk of infection in animal models *(20)*. In a particularly informative study, resistance to *Streptococcus pneumoniae* in mice was decreased by morphine, which was associated with decreased production of TNF-α, IL-1, IL-6, macrophage inflammatory protein (MIP)-2, and KC in the lungs *(33)*. However, some interesting exceptions have been documented. For example, the administration of morphine by timed-release pellets had an adverse effect on a variety of immune system parameters in mice, but morphine actually enhanced resistance to a challenge dose of live *Listeria monocytogenes (34)*. It was demonstrated that the implantation of morphine pellets caused a substantial and prolonged increase in serum corticosterone and an associated increase in the absolute number of neutrophils in the blood. It is likely that this increase in neutrophils explains the increased resistance to *L. monocytogenes*, which is heavily dependent on neutrophils early in the disease process *(35)*.

Although numerous studies have shown that cytokines play a key role in both innate and acquired immunity, the effects of partial suppression of the response of several cytokines on resistance to infection have not been conclusively determined. Thus, it is likely that the suppression of cytokine responses caused by opioids is partly responsible for the observed decreases in resistance to infection, but the exact portion of the alteration in resistance to infection mediated by changes in cytokine responses is not known. This is a problem well suited to the new discipline of systems biology, which "studies biological systems by systematically perturbing them (biologically, genetically, or chemically); monitoring the gene, protein, and informational pathway responses; integrating these data; and ultimately, formulating mathematical models that describe the structure of the system and its response to individual perturbations" *(36)*. This will require the use of high-throughput methods to comprehensively evaluate the changes caused by opioids in the complex events associated with cytokine responses and resistance to infection.

3.3. Effects of Opioids on the Immune System in Humans

There is evidence that opioid addiction increases the risk of infectious disease in human beings *(37)*. However, there has been concern that this increased risk of infection is secondary to the use of nonsterile needles or syringes for intravenous administration of opioids or to sleep disruption *(38)* or nutritional deficits *(39)* in opioid abusers. Thus, the role of immunosuppression in decreased resistance to infection has not been definitively established. Administration of opioids for pain relief or anesthesia can suppress pro-immune cytokine responses under some circumstances. For example, Sufentanil decreases perioperative IL-6 induced by surgery *(40)*. κ Opioid agonists decrease proinflammatory cytokine responses and have been proposed as therapeutic agents for inflammatory diseases such as arthritis *(31)*. Fentanyl decreases the fever response induced by exogenous IL-2 *(41)*.

Morphine treatment of human subjects for 36 h inhibits natural-killer (NK) and cytokine-activated NK cell activity *(42)*. Heroin addicts have decreased MCP-1 messenger ribonucleic acid (mRNA) expression in peripheral blood monocytes *(43)*. A strong association has been reported between postsurgical infections and the amount of opioid delivered by patient-controlled analgesia pumps. This association could not be explained by a number of possible confounding variables *(44)*, and the authors suggest that opioid-mediated immunosuppression is the most likely explanation for the findings. However, results from other studies do not indicate meaningful immunosuppression induced by opioids. For example, anesthesia that includes opioids is associated with increased production of IL-1β and TNF-α and decreased IL-10 upon stimulation of peripheral blood mononuclear cells ex vivo *(45)*. Similarly, it has been reported that heroin addicts have increased cytokine responses that normalize during subsequent methadone therapy *(46)*. High-dose fentanyl has no effect on postsurgery cytokine and chemokine production *(47)*. There was no obvious suppression of delayed type hypersensitivity responses (which are dependent on Th1 cytokines) in heroin addicts *(48)*. The basis for these conflicting results is not clear. However, on the basis of the animal studies mentioned previously, it is clear that the effects of opioids are not universally immunosuppressive or immunoenhancing. Rather, the effect depends on a number of factors related to timing of dosing and immunological challenge, the nature of the infecting microorganism, and other variables.

Withdrawal from opioids can also decrease cytokine production by macrophages *(49)*. Obviously, this is not a direct effect of opioids, but it is probably a results of the stress associated with withdrawal.

3.4. Mechanisms of Opioid-Induced Immunomodulation

A major issue with regard to the mechanism of opioid action on the immune system is the role of direct effects mediated by opioid receptors on cells of the immune system vs indirect effects mediated through opioid receptors on other cells or tissues, particularly those in the brain. Evidence also exists that opioids can affect the immune system in ways that are not dependent on classical opioid receptors *(50)*, so this possibility must be considered as well. Indirect effects are clearly involved in some of the effects of opioids on the immune system in rodent models. For example, suppression of NK cell activity and thymic atrophy in mice implanted with 75 mg of timed-release morphine pellets are mediated by increased corticosterone concentrations *(51,52)*. These effects can be inhibited by naltrexone, suggesting that classical opioid receptors are involved in this effect, but presumably these receptors are located in the hypothalamus, where the signal to increase corticosterone release from the adrenals usually originates. The role of indirect effects of opioids in altering cytokine responses is not as clear. One of the most thorough studies in this regard indicates that opioid induced corticosterone mediates part of the suppression of IFN-γ and IL-2, and sympathetic nervous system responses mediate an additional portion of the suppression of IFN-γ produced by spleen cells in response to mitogens ex vivo *(22)*.

Direct suppression by opioids of antibody responses have been reported, and this suppression could be prevented by adding IL-6 to the culture system, suggesting that decreased cytokine production may be involved *(53)*. However, the situation is apparently more complex than originally suspected. We reported that opioid agonists did not have significant direct effects on antibody responses in vitro, but they did in vivo *(54)*. Using a virtually identical experimental system, another group had previously reported that opioid agonists directly inhibit antibody responses in vitro and that this inhibition could be prevented by opioid antagonists *(55)*. In an effort to understand these contradictory results, this group systematically considered variables that might have accounted for the disparity. They found that both studies used C57Bl/6 × C3H F1 mice, but the mice came from different suppliers: Charles River Labs and Jackson Labs. The progenitors of the Jackson and Charles River strains were separated 45 yr ago, and the Jackson Labs C3H line (C3H/HeJ) is well known for its hyporesponsiveness to LPS as compared with other C3H lines. When Eisenstein and colleagues *(56)* evaluated both strains, they found that opioids did not directly affect antibody responses in vitro using cells from the Charles River line, but they did directly suppress antibody responses by cells from the Jackson Labs line. This apparently explains the previously mentioned contradictory results, because we had used C3H/HeN (Charles River Labs)-derived mice and the Eisenstein group had used C3H/HeJ (Jackson Labs)-derived mice. The implications of these findings are interesting. If functional opioid receptors are expressed on the lymphocytes or macrophages of some mouse strains and not others, without obvious alterations in immune status, what does this imply with regard to the importance of these receptors? Comparable strain differences in cytokine responses have been documented *(23)*. Thus, it seems unwise at present to generalize with regard to the relative importance of direct and indirect mechanisms of action of opioids, because it is not known whether this unusual strain-dependence applies to other species.

The molecular mechanisms by which receptor-mediated immunomodulation is induced by opioids have not been extensively investigated. However, there are indications that some opioids can act through the δ-opioid receptor (a G protein-coupled receptor) to activate mitogen-activated protein (MAP) kinases in a human T-cell tumor line *(57)*, which can affect cytokine expression *(58)*. However, a different pattern of effects was reported in T-cells activated by anti-CD3 and CD28 antibodies, which mimic an actual immunological stimulus. Opioids act to decrease activation of the IFN-γ promoter by mechanisms that include increasing cellular cAMP, decreasing MAP kinase activation, and decreasing the activation of the transcription factors NF-κB and nuclear factor of activated T-cells *(6)*. The increase in cAMP has been observed in most reports, and it may have common effects on cytokine expression, when the cells are exposed to similar stimuli. It is interesting that cannabinoids also increase cellular cAMP and share some of the effects of opioids on cytokine production (discussed in later section of this chapter). Opioid agonists also increase intracellular Ca^{2+} in human B-cell lines, suggesting another mechanism of signaling that would be expected to affect the expression of cytokines *(59)*. However, considering that it has been known for more than 20 yr that opioids affect cytokine production in the immune system, surprisingly little is known of the mechanisms by which this occurs.

Furthermore, it is not entirely clear whether most of the effects of commonly abused opioids on cytokine responses are mediated directly through receptors on immune system cells, indirectly through opioid receptors on other cell types, or through nonreceptor-mediated mechanisms. Evidence for all of these possibilities has been reported for various immunological end points, but few studies have addressed all these options with regard to cytokine production. The well-characterized pharmacologically defined receptors for opioids are designated μ, κ, and δ. Most of the opioids that are used by opioid-dependent persons (e.g., morphine) act mostly or entirely through the μ receptor *(60)*. The most commonly used antagonists (naltrexone and naloxone) inhibit the action of opioids at all of these receptors. The development of transgenic (knockout) mice lacking one or more opioid receptors has been useful in this research, but these mice have only been used in limited ways to address the effects of opioids on cytokine production. In one study, morphine-induced suppression of IL-1 and IL-6 production by macrophages was similar in μ-receptor knockout mice and wild-type mice *(61)*, suggesting that μ-receptors on macrophages are not critical in the suppression of IL-1 and IL-6 production by opioids in vivo. In the same study, production of TNF-α by macrophages was suppressed by morphine in normal, but not μ-receptor knockout mice, suggesting that μ receptors are involved in this action. However, in another study by the same group, using very similar methods, morphine increased LPS-induced production of TNF-α and

IL-1β *(22)*. This increase was not observed in μ-receptor knockout mice. The actual role of μ-receptors on production of proinflammatory cytokines by macrophages will not be clear until these contradictory findings are resolved.

One of the more interesting mechanisms by which the immune system and the opioid system interact is heterologous desensitization of receptors. Both opioid and chemokine receptors are G protein-coupled seven trans-membrane domain proteins *(62)*. The endogenous opioid ligands (enkephalins and endorphins) together with cytokines and chemokines apparently interact to form a network controlling the association between pain and inflammation. Recent evidence suggests that the pain associated with inflammation is partly due to the desensitization of opioid receptors by chemokines *(62)*. The reciprocal relationship also exists: opioid agonists can decrease responses of cells to chemokines by desensitizing chemokine receptors *(62–65)*, and this is probably one of the mechanisms by which opioids suppress inflammation. The precise mechanisms for this desensitization are not clear, but signaling seems to be involved and internalization of receptors has been reported for opioid receptors *(63)* but not CCR5 (chemokine) receptor *(66)*.

4. CANNABINOIDS

4.1. Background, In Vitro Effects, and In Vivo Effects in Animal Models

Cannabinoids, including the major active ingredient of marijuana, Δ-9-tetrahydrocannabinol (THC), have a broad range of immunomodulatory effects in vitro and in animal models. Cannabinoid CB1 receptors are located primarily in the central nervous system, but they also are present on cells in the periphery, including cells of the immune system *(67)*. Cannabinoid CB2 receptors are located exclusively in the periphery and are also present on cells of the immune system *(67)*. The effects of natural cannabinoid agonists administered exogenously, THC or synthetic analogs, and by marijuana smoke generally are reported to be immunosuppressive or anti-inflammatory *(68–74)*. However, antagonists of the CB1 cannabinoid receptor can decrease inflammation in vivo in an animal model *(75)*. This might seem to suggest that natural cannabinoid ligands at concentrations attained in vivo may have proinflammatory effects, at least in some models of inflammation. However, this is conclusion is not supported by results from CB1 receptor knockout mice, which are more prone to damaging inflammatory responses than normal mice *(76)*. Thus, the reported decreased in inflammation caused by a CB1 antagonist may simply reflect partial agonist activity at a high dosage. It also should be noted that the effect of a particular cannabinoid ligand on production of a particular cytokine can depend heavily on the intensity of the activation stimulus. For example, IL-2 production is enhanced in T- cells treated with suboptimal-activating stimuli but suppressed in T-cells treated with optimal activation stimuli *(77)*. This observation may explain some of the apparently contradictory reports in the literature on this subject.

Cannabinoids can suppress production of IL-2 and IFN-γ by T-cells *(78)* and the production of TNF-α, IL-12, and IL-6 by macrophages *(73,79,80)*. In contrast, IL-10 production tends to be upregulated by cannabinoids *(79)*. In addition, there are indications that the suppression of T-cell cytokines is not indiscriminate, but may primarily involve Th1 cytokines *(78)*.

4.2. Effects of Cannabinoids in Human Subjects

Recent results indicate a similar range of effects in human subjects *(81–83)*. In particular, the response to T-cell mitogens and production of IL-2 is decreased in heavy marijuana users compared with occasional users or nonusers. In contrast, concentrations of IL-10 and TGF-β, which tend to be immunosuppressive or anti-inflammatory are increased *(81)*. However, further immunosuppression in patients with immune dysfunction caused by HIV-1 infection or multiple sclerosis was not observed after cannabinoid exposure *(84–86)*.

4.3. Mechanisms of Cannabinoid Action in the Immune System

The mechanisms by which cannabinoids act on the immune system have not been completely elucidated, but evidence has been presented for both receptor mediated and nonreceptor-mediated mechanisms. In addition, it has been reported that psychoactive cannabinoids can induce a neuroendocrine stress response *(87)*, and such a response has been shown in other systems to alter cytokine production. Cannabinoid receptors are G protein-coupled receptors, so it would be expected that cannabinoids would act by altering signaling events typically induced by activation of such receptors. Experimental results generally have been consistent with this expectation *(71,88)*. For example, cannabinoids can act to decrease cAMP concentrations during cellular activation leading to decreased activity of cAMP-dependent transcription factors *(89)*. However, other signaling events are also inhibited in T-lymphocytes by cannabinoids, such as activation of extracellular signal-regulated kinases and of the transcription factors activator protein (AP)-1 and NF-κB *(71,72,90)*. This effect may be relatively specific for T-cells, or it may depend on the nature of the activating stimulus, because cannabinoid agonists alone can activate ERK in some cell types *(91–93)*. Increases in intracellular Ca^{2+} and NF-AT mediated by CB2 receptors on mouse spleen cells and a human T-cell line also have been reported *(94)*. Enhanced production of IL-2 can be induced by suboptimally stimulated T-cells, and this effect is apparently CB1 and CB2-independent *(67)*. Thus, almost every conceivable category of mechanism (receptor dependent and receptor independent; direct effects on immune system cells and indirect effects mediated by the neuroendocrine stress response) has either been documented or remains a possibility. Clearly, further work is needed to delineate the relative importance of each mechanism and to understand which ones are induced by particular scenarios of cannabinoid exposure. Although no anti-inflammatory effects were noted in a clinical trial of Δ-9-THC for Mediterranean fever *(95)*, the anti-inflammatory properties of some synthetic cannabinoids in animal models have prompted a proposal to develop them as therapeutic agents for inflammatory-mediated illnesses such as hepatitis *(70)*.

5. COCAINE

5.1. Introduction and In Vitro Effects

Cocaine acts on the central nervous system primarily by preventing the reuptake of norepinephrine and dopamine from synapses and thereby prolonging or intensifying the effects of these neurotransmitters *(60,96)*. However, it also acts in the central nervous system and in peripheral tissues by binding to the intracellular sigma receptor *(97,98)*. Cocaine also induces a brief neuroendocrine stress response (with elevated concentrations of glucocorticoids in the blood) in rodent *(99,100)* and nonhuman primate models *(101)* as well as in human subjects *(102,103)*.

Direct effects of cocaine on cytokine production by immune system cells in vitro have been documented. Cocaine at relevant concentrations (10–200 ng/mL) suppresses IL-2-mediated expression of IFN-γ and IL-8 at the mRNA and protein level in human peripheral blood mononuclear cells *(104)*. Similarly, cocaine at relevant concentrations inhibits Concanavalin A-induced production of IL-2, IL-4, IL-5, IL-10, and IFN-γ by mouse spleen cells and LPS-induced production of IL-1β, IL-6, and TNF-α by mouse macrophages *(105)*. Other investigators have reported similar effects, but the concentrations of cocaine used were more than 1000 times greater than concentrations that can occur in animals or human beings. Suppressed production of IFN-γ and IL-10 by peripheral blood mononuclear cells from dependent persons has been reported using cocaine concentrations that are approx 10-fold greater than concentrations typically reported in vivo *(106)*. Cocaine also increases TGF-β production, and this generally immunosuppressive cytokine may explain some of its effects *(107)*.

5.2. Effects of Cocaine in Animal Models

The effects cocaine on cytokine production in vivo in animal models are more complex. Acute or daily cocaine exposure for 7 d in mice decreases the production of IL-2, IFN-γ, and IL-4 by mitogen-stimulated spleen cells ex vivo *(108)*. Similar results have been reported for five daily doses of cocaine *(109)*. However, neurological sensitization to the effects of cocaine by intermittent dosing caused an increase in IFN-γ production *(110)*. Chronic exposure of rats to cocaine is associated with decreased production of IL-2 and TNF-α and increased production of IL-10 and TGF-β upon stimulation of spleen cells with mitogen ex vivo *(111)*. Cocaine-mediated increases in IL-10 production have been identified as the mechanism by which cocaine can decrease resistance to cancer *(112)*. These results would generally be consistent with an overall anti-inflammatory or immunosuppressive effect of cocaine. However, acute administration of cocaine enhances one immune parameter (T-dependent antibody response) by a mechanism that depends on a brief increase in serum corticosterone *(99,100)*. Considering the generally accepted immunosuppressive properties of glucocorticoids, this may seem paradoxical. However, it is becoming clear that brief or low-level increases in glucocorticoids can enhance some immune functions, whereas prolonged or intense increases generally suppress immune functions *(113,114)*.

5.3. Effects of Cocaine in Human Subjects

Cocaine dependence is associated with an increased risk of infectious diseases in human subjects *(115,116)*. A causative role for cocaine-mediated immunosuppression has not been definitively established, but evidence for an immune system-related mechanism has been presented for the cocaine-induced increase in susceptibility to HIV-1 in the brain *(116)*. In addition, two intriguing studies suggest suppression of cytokine responses known to be important in resistance to infection. Alveolar macrophages from smokers of crack cocaine exhibit suppression of cytokine mediated antimicrobial mechanisms such as nitric oxide production *(117)*. This was not the case for tobacco smoke, indicating that the effects of cocaine do not reflect only the nonspecific action of particles associated with smoking. In another study, indwelling intravenous catheters were used to stimulate an innate immune response. Subjects receiving 0.4 mg/kg cocaine intravenously had reduced IL-6 levels at 2 h, possibly caused by increased release of cortisol *(118)*. These results suggest that cocaine-induced immunosuppression may be responsible for decreased resistance to infection. However, enhancement of innate immunity after cocaine administration in human subjects also has been reported *(119)*. Ten to 45 min after a single dose of cocaine by inhalation or by intravenous administration, neutrophils were activated and were better able to kill *Staphylococcus aureus* than neutrophils from subjects treated with placebo. It is possible that this corresponds to the brief, glucocorticoid-mediated enhancement of immune function reported in rodents *(120)* and that direct effects of cocaine would suppress these same parameters at later times after dosing.

It should be noted that not all studies have documented changes in cytokine production in a manner that would be consistent with immunosuppression. For example, the administration of cocaine to cocaine-dependent human subjects increases ex vivo production of IFN-γ by peripheral blood mononuclear cells *(106)*. Similarly, when persons who regularly smoked crack cocaine were asked to abstain for 8 h and cocaine then administered under controlled conditions, this was associated with increased activation of neutrophils and IL-8 production *(119)*. However, these findings may result more from the amelioration of the stress response that often is associated with withdrawal from cocaine *(121)* than from direct action of cocaine. Similar findings also have been observed in a few studies in which cells were exposed to cocaine in vitro. In one study, cocaine increased the production of type I interferons in culture by L929 cells and macrophages *(122)*. In another, it activated the production of TNF-α by brain endothelial cells in vitro *(123)*. However, both of these cytokines can be induced by very small quantities of contaminants such as lipopolysaccharide.

5.4. Mechanisms of Cocaine-Mediated Immunomodulation

Few studies have been done to specifically delineate the mechanisms by which cocaine affects cytokine production. As already mentioned, cocaine-induced elevation in serum corticosterone concentrations seem to be responsible for its enhancement of antibody production and alteration of Th1/Th2 cytokine balance *(99,100)*. This may seem paradoxical in view of the generally accepted immunosuppressive properties of glucocorticoids. However, we have previously pointed out that the effects of glucocorticoids on the immune system are exquisitely dependent on the concentration and duration of exposure *(113)*. Thus, the results reported for cocaine are consistent with immunoenhancement caused by short-term restraint *(120)*. However, these results suggest that the generally suppressive effects of cocaine on other cytokine-related immune functions are probably not secondary to the stress response, because such suppressive effects require intense or prolonged exposure to stressors *(113)*. Receptor-mediated mechanisms for cocaine's effects on the immune system have been considered, and there are indications that the binding of sigma receptors may be involved. For example, cocaine increases IL-10 production in mice by a sigma receptor-dependent mechanism (demonstrated by the action of a sigma receptor antagonist). This increase in IL-10 production is responsible for decreased resistance to exogenous tumor cells in this model *(112)*. It should also be noted when considering mechanisms, that many cocaine abusers also consume ethanol. This leads to the production of a metabolite, cocaethylene, which inhibits IL-2 production by activated peripheral blood mononuclear cells in vitro at very low concentrations *(124)*. Thus, drug interactions and effects of metabolites may part of the mechanism of action of cocaine on the immune system. The action of cocaine on catecholamine metabolism has not been carefully investigated and represents another potential mechanism by which it could affect the immune system.

6. ETHANOL

6.1. Background and In Vitro Effects of Ethanol

The association between excessive ethanol consumption and increased risk of infection has been recognized since colonial times in America *(125)*. The risk has been quantified in later studies, and a substantial proportion of cases of certain infectious diseases seem to be related to alcohol abuse *(126)*. Subsequent studies in animals and in human subjects have revealed that both acute and chronic ethanol consumption are associated with decreased innate and acquired immunity *(126)*. It is likely that immunosuppression is a mechanism by which ethanol suppresses resistance to infection, but direct evidence for this is limited.

Addition of ethanol to immune system cells in culture adversely affects a wide range of immune functions. Relevant concentrations of ethanol for such studies must be considered, and a concentration as high as approx 0.5% (w/v) (approx 108 m*M*) would be considered by some investigators too high to be relevant. This number is based primarily on the fact that concentrations in this range in the blood may be associated with a fatal outcome in humans *(127)*. However, it also should be considered that blood ethanol concentrations in this range have been measured in patients judged by emergency room staff to be sober, although they admitted to recent ethanol consumption *(128)*. This probably requires the development of tolerance, which would be associated with chronic ethanol consumption. Thus, it seems reasonable to include concentrations of ethanol up to 0.5% (w/v) in experiments designed to evaluate the effects of ethanol on immune system cells in vitro. Significant suppression of the production of pro-inflammatory cytokines and significant enhancement of IL-10 production by peripheral blood mononuclear cells in vitro have been reported at much lower ethanol concentrations (0.1% w/v [129,130]). Ethanol at relevant concentrations inhibits TNF-α-induced activation of endothelial cells as indicated by the production of key cytokines and chemokines as well as cellular adhesion molecules (131). Production of TNF-α and IFN-γ (Th0 and Th1 cytokines, respectively) induced by Concanavalin A (a T-lymphocyte mitogen) in spleen cell cultures is suppressed by ethanol (132). Although it is clear that ethanol can directly affect cytokine production in vitro, the role of its major metabolites (acetaldehyde and acetate) has not been examined in this regard. Thus, it remains possible that the effects of ethanol on cytokine production in vivo represent a combination of direct effects of ethanol, direct effects of metabolites, and indirect effects.

6.2. Effect of Ethanol on the Immune System in Animal Models

Animal models for chronic and acute ethanol exposure generally have indicated effects on cytokine expression that would be consistent with the generally reported decreases in resistance to infection, that is, suppression of the production of proinflammatory cytokines (e.g., IL-12) and enhancement of anti-inflammatory or immunosuppressive cytokines (e.g., IL-10 [133–135]). However, it should be noted that there is a considerable body of data from investigators seeking to explain alcoholic liver disease suggesting that chronic ethanol exposure can, under some circumstances, enhance the production of pro-inflammatory cytokines. This would be expected, considering the key role that inflammation seems to play in alcoholic liver disease. However, the paradox of decreased resistance to infection and increased inflammatory disease associated with the same treatment has not yet been resolved. Possible resolutions to this paradox will be mentioned in the following discussion of results from animal models.

Host resistance to intracellular (136,137) and extracellular (133,138,139) bacteria, viruses (140,141), and other classes of pathogens (142) typically are decreased in rodents treated with ethanol. None of these studies compared acute and chronic ethanol exposure, but some involved acute and some involved chronic exposure, suggesting that both modes can decrease host resistance. It should be noted that results from a few studies indicate no significant change in resistance to infection. Both decreased resistance (136) and normal resistance (143) to Listeria monocytogenes have been reported by different groups, and the basis for these differing results is not clear. Although a review of the published literature suggests that the evidence for decreased resistance is overwhelming, it seems likely that some studies in which the results were negative (i.e., ethanol did not alter host resistance) probably have not been published. The example just cited of conflicting results with *L. monocytogenes* illustrates that host resistance and the effects of ethanol are both complex. It is likely that there are a number of factors not yet elucidated that influence both the effects of ethanol on the immune system and the ability of the immune system to clear infectious agents.

Innate immunity and inflammation, antibody responses, and T-cell-mediated responses generally are suppressed by acute or chronic ethanol treatment in animal models, and key cytokines associated with these responses are similarly affected. For example, the toll-like receptor mediated induction of pro-inflammatory cytokines and chemokines (IL-12, IL-6, and CXCL9) by peritoneal macrophages in mice is suppressed by acute ethanol administration, whereas the production of the generally anti-inflammatory cytokine IL-10 is substantially increased *(133,134)*. In the same studies, the expression of other cytokine genes (e.g., IL-15, IL-1, IFN-γ, and IFN-α) also was suppressed as was the expression of most chemokine genes. Interferon-gamma production and resultant T-cell-mediated delayed-type hypersensitivity responses are suppressed by chronic ethanol administration in mice *(144)*. Similarly, the T-cell response to hepatitis C core protein is inhibited by chronic ethanol exposure in mice, and this can be prevented by co-treatment with mouse IL-2 in vivo *(145)*. Acute exposure to ethanol decreases the antibody response to a complex T-dependent antigen (sheep erythrocytes) and also decreases the production of IL-2 and IL-4 in response to this antigen *(146)*. The effects of chronic ethanol treatment on antibody and responses (and related cytokine responses) have not been entirely consistent and may vary depending on the antigen used. For example, the humoral response to the hepatitis C nonstructural protein, NS5, was suppressed by chronic exposure to ethanol in a liquid diet preparation. However, coadministration of murine IL-2 partially prevented the suppression of the B-cell response *(147)*. Using a synthetic amino acid copolymer as antigen, another group of investigators has consistently reported no effect or enhanced antibody responses *(148)*.

Mice have been used extensively in studies of the effects of ethanol on the immune system because of the excellent availability of reagents, transgenic strains, and methods to evaluate all aspects of immune function. However, until recently most studies of the effects of chronic ethanol consumption have involved ethanol administration in a liquid diet developed for rats as the sole source of food and water *(149)*. Rats consuming ethanol in this diet maintain a normal or near-normal rate of growth, whereas most mouse strains have a decreased rate of growth and/or loss of body weight. Ethanol is such an aversive stimulus that mice will not consume a sufficient diet to maintain normal nutrition. A particularly important study demonstrated that a substantial portion of the effects of chronic ethanol consumption on lymphocyte depletion in the spleen, thymus, and gut-associated lymphoid tissue is mediated by the stress response associated with this undernutrition *(150)*. Because many of the studies cited in this chapter involved the use of this liquid diet, it would probably be worthwhile to repeat many of these studies using a more recent approach that has been adopted by a number of investigators. Several years ago, Meadows and colleagues *(151)* exhaustively documented a method for administration of ethanol chronically by providing mice with a 20% ethanol solution as the sole source of water. They found that this method permitted consumption of relevant quantities of ethanol, provided adequate nutrition and hydration, and did not induce a meaningful stress response. A number of investigators have now adopted this method *(141,152)*.

6.3. Effects of Ethanol on Human Subjects

Effects of ethanol on cytokine production by human cells in culture and in humans exposed to ethanol under experimental conditions or during the course of alcohol dependence have been reported. One of the more consistent observations has been a decrease in proinflammatory cytokine and chemokine production and decreased migration of leukocytes to sites of inflammation or infection. The situation for acute exposure is particularly convincing. In persons who consume alcohol in the laboratory, decreased migration of neutrophils to a site of inflammation on the skin has been documented *(153)*. Cytokine and chemokine production by peripheral blood monocytes from such persons is decreased *(154)*. Comparable exposure to ethanol in vitro causes comparable decreases in the production of proinflammatory cytokines and increased production of IL-10 *(155,156)*. The situation is apparently more complex in persons who are dependent on alcohol. An increased concentration of IL-10 in the serum was associated with chronic alcohol consumption in patients undergoing surgery for cancer of the upper aerodigestive tract *(157)*. The ratio of IL-10 to IL-6 also was increased in the alcoholic patients, suggesting suppression of normal postsurgical inflammation in the alcoholic patients. Both of these parameters were significantly associated with an increased risk of surgical wound infection and pneumonia in the alcoholic patients *(157,158)*. However, patients with alcoholic liver disease exhibited increased concentrations of TNF-α and IL-6 in the plasma *(159)*. Of interest, these increases were not associated with the degree of liver disease *per se* but were closely associated with acute excessive ethanol consumption and resultant increases in circulating lipopolysaccharide, suggesting that ethanol may continue to suppress innate immunity and inflammation in chronic alcoholics as it does following occasional acute ethanol consumption. However, acute excessive ethanol consumption apparently also causes an increase in LPS (presumably by decreasing the permeability of the gut to Gram-negative bacteria), and the proinflammatory properties of lipopolysaccharide are not completely abrogated by the anti-inflammatory action of ethanol. At least this working hypothesis would serve to explain the apparent paradox of the generally anti-inflammatory effects of ethanol and the involvement of inflammation in the pathogenesis of alcoholic liver disease. An apparent rebound in sensitivity to lipopolysaccharide after the suppression of responsiveness by ethanol *(160)* as well as other mechanisms may also be involved.

A similarly complicated situation apparently occurs with regard to cytokines associated with acquired immune responses. There are indications that cell-mediated immunity and associated cytokine responses are diminished in human beings who chronically consume excessive quantities of ethanol *(157,158)*. This is associated with decreased production of IFN-γ and IL-6 and increased production of IL-10 as well as an increased risk of postoperative infections. It is important to note that these studies involved persons who were actively drinking and still had detectable blood ethanol concentrations before surgery.

Alcohol-dependent persons without liver disease typically have peripheral blood lymphocytes with an activated phenotype *(161)*, and lymphocytes from alcohol-dependent persons with liver disease exhibit increased production of a number of cytokines upon ex vivo stimulation *(162)*. Recent results suggest that both T-cell-mediated *(163–165)* and innate immune mechanisms *(166–169)* are involved in liver damage associated with chronic ethanol consumption. The differences in immune system parameters in alcohol-dependent persons without liver disease as compared with those with liver disease strongly suggest that many of the immunological effects in the latter are not caused by ethanol *per se* but that they are secondary to liver damage. It had been proposed that liver dysfunction contributes to increased transit of lipopolysaccharide from the gut and that this explains much of the immune system activation observed in alcohol-dependent persons with liver disease. However, careful analysis of lipopolysaccharide concentrations revealed that elevated levels were associated more closely with recent alcohol consumption than with liver dysfunction or damage *(159)*. In any case, it should be emphasized that the ongoing inflammatory changes in the liver and the activated phenotype of lymphocytes in the blood are not associated with increased immunity to infection or cancer. In fact, alcoholics with liver disease have diminished T-cell-mediated immune responses and resistance to infection *(170)*. Thus, immune activation in alcohol-dependent persons with liver disease probably represents nonspecific *(126)* or perhaps even autoimmune *(171)* activity of the immune system, which does not contribute to resistance to infection and may interfere with responses needed for normal resistance.

6.4. Mechanisms of Ethanol-Mediated Immunomodulation

The mechanisms by which ethanol may act to alter cytokine production include direct and indirect effects. Ethanol induces a neuroendocrine stress response in rodents *(172,173)*, nonhuman primates *(174)*, and human beings *(175,176)*. A review of the literature suggests that relatively high blood ethanol levels are required (approx 0.2% or greater) and that increased cortisol concentrations have been reported in most studies in which this blood level was reached or exceeded *(177)*. Glucocorticoid hormones such as cortisol in humans and corticosterone in rodents typically suppress the production of proinflammatory cytokines as well as some of the cytokines involved in specific immune responses *(178–180)*. Glucocorticoids also can increase the production of anti-inflammatory or immunosuppressive cytokines such as IL-10 *(181)*. By administering corticosterone in a manner designed to yield a comparable area under the corticosterone concentration vs time curve as measured in mice treated with ethanol, we demonstrated that ethanol-induced corticosterone is sufficient to suppress the production of TNF-α for 1 h, but not for the whole period of time (2.5 h) for which ethanol altered TNF-α production *(182)*. In other studies that have not yet been published, we observed that a glucocorticoid receptor antagonist (mifeprisone) did not diminish the effects of ethanol on cytokine production induced by lipopolysaccharide in mice. Thus, it is not clear to what extent the effects of glucocorticoids or other stress mediators are responsible for the effects of ethanol on cytokine production in human beings. An important factor in this regard may be the drinking pattern of each individual. In rodents, continuous exposure to ethanol for 7 d leads to habituation or tolerance with regard to the stress response *(183)*. However, this has not been uniformly noted in alcohol-dependent human subjects. One study indicated no increase in serum cortisol in alcohol-dependent subjects with high blood ethanol concentrations *(184)*, but another indicated that such increases do occur *(175)*. A possible explanation for this difference comes from a report indicating that alcohol-dependent persons do not all drink continuously. Approximately 50% are better characterized as binge, episodic, or sporadic drinkers than steady drinkers *(185)*. Steady drinking would be expected to induce a more uniform state of tolerance of habituation that would decrease the stress response to ethanol, whereas the other drinking patterns might not induce such a state.

Another issue that has probably been insufficiently studied with regard to the mechanisms by which ethanol consumption affects cytokine responses is the role of its major metabolites, acetaldehyde and acetate. Both enhancing and inhibitory effects on TNF-α production have been reported in rodent models *(166,186)*.

An emerging body of evidence suggests an additional mechanism by which ethanol may affect cytokine production as well as a variety of other cellular functions: alteration of cellular signaling. In a series of particularly informative studies, Szabo and colleagues (154) have demonstrated that ethanol decreases the production of proinflammatory cytokines and chemokines by human monocytes in culture. Very similar effects were noted when the exposure to ethanol resulted from ingestion of ethanol under controlled conditions to achieve blood ethanol concentrations of approx 0.1% (w/v). Further studies using an in vitro system revealed that ethanol interferes with the activation of NF-κB induced by lipopolysaccharide (through TLR4 *[156]*). Adenyl cyclase activation in human lymphocyte membranes in vitro is suppressed in a manner that correlates with blood ethanol concentration following ethanol consumption by ethanol-dependent human subjects *(187)*. Recent studies in rodents in vivo and ex vivo support the idea that ethanol disrupts signaling through all TLRs. This inhibition is associated with decreased activation of the MAP kinases and of IRAK-1, which is activated early in the TLR signaling pathways. Although indirect mechanisms of action have not been entirely ruled out, and the role of metabolites is not clear, we found no evidence for the involvement of corticosterone in the inhibition of TLR3 mediated cytokine production by acute ethanol administration in mice *(188)*. We *(133)* and others *(189)* have speculated that ethanol may modulate signaling by disrupting lipid rafts, which are involved in signal transduction for TLR and other receptors important for immune function. However, no definitive experimental evidence has been reported in support of this speculation.

7. NICOTINE

7.1. Background, In Vitro Effects, and In Vivo Effects of Nicotine in Animal Models

Nicotine is one of more than 3900 chemicals found in cigarette smoke *(190)*, and it is also found at potentially toxic levels in raw tobacco, smokeless tobacco products, and pesticide preparations *(191)*. Nicotine is known to be the addictive component of tobacco and also has been implicated in causing some of tobacco's immunomodulatory effects *(192)*. Research on the immunological effects of tobacco has focused on nicotine because of its highly addictive properties, widespread use, and relatively simple and well-established initial mechanism of action. Many of the immunomodulatory effects of nicotine have been suggested to result from nicotine's direct contact with nicotinic acetylcholine receptors on immune system cells, whereas some effects may be secondary to an increase of catecholamines and glucocorticoids caused by nicotine's action on the central nervous system *(193,194)*. A number of reports indicate that cytokine production can be affected by nicotine.

When evaluating the body of literature on how nicotine modulates cytokine production, the concentration or dosage of nicotine used is important. In a number of published studies, the concentrations of nicotine used in vitro were orders of magnitude greater than concentrations found in average tobacco users. Somewhat-higher concentrations of nicotine than measured in smokers may be relevant with regard to cases of overt toxicity caused by exposure to nicotine-based pesticides or occupational exposure to tobacco leaves. However, the relevance of substantially greater concentrations, particularly in vitro, is questionable. The average nicotine concentration in the blood of smokers is 33 ng/mL *(195)*. A similar value has been reported for persons who use smokeless tobacco *(196,197)*. Several reports indicate that nicotine at concentrations greater than 10 µg/mL in vitro suppresses the expression of cytokines induced by LPS or anti-CD3 antibodies. However, these concentrations are orders of magnitude greater than lethal concentrations for animals, and they are also orders of magnitude greater than the binding affinity for nicotinic acetylcholine receptors. This does not exclude the possibility that these results are relevant. For example, increased concentrations of nicotine may be required to compensate for downregulation of nicotinic receptors in vitro or on cell lines as compared with normal cells. It has also been suggested that local nicotine concentrations in the oral cavity of smokeless tobacco users and in the lungs of smokers is probably substantially greater than in the plasma *(198)*. However, the relevance of results with high concentrations of nicotine is not certain unless they have been corroborated in vivo with doses of nicotine that can be realistically tolerated.

Although it is commonly stated that immune system cells have both muscarinic and nicotinic acetylcholine receptors, we have noted previously that the evidence presented in many studies is insufficient to support this conclusion *(199)*. For example, some studies present binding curves that do not reach a plateau (thus not clearly demonstrating saturable binding, which is characteristic of bona fide receptors). Other studies indicate binding constants that are orders of magnitude less than reported in numerous studies of these receptors in the nervous system. However, convincing evidence has recently emerged indicating that at least some cell types in the immune system do indeed respond to nicotine through bona fide nicotinic acetylcholine receptors. For example, the LPS-induced production of TNF-α by a mouse macrophage cell line is inhibited by nicotine at 100 ng/mL in vitro, and this inhibition is prevented in cells treated with anti-sense RNA to prevent expression of the α-7 subunit of the nicotinic acetylcholine receptor *(200)*. Failure of vagus nerve stimulation to suppress LPS-induced TNF-α production in vivo in mice (which is thought to be mediated by nicotinic acetylcholine receptors) was observed in knockout mice lacking the α-7 subunit of the nicotinic acetylcholine receptor *(6)*. However, definitive characterization of pharmacologically defined receptors on cells of the immune system still lags behind the identification of the apparent functional relevance of these receptors.

Significant effects of nicotine on cytokine responses have been reported at relevant nicotine concentrations (100 ng/mL) in mouse macrophages treated with live *Legionella pneumophila (201)*. The production of TNF-α was significantly decreased as was inhibition of bacterial growth. There was some indication of inhibition of other cytokines (e.g., IL-6 and IL-12) at relevant concentrations of nicotine, but substantial inhibition required higher concentrations. It has also been reported that chronic exposure to nicotine in mice inhibits cellular signaling induced through the T-cell receptor for antigen *(202)*. This would very likely inhibit T-cell cytokine production as well, although this was not formally evaluated. Paradoxically, enhancement of dendritic cell function and associated activation of Th1 cytokine production has been reported in one study of dendritic cells exposed to relevant concentrations of nicotine *(203)*. This may simply indicate that the action of nicotine in the immune system is strictly dependent on the nature of the immune stimulus and on the cell type involved. However, the findings of both suppression and enhancement of cytokine production is reminiscent of the effects of nicotine on two common forms of inflammatory bowel disease.

7.2. Effects of Nicotine on the Immune System in Human Subjects

In human subjects, a number of studies of the effects of nicotine on the immune system have focused on apparently beneficial effects of nicotine in patients with ulcerative colitis and adverse effects in patients with Crohn's disease *(204)*. Interestingly, the pattern of pro-inflammatory cytokine (IL-1) and chemokine (IL-8) expression is similarly decreased in smokers in both diseases *(205,206)*. Thus, the basis for the differing effects on disease progression is not known. However, one possibility is suggested by the work of Aicher et al., which has already been mentioned *(203)*. If Crohn's disease is triggered primarily by a specific immune response involving dendritic cells, nicotine may enhance it, whereas the progression of ulcerative colitis might be more related to inflammatory stimuli such as endotoxin. In this case, nicotine would be expected to inhibit progression.

7.3. Mechanisms of Nicotine's Immunomodulatory Effects

Recent studies in mice have suggested a novel mechanism that might be effective in nicotine in treatment of sepsis to prevent multiple organ failure and lethal septic shock. Several investigators have determined that a protein first identified as a DNA-binding protein, HMGB-1 (high mobility group protein B1) also acts as a cytokine and is an important mediator of organ failure and death late in the course of septic shock *(207,208)*. Nicotine prevents the expression of this gene induced by LPS or by sepsis, and it decreases lethality even when administered after the initiation of infection *(200)*. Therapeutic agents with this capability have been elusive in animal models and human subjects. The results also demonstrate that the action of nicotine in this system depends on the α-7 subunit of the nicotinic receptor, and these results are consistent with previous reports that stimulation of the vagus nerve (which presumably leads to systemic release of acetylcholine) also can be protective in sepsis *(6)*. These findings represent one of the clearest demonstrations that endogenous as well as exogenous ligands for neurotransmitter receptors can affect immune or inflammatory response in quantitatively important ways and are not always associated merely with "fine tuning" of responses.

The ability of peripheral blood monocytes to produce cytokines has also been evaluated after treatment with Gram-negative bacteria like *Legionella pneumophila*. In one study *Legionella pneumophila* was added to macrophages in vitro to stimulate cytokine production. Nicotine was then added to examine its effects on cytokine levels. There were significant decreases in antimicrobial activity and in the production of IL-6, IL-12, and TNF-α *(201)*. In another study, treatment with nicotine (1 μM) significant decreased LPS induced TNF-α production by human peripheral blood monocytes *(6)*. This same pattern of cytokine modulation caused by nicotine treatment has also been seen in studies using peritoneal macrophages. In these studies, 5 pM of nicotine treatment causes as much as a 35% decrease in LPS-induced TNF-α production *(6)*.

Taken together, the reports described in previous sections suggest that proinflammatory cytokine production by T-cells, dendritic cells, and macrophages is robustly inhibited by nicotine administration under some circumstances. Nicotine's inhibitory effects on the production of proinflammatory cytokines by stimulated cells are thought to be mediated through nicotine's direct contact with nicotinic acetylcholine receptors on immune cells. However, the nicotine-induced signaling events that affect immune responses and the mechanisms by which this occurs have not been determined. It also should be noted that nicotine induces a stress response characterized by increased endogenous glucocorticoid concentrations *(209)*, and this may constitute an indirect anti-inflammatory mechanism that also contributes to the effects of nicotine.

8. CONCLUSIONS

Although there are studies that indicate exceptions, the overall weight of the evidence suggests that all the drugs discussed in this chapter can adversely affect the immune system and that cytokine dysregulation is an important part of this. In most cases, the effects of the drugs in human subjects are analogous to those in animals. The combination of animal studies, studies with human subjects, and studies using immune system cells in vitro have revealed both indirect and direct (receptor-mediated and nonreceptor-mediated) mechanisms by which these drugs affect the immune system. However, detailed studies of the alterations in cellular signaling responsible for altered immune functions have only been conducted for cannabinoids. Considering that therapeutic agents for modulating cellular signaling are currently under development, understanding the cellular signaling changes associated with immune dysfunction should be a major goal of future research. This may allow intervention to prevent adverse immunological affects associated with abuse of these drugs or with legitimate opioid analgesia.

ACKNOWLEDGMENTS

The author wishes to thank Dr. Pamela Hebért and Mr. Carlton Schwab for assistance in preparing this manuscript.

REFERENCES

1. Leonard WJ. Type I cytokines and interferons and their receptors. In: Paul WE, ed. Fundamental Immunology. Philadelphia: Lippincott Williams & Wilkins; 2003:701–47.
2. Murphy PM. Chemokines. In: Paul WE, ed. Fundamental Immunology. Philadelphia: Lippincott Williams & Wilkins; 2003:801–840.
3. Andrews P. Cocaethylene toxicity. J Addict Dis 1997;16:75–84.
4. House RV, Thomas PT, Bhargava HN. In vitro exposure to peptidic delta opioid receptor antagonists results in limited immunosuppression. Neuropeptides 1997;31:89–93.
5. Wetzel MA, Steele AD, Eisenstein TK, Adler MW, Henderson EE, Rogers TJ. Mu-opioid induction of monocyte chemoattractant protein-1, RANTES, and IFN-gamma-inducible protein-10 expression in human peripheral blood mononuclear cells. J Immunol 2000;165:6519–6524.

6. Wang H, Yu M, Ochani M, et al. Nicotinic acetylcholine receptor alpha7 subunit is an essential regulator of inflammation. Nature 2003;421:384–388.
7. Pruett SB, Obiri N, Kiel JL. Involvement and relative importance of at least two distinct mechanisms in the effects of 2-mercaptoethanol on murine lymphocytes in culture. J Cell Physiol 1989;141:40–45.
8. Bhat RS, Bhaskaran M, Mongia A, Hitosugi N, Singhal PC. Morphine-induced macrophage apoptosis: oxidative stress and strategies for modulation. J Leukoc Biol 2004;75:1131–1138.
9. Patel K, Bhaskaran M, Dani D, Reddy K, Singhal PC. Role of heme oxygenase-1 in morphine-modulated apoptosis and migration of macrophages. J Infect Dis 2003;187:47–54.
10. Sei Y, Yoshimoto K, McIntyre T, Skolnick P, Arora PK. Morphine-induced thymic hypoplasia is glucocorticoid-dependent. J Immunol 1991;146:194–198.
11. House RV, Thomas PT, Bhargava H. In vitro evaluation of fentanyl and meperidine for immunomodulatory activity. Immunol Lett 1995;46:117–124.
12. Thomas PT, Bhargava HN, House RV. Immunomodulatory effects of in vitro exposure to morphine and its metabolites. Pharmacology 1995;50:51–62.
13. Suzuki S, Chuang AJ, Chuang LF, Doi RH, Chuang RY. Morphine promotes simian acquired immunodeficiency syndrome virus replication in monkey peripheral mononuclear cells: induction of CC chemokine receptor 5 expression for virus entry. J Infect Dis 2002;185:1826–1829.
14. Homan JW, Steele AD, Martinand-Mari C, et al. Inhibition of morphine-potentiated HIV-1 replication in peripheral blood mononuclear cells with the nuclease-resistant 2–5A agonist analog, 2–5A(N6B). J Acquir Immune Defic Syndr 2002;30:9–20.
15. Mou L, Lankford-Turner P, Leander MV, Bissonnette RP, Donahoe RM, Royal W. RXR-induced TNF-alpha suppression is reversed by morphine in activated U937 cells. J Neuroimmunol 2004;147:99–105.
16. Donahoe RM, Vlahov D. Opiates as potential cofactors in progression of HIV-1 infections to AIDS. J Neuroimmunol 1998;83:77–87.
17. Nyland SB, Specter S, Ugen KE. Morphine effects on HTLV-I infection in the presence or absence of concurrent HIV-1 infection. DNA Cell Biol 1999;18:285–291.
18. Krol A, Flynn C, Vlahov D, Miedema F, Coutinho RA, van Ameijden EJ. New evidence to reconcile in vitro and epidemiologic data on the possible role of heroin on CD4+ decline among HIV-infected injecting drug users. Drug Alcohol Depend 1999;54:145–154.
19. McCarthy L, Wetzel M, Sliker JK, Eisenstein TK, Rogers TJ. Opioids, opioid receptors, and the immune response. Drug Alcohol Depend 2001;62:111–123.
20. Vallejo R, De Leon-Casasola O, Benyamin R. Opioid therapy and immunosuppression: a review. Am J Ther 2004;11:354–365.
21. Bussiere JL, Adler MW, Rogers TJ, Eisenstein TK. Cytokine reversal of morphine-induced suppression of the antibody response. J Pharmacol Exp Ther 1993;264:591–597.

22. Wang J, Charboneau R, Balasubramanian S, Barke RA, Loh HH, Roy S. The immunosuppressive effects of chronic morphine treatment are partially dependent on corticosterone and mediated by the mu-opioid receptor. J Leukoc Biol 2002;71:782–790.

23. Eisenstein TK, Bussiere JL, Rogers TJ, Adler MW. Immunosuppressive effects of morphine on immune responses in mice. Adv Exp Med Biol 1993;335:41–52.

24. Ho BT, Lu JG, Huo YY, et al. The opioid mechanism of interferon-alpha action. Anticancer Drugs 1994;5:90–94.

25. Saphier D. Neuroendocrine effects of interferon-alpha in the rat. Adv Exp Med Biol 1995;373:209–218.

26. Sacerdote P. Effects of in vitro and in vivo opioids on the production of IL-12 and IL-10 by murine macrophages. Ann NY Acad Sci 2003;992:129–140.

27. Peng X, Mosser DM, Adler MW, Rogers TJ, Meissler JJ Jr., Eisenstein TK. Morphine enhances interleukin-12 and the production of other pro-inflammatory cytokines in mouse peritoneal macrophages. J Leukoc Biol 2000;68:723–728.

28. Nelson CJ, Lysle DT. Morphine modulation of the contact hypersensitivity response: characterization of immunological changes. Clin Immunol 2001;98:370–377.

29. Limiroli E, Gaspani L, Panerai AE, Sacerdote P. Differential morphine tolerance development in the modulation of macrophage cytokine production in mice. J Leukoc Biol 2002;72:43–48.

30. Pacifici R, di Carlo S, Bacosi A, Pichini S, Zuccaro P. Pharmacokinetics and cytokine production in heroin and morphine-treated mice. Int J Immunopharmacol 2000;22:603–614.

31. Walker JS. Anti-inflammatory effects of opioids. Adv Exp Med Biol 2003;521:148–160.

32. Philippe D, Dubuquoy L, Groux H, et al. Anti-inflammatory properties of the mu opioid receptor support its use in the treatment of colon inflammation. J Clin Invest 2003;111:1329–1338.

33. Wang J, Barke RA, Charboneau R, Roy S. Morphine impairs host innate immune response and increases susceptibility to *Streptococcus pneumoniae* lung infection. J Immunol 2005;174:426–434.

34. LeVier DG, McCay JA, Stern ML, et al. Immunotoxicological profile of morphine sulfate in B6C3F1 female mice. Fundam Appl Toxicol 1994;22: 525–542.

35. Conlan JW, North RJ. *Listeria monocytogenes*, but not *Salmonella typhimurium*, elicits a CD18-independent mechanism of neutrophil extravasation into the murine peritoneal cavity. Infect Immun 1994;62:2702–2706.

36. Ideker T, Galitski T, Hood L. A new approach to decoding life: systems biology. Annu Rev Genomics Hum Genet 2001;2:343–372.

37. Louria DB, Hensle T, Rose J. The major medical complications of heroin addiction. Ann Intern Med 1967;67:1–22.

38. Kay DC, Pickworth WB, Neider GL. Morphine-like insomnia from heroin in nondependent human addicts. Br J Clin Pharmacol 1981;11:159–169.

39. el-Nakah A, Frank O, Louria DB, Quinones MA, Baker H. A vitamin profile of heroin addiction. Am J Public Health 1979;69:1058–1060.
40. Akural EI, Salomaki TE, Bloigu AH, et al. The effects of pre-emptive epidural sufentanil on human immune function. Acta Anaesthesiol Scand 2004;48: 750–755.
41. Negishi C, Lenhardt R, Ozaki M, et al. Opioids inhibit febrile responses in humans, whereas epidural analgesia does not: an explanation for hyperthermia during epidural analgesia. Anesthesiology 2001;94:218–222.
42. Yeager MP, Colacchio TA, Yu CT, et al. Morphine inhibits spontaneous and cytokine-enhanced natural killer cell cytotoxicity in volunteers. Anesthesiology 1995;83:500–508.
43. Song LC, Song C, Zeng J, Xiong ML, Lou N. Reduction in monocyte chemoattractant protein-1 mRNA expression in peripheral blood mononuclear cells of diamorphine addicts. Acta Pharmacol Sin 2002;23:336–338.
44. Horn SD, Wright HL, Couperus JJ, et al. Association between patient-controlled analgesia pump use and postoperative surgical site infection in intestinal surgery patients. Surg Infect (Larchmt) 2002;3(2):109–118.
45. Brand JM, Frohn C, Luhm J, Kirchner H, Schmucker P. Early alterations in the number of circulating lymphocyte subpopulations and enhanced proinflammatory immune response during opioid-based general anesthesia. Shock 2003;20:213–217.
46. Zajicova A, Wilczek H, Holan V. The alterations of immunological reactivity in heroin addicts and their normalization in patients maintained on methadone. Folia Biol (Praha) 2004;50:24–28.
47. Brix-Christensen V, Tonnesen E, Sorensen IJ, Bilfinger TV, Sanchez RG, Stefano GB. Effects of anaesthesia based on high versus low doses of opioids on the cytokine and acute-phase protein responses in patients undergoing cardiac surgery. Acta Anaesthesiol Scand 1998;42:63–70.
48. Heathcote J, Taylor KB. Immunity and nutrition in heroin addicts. Drug Alcohol Depend 1981;8:245–255.
49. Rahim RT, Meissler JJ, Zhang L, Adler MW, Rogers TJ, Eisenstein TK. Withdrawal from morphine in mice suppresses splenic macrophage function, cytokine production, and costimulatory molecules. J Neuroimmunol 2003;144:16–27.
50. Bussiere JL, Adler MW, Rogers TJ, Eisenstein TK. Differential effects of morphine and naltrexone on the antibody response in various mouse strains. Immunopharmacol Immunotoxicol 1992;14:657–673.
51. Freier DO, Fuchs BA. A mechanism of action for morphine-induced immunosuppression: corticosterone mediates morphine-induced suppression of natural killer cell activity. J Pharmacol Exp Ther 1994;270:1127–1133.
52. Fuchs BA, Pruett SB. Morphine induces apoptosis in murine thymocytes in vivo but not in vitro: involvement of both opiate and glucocorticoid receptors. J Pharmacol Exp Ther 1993;266:417–423.
53. Bussiere J, Adler M, Rogers T, Eisenstein T. Interleukin-6 reverses morphine-induced suppression of the antibody response. NIDA Research Monograph Series 1992;119.

54. Pruett SB, Han YC, Fuchs BA. Morphine suppresses primary humoral immune responses by a predominantly indirect mechanism. J Pharmacol Exp Ther 1992;262:923–928.
55. Taub DD, Eisenstein TK, Geller EB, Adler MW, Rogers TJ. Immumomodulatory activity of μ-and κ-selective opioid agonists. Proc Natl Acad Sci USA 1991;88:360–364.
56. Eisenstein TK, Meissler JJ, Jr., Rogers TJ, Geller EB, Adler MW. Mouse strain differences in immunosuppression by opioids in vitro. J Pharmacol Exp Ther 1995;275:1484–1489.
57. Hedin KE, Bell MP, Huntoon CJ, Karnitz LM, McKean DJ. Gi proteins use a novel beta gamma- and Ras-independent pathway to activate extracellular signal-regulated kinase and mobilize AP-1 transcription factors in Jurkat T lymphocytes. J Biol Chem 1999;274:19,992–20,001.
58. Shahabi NA, Daaka Y, McAllen K, Sharp BM. Delta opioid receptors expressed by stably transfected jurkat cells signal through the map kinase pathway in a ras-independent manner. J Neuroimmunol 1999;94:48–57.
59. Heagy W, Shipp MA, Finberg RW. Opioid receptor agonists and Ca2+ modulation in human B cell lines. J Immunol 1992;149:4074–4081.
60. O'Brien CP. Drug addiction and drug abuse. In: Hardman JG, Limbird LE, eds. Goodman and Gilman's The Pharmacological Basis of Therapeutics. New York: McGraw-Hill; 2001, pp. 621–642.
61. Roy S, Barke RA, Loh HH. Mu-opioid receptor knockout mice: role of mu-opioid receptor in morphine mediated immune functions. Brain Res Mol Brain Res 1998;61:190–194.
62. Szabo I, Chen XH, Xin L, et al. Heterologous desensitization of opioid receptors by chemokines inhibits chemotaxis and enhances the perception of pain. Proc Natl Acad Sci USA 2002;99:10,276–10,281.
63. Zhang N, Rogers TJ, Caterina M, Oppenheim JJ. Proinflammatory chemokines, such as C-C chemokine ligand 3, desensitize mu-opioid receptors on dorsal root ganglia neurons. J Immunol 2004;173:594–599.
64. Zhang N, Hodge D, Rogers TJ, Oppenheim JJ. Ca2+-independent protein kinase Cs mediate heterologous desensitization of leukocyte chemokine receptors by opioid receptors. J Biol Chem 2003;278:12,729–12,736.
65. Rogers TJ, Steele AD, Howard OM, Oppenheim JJ. Bidirectional heterologous desensitization of opioid and chemokine receptors. Ann NY Acad Sci 2000; 917:19–28.
66. Szabo I, Wetzel MA, Zhang N, et al. Selective inactivation of CCR5 and decreased infectivity of R5 HIV-1 strains mediated by opioid-induced heterologous desensitization. J Leukoc Biol 2003;74:1074–1082.
67. Jan TR, Rao GK, Kaminski NE. Cannabinol enhancement of interleukin-2 (IL-2) expression by T cells is associated with an increase in IL-2 distal nuclear factor of activated T cell activity. Mol Pharmacol 2002;61:446–454.
68. Smith SR, Denhardt G, Terminelli C. The anti-inflammatory activities of cannabinoid receptor ligands in mouse peritonitis models. Eur J Pharmacol 2001;432:107–119.

69. Juttler E, Potrovita I, Tarabin V, et al. The cannabinoid dexanabinol is an inhibitor of the nuclear factor-kappa B (NF-kappaB). Neuropharmacology 2004;47: 580–592.
70. Lavon I, Sheinin T, Meilin S, et al. A novel synthetic cannabinoid derivative inhibits inflammatory liver damage via negative cytokine regulation. Mol Pharmacol 2003;64:1334–1341.
71. Herring AC, Faubert Kaplan BL, Kaminski NE. Modulation of CREB and NF-kappaB signal transduction by cannabinol in activated thymocytes. Cell Signal 2001;13:241–250.
72. Faubert Kaplan BL, Kaminski NE. Cannabinoids inhibit the activation of ERK MAPK in PMA/Io-stimulated mouse splenocytes. Int Immunopharmacol 2003;3:1503–1510.
73. Facchinetti F, Del Giudice E, Furegato S, Passarotto M, Leon A. Cannabinoids ablate release of TNFalpha in rat microglial cells stimulated with lypopolysaccharide. Glia 2003;41:161–168.
74. Coffey RG, Yamamoto Y, Snella E, Pross S. Tetrahydrocannabinol inhibition of macrophage nitric oxide production. Biochem Pharmacol 1996;52:743–751.
75. Croci T, Landi M, Galzin AM, Marini P. Role of cannabinoid CB1 receptors and tumor necrosis factor-alpha in the gut and systemic anti-inflammatory activity of SR 141716 (rimonabant) in rodents. Br J Pharmacol 2003;140:115–122.
76. Massa F, Marsicano G, Hermann H, et al. The endogenous cannabinoid system protects against colonic inflammation. J Clin Invest 2004;113:1202–1209.
77. Jan TR, Kaminski NE. Role of mitogen-activated protein kinases in the differential regulation of interleukin-2 by cannabinol. J Leukoc Biol 2001;69:841–849.
78. Klein TW, Newton C, Larsen K, et al. Cannabinoid receptors and T helper cells. J Neuroimmunol 2004;147:91–94.
79. Smith SR, Terminelli C, Denhardt G. Effects of cannabinoid receptor agonist and antagonist ligands on production of inflammatory cytokines and anti-inflammatory interleukin-10 in endotoxemic mice. J Pharmacol Exp Ther 2000;293:136–150.
80. Chang YH, Lee ST, Lin WW. Effects of cannabinoids on LPS-stimulated inflammatory mediator release from macrophages: involvement of eicosanoids. J Cell Biochem 2001;81:715–723.
81. Pacifici R, Zuccaro P, Pichini S, et al. Modulation of the immune system in cannabis users. JAMA 2003;289:1929–1931.
82. Roth MD, Baldwin GC, Tashkin DP. Effects of delta-9-tetrahydrocannabinol on human immune function and host defense. Chem Phys Lipids 2002;121: 229–239.
83. Tashkin DP, Baldwin GC, Sarafian T, Dubinett S, Roth MD. Respiratory and immunologic consequences of marijuana smoking. J Clin Pharmacol 2002;42(11 Suppl):71S–81S.
84. Bredt BM, Higuera-Alhino D, Shade SB, Hebert SJ, McCune JM, Abrams DI. Short-term effects of cannabinoids on immune phenotype and function in HIV-1-infected patients. J Clin Pharmacol 2002;42(11 Suppl):82S–9S.

85. Abrams DI, Hilton JF, Leiser RJ, et al. Short-term effects of cannabinoids in patients with HIV-1 infection: a randomized, placebo-controlled clinical trial. Ann Intern Med 2003;139:258–266.
86. Killestein J, Hoogervorst EL, Reif M, et al. Immunomodulatory effects of orally administered cannabinoids in multiple sclerosis. J Neuroimmunol 2003; 137:140–143.
87. Johnson KM, Dewey WL, Ritter KS, Beckner JS. Cannabinoid effects on plasma corticosterone and uptake of 3H-corticosterone by mouse brain. Eur J Pharmacol 1978;47:303–310.
88. Herring AC, Koh WS, Kaminski NE. Inhibition of the cyclic AMP signaling cascade and nuclear factor binding to CRE and kB elements by cannabinol, a minimally CNS-active cannabinoid. Biochem Pharmacol 1998;55:1013–1023.
89. Herring AC, Kaminski NE. Cannabinol-mediated inhibition of nuclear factor-kB, cAMP response element-binding protein, and interleukin-2 secretion by activated thymocytes. J Pharmacol Exp Ther 1999;291:1156–1163.
90. Faubert BL, Kaminski NE. AP-1 activity is negatively regulated by cannabinol through inhibition of its protein components, c-fos and c-jun. J Leukoc Biol 2000;67:259–266.
91. Davis MI, Ronesi J, Lovinger DM. A predominant role for inhibition of the adenylate cyclase/protein kinase A pathway in ERK activation by cannabinoid receptor 1 in N1E-115 neuroblastoma cells. J Biol Chem 2003;278:48,973–48,780.
92. Alberich Jorda M, Rayman N, Tas M, et al. The peripheral cannabinoid receptor Cb2, frequently expressed on AML blasts, either induces a neutrophilic differentiation block or confers abnormal migration properties in a ligand-dependent manner. Blood 2004;104:526–534.
93. Galve-Roperh I, Rueda D, Gomez del Pulgar T, Velasco G, Guzman M. Mechanism of extracellular signal-regulated kinase activation by the CB(1) cannabinoid receptor. Mol Pharmacol 2002;62:1385–1392.
94. Rao GK, Zhang W, Kaminski NE. Cannabinoid receptor-mediated regulation of intracellular calcium by delta(9)-tetrahydrocannabinol in resting T cells. J Leukoc Biol 2004;75:884–892.
95. Holdcroft A, Smith M, Jacklin A, et al. Pain relief with oral cannabinoids in familial Mediterranean fever. Anaesthesia 1997;52:483–486.
96. Hoffman BB, Taylor P. Neurotransmission. In: Hardman JG, Limbird LE, eds. Goodman and Gilman's The Pharmacological Basis of Therapeutics. New York: McGraw-Hill; 2001, p. 144.
97. Matsumoto RR, McCracken KA, Pouw B, Zhang Y, Bowen WD. Involvement of sigma receptors in the behavioral effects of cocaine: evidence from novel ligands and antisense oligodeoxynucleotides. Neuropharmacology 2002;42:1043–1055.
98. Gardner B, Zhu LX, Roth MD, Tashkin DP, Dubinett SM, Sharma S. Cocaine modulates cytokine and enhances tumor growth through sigma receptors. J Neuroimmunol 2004;147:95–98.

 99. Stanulis ED, Matulka RA, Jordan SD, Rosecrans JA, Holsapple MP. Role of corticosterone in the enhancement of the antibody response after acute cocaine administration. J Pharmacol Exp Ther 1997;280:284–291.
100. Stanulis ED, Jordan SD, Rosecrans JA, Holsapple MP. Disruption of Th1/Th2 cytokine balance by cocaine is mediated by corticosterone. Immunopharmacology 1997;37:25–33.
101. Broadbear JH, Winger G, Woods JH. Self-administration of fentanyl, cocaine and ketamine: effects on the pituitary-adrenal axis in rhesus monkeys. Psychopharmacology (Berl) 2004;176:398–406.
102. Heesch CM, Negus BH, Keffer JH, Snyder RW 2nd, Risser RC, Eichhorn EJ. Effects of cocaine on cortisol secretion in humans. Am J Med Sci 1995;310:61–64.
103. Baumann MH, Gendron TM, Becketts KM, Henningfield JE, Gorelick DA, Rothman RB. Effects of intravenous cocaine on plasma cortisol and prolactin in human cocaine abusers. Biol Psychiatry 1995;38:751–755.
104. Mao JT, Huang M, Wang J, Sharma S, Tashkin DP, Dubinett SM. Cocaine down-regulates IL-2-induced peripheral blood lymphocyte IL-8 and IFN-gamma production. Cell Immunol 1996;172:217–223.
105. Wang Y, Huang DS, Watson RR. In vivo and in vitro cocaine modulation on production of cytokines in C57BL/6 mice. Life Sci 1994;54:401–411.
106. Gan X, Zhang L, Newton T, et al. Cocaine infusion increases interferon-gamma and decreases interleukin-10 in cocaine-dependent subjects. Clin Immunol Immunopathol 1998;89:181–190.
107. Chao CC, Molitor TW, Gekker G, Murtaugh MP, Peterson PK. Cocaine-mediated suppression of superoxide production by human peripheral blood mononuclear cells. J Pharmacol Exp Ther 1991;256:255–258.
108. Di Francesco P, Marini S, Pica F, Favalli C, Tubaro E, Garaci E. In vivo cocaine administration influences lymphokine production and humoral immune response. Immunol Res 1992;11:74–79.
109. Casalinuovo IA, Gaziano R, Di Francesco P. Cytokine pattern secretion by murine spleen cells after inactivated *Candida albicans* immunization. Effect of cocaine and morphine treatment. Immunopharmacol Immunotoxicol 2000;22:35–48.
110. Kubera M, Filip M, Basta-Kaim A, et al. The effect of cocaine sensitization on mouse immunoreactivity. Eur J Pharmacol 2004;483:309–315.
111. Pacifici R, Fiaschi AI, Micheli L, et al. Immunosuppression and oxidative stress induced by acute and chronic exposure to cocaine in rat. Int Immunopharmacol 2003;3:581–592.
112. Zhu LX, Sharma S, Gardner B, et al. IL-10 mediates sigma 1 receptor-dependent suppression of antitumor immunity. J Immunol 2003;170:3585–3591.
113. Pruett SB. Quantitative aspects of stress-induced immunomodulation. Int Immunopharmacol 2001;1:507–520.
114. Dhabhar FS. Stress-induced enhancement of cell-mediated immunity. Ann NY Acad Sci 1998;840:359–372.

115. Tashkin DP. Evidence implicating cocaine as a possible risk factor for HIV infection. J Neuroimmunol 2004;147:26–27.

116. Nair MP, Schwartz SA, Mahajan SD, et al. Drug abuse and neuropathogenesis of HIV infection: role of DC-SIGN and IDO. J Neuroimmunol 2004;157:56–60.

117. Shay AH, Choi R, Whittaker K, et al. Impairment of antimicrobial activity and nitric oxide production in alveolar macrophages from smokers of marijuana and cocaine. J Infect Dis 2003;187:700–704.

118. Halpern JH, Sholar MB, Glowacki J, Mello NK, Mendelson JH, Siegel AJ. Diminished interleukin-6 response to proinflammatory challenge in men and women after intravenous cocaine administration. J Clin Endocrinol Metab 2003;88:1188–1193.

119. Baldwin GC, Buckley DM, Roth MD, Kleerup EC, Tashkin DP. Acute activation of circulating polymorphonuclear neutrophils following in vivo administration of cocaine. A potential etiology for pulmonary injury. Chest 1997;111: 698–705.

120. Dhabhar FS, McEwen BS. Stress-induced enhancement of antigen-specific cell-mediated immunity. J Immunol 1996;156:2608–2615.

121. Contoreggi C, Herning RI, Koeppl B, et al. Treatment-seeking inpatient cocaine abusers show hypothalamic dysregulation of both basal prolactin and cortisol secretion. Neuroendocrinology 2003;78:154–162.

122. Grattendick K, Jansen DB, Lefkowitz DL, Lefkowitz SS. Cocaine causes increased type I interferon secretion by both L929 cells and murine macrophages. Clin Diagn Lab Immunol 2000;7:245–250.

123. Lee YW, Hennig B, Fiala M, Kim KS, Toborek M. Cocaine activates redox-regulated transcription factors and induces TNF-alpha expression in human brain endothelial cells. Brain Res 2001;920:125–133.

124. Chiappelli F, Kung MA, Villanueva P, Lee P, Frost P, Prieto N. Immunotoxicity of cocaethylene. Immunopharmacol Immunotoxicol 1995;17:399–417.

125. Rush B. An enquiry into the effects of ardent spirits upon the human body and mind with an account of the means of preventing and of the remedies for curing them (reprinted). Q J Stud Alcohol 1943;4:325–341.

126. Cook RT. Alcohol abuse, alcoholism, and damage to the immune system—a review. Alcohol Clin Exp Res 1998;22:1927–1942.

127. Poikolainen K. Estimated lethal ethanol concentrations in relation to age, aspiration, and drugs. Alcohol Clin Exp Res 1984;8:223–225.

128. Urso T, Gavaler JS, Van Thiel DH. Blood ethanol levels in sober alcohol users seen in an emergency room. Life Sci 1981;28:1053–1056.

129. Mandrekar P, Catalano D, Girouard L, Szabo G. Human monocyte IL-10 production is increased by acute ethanol treatment. Cytokine 1996;8:567–577.

130. Szabo G, Mandrekar P, Catalano D. Inhibition of superantigen-induced T cell proliferation and monocyte IL-1 beta, TNF-alpha, and IL-6 production by acute ethanol treatment. J Leukoc Biol 1995;58:342–350.

131. Saeed RW, Varma S, Peng T, Tracey KJ, Sherry B, Metz CN. Ethanol blocks leukocyte recruitment and endothelial cell activation in vivo and in vitro. J Immunol 2004;173:6376–6383.

132. Chen GJ, Huang DS, Watzl B, Watson RR. Ethanol modulation of tumor necrosis factor and gamma interferon production by murine splenocytes and macrophages. Life Sci 1993;52:1319–1326.
133. Pruett SB, Zheng Q, Fan R, Matthews K, Schwab C. Ethanol suppresses cytokine responses induced through Toll-like receptors as well as innate resistance to *Escherichia coli* in a mouse model for binge drinking. Alcohol 2004; 33:147–155.
134. Pruett SB, Schwab C, Zheng Q, Fan R. Suppression of innate immunity by ethanol: a global perspective and a new mechanism beginning with inhibition of signaling through Toll-like receptors 3 and 4. J Immunol 2004;173: 2715–2724.
135. Mason CM, Dobard E, Kolls JK, Nelson S. Ethanol and murine interleukin (IL)-12 production. Alcohol Clin Exp Res 2000;24:553–559.
136. Saad AJ, Domiati-Saad R, Jerrells TR. Ethanol ingestion increases susceptibility of mice to *Listeria monocytogenes*. Alcohol Clin Exp Res 1993;17:75–85.
137. Sibley D, Jerrells TR. Alcohol consumption by C57BL/6 mice is associated with depletion of lymphoid cells from the gut-associated lymphoid tissues and altered resistance to oral infections with *Salmonella typhimurium*. J Infect Dis 2000;182:482–489.
138. Boe DM, Nelson S, Zhang P, Bagby GJ. Acute ethanol intoxication suppresses lung chemokine production following infection with *Streptococcus pneumoniae*. J Infect Dis 2001;184:1134–1142.
139. Greenberg SS, Zhao X, Hua L, Wang JF, Nelson S, Ouyang J. Ethanol inhibits lung clearance of *Pseudomonas aeruginosa* by a neutrophil and nitric oxide-dependent mechanism, in vivo. Alcohol Clin Exp Res 1999;23:735–744.
140. Cotte J, Forestier F, Quero AM, Bourrinet P, German A. The effect of alcohol ingestion on the susceptibility of mice to viral infections. Alcohol Clin Exp Res 1982;6:239–246.
141. Clemens DL, Jerrells TR. Ethanol consumption potentiates viral pancreatitis and may inhibit pancreas regeneration: preliminary findings. Alcohol 2004; 33:183–189.
142. Steven WM, Kumar SN, Stewart GL, Seelig LL Jr. The effects of ethanol consumption on the expression of immunity to *Trichinella spiralis* in rats. Alcohol Clin Exp Res 1990;14:87–91.
143. Salerno JA, Waltenbaugh C, Cianciotto NP. Ethanol consumption and the susceptibility of mice to *Listeria monocytogenes* infection. Alcohol Clin Exp Res 2001;25:464–472.
144. Ostrovidov S, Howard LM, Ikeda M, Ikeda A, Waltenbaugh C. Restoration of ethanol-compromised T(h)1 responses by sodium orthovanadate. Int Immunol 2002;14:1239–1245.
145. Geissler M, Geisen A, Wands JR. Inhibitory effects of chronic ethanol consumption on cellular immune responses to hepatitis C virus core protein are reversed by genetic immunizations augmented with cytokine-expressing plasmids. J Immunol 1997;159:5107–5113.

146. Han Y-C, Pruett SB. Mechanisms of ethanol-induced suppression of a primary antibody response in a mouse model for binge drinking. J Pharmacol Exp Ther 1995;275:950–957.

147. Encke J, Wands JR. Ethanol inhibition: the humoral and cellular immune response to hepatitis C virus NS5 after genetic Immunization. Alcohol Clin Exp Res 2000;24:1063–1069.

148. Waltenbaugh C, Mikszta J, Ward H, Hsiung L. Alteration of copolymer-specific humoral and cell-mediated immune responses by ethanol. Alcohol Clin Exp Res 1994;18:1–7.

149. Lieber CS, DeCarli LM. The feeding of ethanol in liquid diets: 1986 update. Alcohol Clin Exp Res 1986;10:550–553.

150. Padgett EL, Sibley DA, Jerrells TR. Effect of adrenalectomy on ethanol-associated changes in lymphocyte cell numbers and subpopulations in thymus, spleen, and gut-associated lymphoid tissues. Int J Immunopharmacol 2000; 22:285–298.

151. Meadows GG, Blank SE, Duncan DD. Influence of ethanol consumption on natural killer cell activity in mice. Alcohol Clin Exp Res 1989;13:476–479.

152. Cook RT, Zhu X, Coleman RA, et al. T-cell activation after chronic ethanol ingestion in mice. Alcohol 2004;33:175–181.

153. Gluckman SJ, MacGregor RR. Effect of acute alcohol intoxication on granulocyte mobilization and kinetics. Blood 1978;52:551–559.

154. Szabo G, Chavan S, Mandrekar P, Catalano D. Acute alcohol consumption attenuates interleukin-8 (IL-8) and monocyte chemoattractant peptide-1 (MCP-1) induction in response to ex vivo stimulation. J Clin Immunol 1999;19:67–76.

155. Girouard L, Mandrekar P, Catalano D, Szabo G. Regulation of monocyte interleukin-12 production by acute alcohol: a role for inhibition by interleukin-10. Alcohol: Clin Exp Res 1998;22:211–216.

156. Mandrekar P, Dolganiuc A, Bellerose G, et al. Acute alcohol inhibits the induction of nuclear regulatory factor kappa B activation through CD14/ toll-like receptor 4, interleukin-1, and tumor necrosis factor receptors: a common mechanism independent of inhibitory kappa B alpha degradation? Alcohol Clin Exp Res 2002;26:1609–1614.

157. Spies CD, von Dossow V, Eggers V, et al. Altered cell-mediated immunity and increased postoperative infection rate in long-term alcoholic patients. Anesthesiology 2004;100:1088–1100.

158. Sander M, Irwin M, Sinha P, Naumann E, Kox WJ, Spies CD. Suppression of interleukin-6 to interleukin-10 ratio in chronic alcoholics: association with postoperative infections. Intensive Care Med 2002;28:285–292.

159. Urbaschek R, McCuskey RS, Rudi V, et al. Endotoxin, endotoxin-neutralizing-capacity, sCD14, sICAM-1, and cytokines in patients with various degrees of alcoholic liver disease. Alcohol Clin Exp Res 2001;25(2):261–268.

160. Yamashina S, Wheeler MD, Rusyn I, Ikejima K, Sato N, Thurman RG. Tolerance and sensitization to endotoxin in Kupffer cells caused by acute ethanol

involve interleukin-1 receptor-associated kinase. Biochem Biophys Res Commun 2000;277:686–690.

161. Cook RT, Garvey MJ, Booth BM, Goeken JA, Stewart B, Noel M. Activated CD-8 cells and HLA DR expression in alcoholics without overt liver disease. J Clin Immunol 1991;11:246–253.

162. Song K, Coleman RA, Alber C, et al. TH1 cytokine response of CD57+ T-cell subsets in healthy controls and patients with alcoholic liver disease. Alcohol 2001;24:155–167.

163. Stewart SF, Vidali M, Day CP, Albano E, Jones DE. Oxidative stress as a trigger for cellular immune responses in patients with alcoholic liver disease. Hepatology 2004;39:197–203.

164. Batey RG, Cao Q, Gould B. Lymphocyte-mediated liver injury in alcohol-related hepatitis. Alcohol 2002;27:37–41.

165. Jerrells TR. Role of activated CD8+ T cells in the initiation and continuation of hepatic damage. Alcohol 2002;27:47–52.

166. Duryee MJ, Klassen LW, Freeman TL, Willis MS, Tuma DJ, Thiele GM. Lipopolysaccharide is a cofactor for malondialdehyde-acetaldehyde adduct-mediated cytokine/chemokine release by rat sinusoidal liver endothelial and kupffer cells. Alcohol Clin Exp Res 2004;28:1931–1938.

167. Enomoto N, Ikejima K, Yamashina S, et al. Kupffer cell sensitization by alcohol involves increased permeability to gut-derived endotoxin. Alcohol Clin Exp Res 2001;25(6 Suppl):51S–4S.

168. Fujimoto M, Uemura M, Nakatani Y, et al. Plasma endotoxin and serum cytokine levels in patients with alcoholic hepatitis: relation to severity of liver disturbance. Alcohol Clin Exp Res 2000;24(4 Suppl):48S–54S.

169. McClain CJ, Hill DB, Song Z, Deaciuc I, Barve S. Monocyte activation in alcoholic liver disease. Alcohol 2002;27:53–61.

170. Snyder N, Bessoff J, Dwyer JM, Conn HO. Depressed delayed cutaneous hypersensitivity in alcoholic hepatitis. Am J Dig Dis 1978;23:353–358.

171. Thiele GM, Freeman TL, Klassen LW. Immunologic mechanisms of alcoholic liver injury. Semin Liver Dis 2004;24:273–287.

172. Thiagarajan AB, Mefford IN, Eskay RL. Single-dose ethanol administration activates the hypothalamic-pituitary-adrenal axis: exploration of the mechanism of action. Neuroendocrinol 1989;50:427–432.

173. Carson EJ, Pruett SB. Development and characterization of a binge drinking model in mice for evaluation of the immunological effects of ethanol. Alcoholism: Clin Exp Res 1996;20:132–138.

174. Barr CS, Newman TK, Lindell S, et al. Early experience and sex interact to influence limbic-hypothalamic-pituitary-adrenal-axis function after acute alcohol administration in rhesus macaques (*Macaca mulatta*). Alcohol Clin Exp Res 2004;28:1114–1119.

175. Mendelson JH, Ogata M, Mello NK. Adrenal function and alcoholism. I. Serum cortisol. Psychosom Med 1971;33:145–157.

176. Kutscher S, Heise DJ, Banger M, et al. Concomitant endocrine and immune alterations during alcohol intoxication and acute withdrawal in alcohol-dependent subjects. Neuropsychobiology 2002;45(3):144–149.

177. Pruett SB, Collier SD, Wu W-J. Ethanol-induced activation of the hypothalamic-pituitary-adrenal axis in a mouse model for binge drinking: role of Ro15-4513-sensitive gamma aminobutyric acid receptors, tolerance, and relevance to humans. Life Sci 1998;63:1137–1146.

178. Barber AE, Coyle SM, Marano MA, et al. Glucocorticoid therapy alters hormonal and cytokine responses to endotoxin in man. J Immunol 1993;150: 1999–2006.

179. Ruzek MC, Pearce BD, Miller AH, Biron CA. Endogenous glucocorticoids protect against cytokine-mediated lethality during viral infection. J Immunol 1999;162:3527–3533.

180. Zhang D, Kishihara K, Wang B, Mizobe K, Kubo C, Nomoto K. Restraint stress-induced immunosuppression by inhibiting leukocyte migration and Th1 cytokine expression during the intraperitoneal infection of *Listeria monocytogenes*. J Neuroimmunol 1998;92:139–151.

181. Swain MG, Appleyard C, Wallace J, Wong H, Le T. Endogenous glucocorticoids released during acute toxic liver injury enhance hepatic IL-10 synthesis and release. Am J Physiol 1999;276:G199–G205.

182. Vinson RB, Carroll JL, Pruett SB. Mechanism of supressed neutrophil mobilization in a mouse model for binge drinking: role of glucocorticoids. Am Physiol Soc 1998;275:R1049–R1057.

183. Eskay RL, Chautard T, Torda T, Hwang D. The effects of alcohol on selected regulatory aspects of the stress axis. NIAAA Res Mono 1993;23:3–19.

184. Merry J, Marks V. Plasma-hydrocortisone response to ethanol in chronic alcoholics. Lancet 1969;1:921–923.

185. Epstein EE, Kahler CW, McCrady BS, Lewis KD, Lewis S. An empirical classification of drinking patterns among alcoholics: binge, episodic, sporadic, and steady. Addict Behav 1995;20:23–41.

186. Nakamura Y, Yokoyama H, Higuchi S, Hara S, Kato S, Ishii H. Acetaldehyde accumulation suppresses Kupffer cell release of TNF-Alpha and modifies acute hepatic inflammation in rats. J Gastroenterol 2004;39:140–147.

187. Pauly T, Dahmen N, Szegedi A, et al. Blood ethanol levels and adenylyl cyclase activity in lymphocytes of alcoholic patients. Biol Psychiatry 1999; 45:489–493.

188. Pruett SB, Fan R, Zheng Q. Acute ethanol administration profoundly alters poly I:C-induced cytokine expression in mice by a mechanism that is not dependent on corticosterone. Life Sci 2003;72:1825–1839.

189. Goral J, Choudhry MA, Kovacs EJ. Acute ethanol exposure inhibits macrophage IL-6 production: role of p38 and ERK1/2 MAPK. J Leukoc Biol 2004;75:553–559.

190. Hoffmann D, Wynder EL. Chemical constituents and bioactivity of tobacco smoke. IARC Sci Publ 1986:145–165.

191. Anthony DC, Montine TJ, Valentine WM, Graham DG. Toxic responses of the nervous system. In: Klaassen CD, ed. Casarett and Doull's Toxicology. New York: McGraw-Hill; 2001, pp. 555–557.

192. Sopori M. Effects of cigarette smoke on the immune system. Nat Rev Immunol 2002;2:372–377.

193. Gilbert DG, Meliska CJ, Williams CL, Jensen RA. Subjective correlates of cigarette-smoking-induced elevations of peripheral beta-endorphin and cortisol. Psychopharmacol 1992;106:275–281.

194. Pomerleau OF, Pomerleau CS. Cortisol response to a psychological stressor and/or nicotine. Pharmacol Biochem Behav 1990;36:211–213.

195. Russell MA, Jarvis M, Iyer R, Feyerabend C. Relation of nicotine yield of cigarettes to blood nicotine concentrations in smokers. Br Med J 1980;280:972–976.

196. Cullen JW, ed. The Health Consequences of Using Smokless Tobacco. Bethesda, Maryland: U.S. Department of Health and Human Services; 1986.

197. Russell MA, Jarvis MJ, Devitt G, Feyerabend C. Nicotine intake by snuff users. Br Med J (Clin Res Ed) 1981;283:814–817.

198. Nouri-Shirazi M, Guinet E. Evidence for the immunosuppressive role of nicotine on human dendritic cell functions. Immunology 2003;109:365–373.

199. Pruett SB, Han Y, Munson AE, Fuchs BA. Assessment of cholinergic influences on a primary humoral immune response. Immunol 1992;77:428–435.

200. Wang H, Liao H, Ochani M, et al. Cholinergic agonists inhibit HMGB1 release and improve survival in experimental sepsis. Nat Med 2004;10:1216–1221.

201. Matsunaga K, Klein TW, Friedman H, Yamamoto Y. Involvement of nicotinic acetylcholine receptors in suppression of antimicrobial activity and cytokine responses of alveolar macrophages to *Legionella pneumophila* infection by nicotine. J Immunol 2001;167:6518–6524.

202. Geng Y, Savage SM, Johnson LJ, Seagrave J, Sopori ML. Effects of nicotine on the immune response. I. Chronic exposure to nicotine impairs antigen receptor-mediated signal transduction in lymphocytes. Toxicol Appl Pharmacol 1995;135:268–278.

203. Aicher A, Heeschen C, Mohaupt M, Cooke JP, Zeiher AM, Dimmeler S. Nicotine strongly activates dendritic cell-mediated adaptive immunity: potential role for progression of atherosclerotic lesions. Circulation 2003;107:604–611.

204. Thomas GA, Rhodes J, Green JT. Inflammatory bowel disease and smoking—a review. Am J Gastroenterol 1998;93:144–149.

205. Arnott ID, Williams N, Drummond HE, Ghosh S. Whole gut lavage fluid interleukin-1beta and interleukin-8 in smokers and non-smokers with Crohn's disease in clinical remission. Dig Liver Dis 2002;34:424–429.

206. Sher ME, Bank S, Greenberg R, et al. The influence of cigarette smoking on cytokine levels in patients with inflammatory bowel disease. Inflamm Bowel Dis 1999;5:73–78.

207. Yang H, Wang H, Tracey KJ. HMG-1 rediscovered as a cytokine. Shock 2001; 15:247–253.
208. Wang H, Yang H, Czura CJ, Sama AE, Tracey KJ. HMGB1 as a late mediator of lethal systemic inflammation. Am J Respir Crit Care Med 2001;164:1768–1773.
209. Mellon RD, Bayer BM. The effects of morphine, nicotine and epibatidine on lymphocyte activity and hypothalmic-pituitary-adrenal axis responses. J Pharm Exp Ther 1999;288:635–642.

9
Preclinical Approaches for the Safety Assessment of Cytokines

Peter T. Thomas and Melissa S. Beck-Westermeyer

SUMMARY

In response to the need to treat disorders such as Crohn's disease, rheumatoid arthritis, allergies, and diabetes, the pharmaceutical industry has pursued the development of cytokines as therapeutic agents. Cytokines possess unique characteristics that are distinctly different than small molecule drugs. These hormone-like proteins possess complex structural features, demonstrate unique binding specificities and, in many cases, share overlapping physiological functions. These characteristics make development of cytokine therapeutics a challenge to drug-development scientists. Preclinical development of these molecules is complicated by the challenge of differentiating between toxicity and exaggerated immunopharmacology and understanding and analyzing the impact of antidrug antibody on the pharmacodynamics and pharmacokinetics of the parent compound. Regulatory agencies have recognized these issues and established guidance documents to address the nonclinical development of biological products. Several examples, including the nonclinical development of interleukin (IL)-4, IL-5 antagonists, IL-6, IL-10, and IL-18 are reviewed in the context of these issues.

Key Words: Cytokines; immunotoxicology; preclinical development; immunogenicity.

1. INTRODUCTION

The beginnings of the understanding of cytokines came from early descriptive studies (1) identifying a substance produced by sensitized lymphocytes that inhibited the migration of nonimmune macrophages from the site of

From: *Methods in Pharmacology and Toxicology: Cytokines in Human Health: Immunotoxicology, Pathology, and Therapeutic Applications*
Edited by: R. V. House and J. Descotes © Humana Press Inc., Totowa, NJ

inflammation. This agent was coined macrophage migration inhibitory factor (MIF). Since that time, during the last 25 years, additional classes of immunoregulatory substances have been described as having antiviral properties (the interferons), growth and functional control of various leukocytes, (the interleukins such as interleukin [IL]-1, IL-2 and, more recently, IL-23). In addition, other hematopoietic and nonhematopoietic growth factors that control nerve and epidermal cell growth and differentiation, among others, have been characterized.

The systematic evolution of cytokine nomenclature is the result, in part, of the disparate origins of the basic research that lead to their description. The product of immunologically sensitized lymphocytes was first described as a lymphokine. The term "cytokine" was proposed in the mid-1970s to reflect the fact that a variety of immune and nonimmune cells produced immunologically active monokines and lymphokines (2). The term "interleukin" was later proposed to describe a group of distinct molecules that regulated communication between leukocytes and other nonhematopoietic and somatic cell types. Presently, many cytokines fall within the interleukin nomenclature. There are, however, several additional molecules that have retained their original functional names (e.g., interferon [IFN]-γ, transforming growth factor-β, leukocyte inhibitory factor, and others).

Cytokines can be classified according to different criteria, including their structural features, ability to bind to various receptor types, and physiological function (3). Hormones and growth factors also share many properties of cytokines. However, there are important differences. Most cytokines are small polypeptide molecules with a molecular weight (MW) of 30,000 Daltons or less. However, some cytokines form higher MW oligomers or heterodimers. For example, the recently described composite cytokine, IL-23, is composed of the IL-12 p40 subunit and a novel p19 protein. IL-23 has been shown to exhibit biological activities similar to and unique from IL-12 (4). The production of cytokines is not constitutive and is usually in response to various stimuli, primarily by those resulting from an immune response. Activity is usually at the level of messenger ribonucleic acid transcription or translation. With the exception of several of the proinflammatory cytokines (e.g., IL-1, TNF, IL-6, IFN-γ), their effective pharmacodynamic range is usually localized. However, unlike most hormones that have limited action and target cell specificity, physiological effects displayed by cytokines are diverse with some targeting hematopoietic cells. Differences aside, it appears that cytokines, polypeptide hormones, and growth factors all serve to facilitate extracellular signaling pathways. Furthermore, many have common structural features. Once bound to a receptor, the signal transduction pathways of many of these molecules appear to be the same. Research into the structure and function of cytokines and their receptors has resulted in classification of these molecules into loose families according to their overall three-dimensional structure *(5,6)*. Grouping of cytokines and their receptors is largely based on primary or higher order structural homologies among the molecules.

Cytokine function is as diverse as the cytokine family itself. Not suprisingly, as part of the pleiotropic nature of these molecules, considerable overlap exists between the cytokines and their intended targets. It is beyond the scope of this chapter to discuss the functional diversity of the cytokine families. Mantovani provided a useful classification of cytokine function as it relates to the fundamental states of immunity and hematopoiesis *(3)*. They include the proinflammatory cytokines involved in innate immunity, cytokines involved in specific immunity, hematopoietic cytokines, as well as cytokines that have anti inflammatory and immunosuppressive activities.

In response to infection, endotoxin, stress, or other acute insult, the initial proinflammatory cytokines, including TNF-α, IL-1, and IL-6, are released as a means of defense. Simply put, this reaction is the hallmark of the innate response of inflammation, and the cytokines involved in the response are, by necessity, pleiotropic in their actions and influence. TNF-α and IL-1 have demonstrable properties to magnify endogenous mechanisms to increase leukocyte cell numbers and recruitment in proximity to the affected area. Meanwhile, IL-6 serves to regulate and intensify the systemic response to the infection by stimulating production of acute phase proteins in the liver. This includes C-reactive protein, and serum amyloid P, which leads to complement activation, increased phagocytic activity through opsonization of bacteria, and a generalized increase in primary, nonspecific immunity *(3)*.

The chemokines also play an important role in nonspecific inflammation *(7)*. This superfamily of small proteins (8–11 kDa) shares structural similarities, including four conserved cysteine residues forming structurally distinct disulfide bonds. Four distinct families (CXC, CC, C, and CX3C) have been described *(8,9)* depending on the location of the first two cysteines in their amino acid sequence. To date, a total of 42 human chemokines have been described. The principal pharmacological action attributed to the chemokines relates to cell trafficking. The four groupings are largely differentiated not only by their structural heterogeneity but also by chemotactic action on different leukocyte cell types. Chemokine–receptor interactions show considerable nonspecificity and redundancy. The mononuclear phagocyte, the most important cell type in innate immunity and also the most evolutionarily conserved, demonstrates the widest response to these proteins.

As a result of an inflammatory response, specific T- and B-cell immunity to the offending agent is developed. Cytokines such as IL-2, IL-4, IL-12, IL-15, and IFN-γ are produced at this time. T-helper 1 (Th1) cells stimulate macrophage activation, opsonization, and phagocytosis; delayed-type hypersensitivity, and promote antibody-dependent cell-mediated cytotoxicity through the production of IFN-γ, IL-2, and TNF. By contrast, Th2 cells produce a complement of cytokines, including IL-4, IL-5, IL-6, IL-10, and IL-13, among others. The humoral immune response, antibody isotype switching, as well as promotion of mucosal immunity and IgA secretion are mediated through the actions of these cells.

The cytokine response patterns that have come to define the Th1-Th2 paradigm have been useful in understanding the mechanisms of a number of disease states as well as designing therapeutic strategies *(10)*. It is believed that an imbalance in the Th1 and Th2 response contributes to diseases such as Crohn's, rheumatoid arthritis, multiple sclerosis, allergies, type 1 diabetes, and some infectious diseases *(3)*. T-cell activation and overproduction of TNF-α, IL-1, and IL-6 may be attributed to the cause of some of the autoimmune disorders and the inflammatory properties described herein *(11)*.

Colony-stimulating factors (CSFs) are a family of acidic glycoproteins that regulate the differentiation and proliferation of a variety of types of hematopoietic progenitor cells. There are well described CSFs that specifically stimulate erythroid and myeloid cell precursors. These include granulocyte colony-stimulating factor (G-CSF), which regulates the survival, proliferation, and differentiation of granulocyte precursors as well as the function of mature neutrophils. Granulocyte-macrophage colony-stimulating factor (GM-CSF) has activity similar to G-CSF except that it is directed toward an earlier myeloid progenitor population. M-CSF on the other hand, specifically stimulates macrophage progenitor cells. In addition to these CSFs, many proteins, including endothelial growth factor and vascular endothelial growth factor, can stimulate the growth of nonhematopoietic cells. Furthermore, cytokines not normally considered growth factors per se are included in this group. They include IL-3, IL-5, IL-6, and stem cell factor, among others.

The actions of this diverse class of molecules include modulation of natural and adoptive immunity, response to neoplasia, cellular differentiation and apoptosis. During the last decade, efforts to leverage the unique characteristics of these molecules and their receptors for the treatment of diseases has lead to the approval of several as therapeutic drugs (Table 1). Furthermore, preclinical and clinical development continues with many of these molecules on a number of disease indications.

Table 1
Growth Factors, Cytokines, and Cytokine
Antagonists Approved for Human Use

Product type	Selected clinical indications
Erythropoetin	Anemia associated with chronic renal failure, HIV infection Anemia associated with cancer chemotherapy
G-CSF	Chemotherapy induced neutropenia Bone marrow transplantation
GM-CSF	Bone marrow transplantation Immunosuppression in acute myelogenous leukemia
Interleukin (IL)-11	Reduced platelet counts following myelosuppressive therapy
Interferon-α	Hairy cell leukemia, malignant melanoma, chronic hepatitis B, C
Interferon-γ1	Chronic granulomatous disease, osteopetrosis, rheumatoid arthritis
Interferon-β	Relapsing multiple sclerosis
Interleukin-2	Renal cell carcinoma, interleukin-4 antagonists, asthma (late stage development)
VEGF antagonists	Metastatic breast cancer
TNF-α antagonists	Rheumatoid arthritis, Crohn's disease
IL-1 receptor antagonists	Rheumatoid arthritis
IL-2 receptor antagonists	Immunosuppression in acute renal transplantation.

G-CSF, granulocyte colony-stimulating factor; GM-CSF, granulocyte-macrophage colony-stimulating factor; TNF, tumor necrosis factor; VEGF, vascular endothelial growth factor.

2. PRECLINICAL SAFETY ASSESSMENT OF CYTOKINE THERAPEUTICS

The purpose of preclinical safety assessment studies is to characterize potential toxicity of new drug candidates and to relate any toxicity to pharmacokinetics and pharmacodynamic action with the intention of selecting safe and effective doses for human clinical studies. All new chemical entities, including cytokine therapeutics, follow this same general strategy. Early in development, repeat-dose pharmacokinetic and toxicokinetic studies typically are performed in relevant species to understand half-life, bioavailability, distribution, clearance, and maximum plasma concentration. These data are later compared with the systemic toxicology findings to ensure that adequate exposure was achieved to support clinical trials. In contrast to small molecules, which are metabolized and excreted by well-characterized Phase I and Phase II drug-metabolizing enzyme pathways, cytokines are degraded by enzymatic proteolysis and the resulting peptides and amino acids are recycled. Therefore, except for evaluating the impact of antidrug antibody on the in vivo pharmacokinetics of the compound, classical ADME studies are usually not relevant. Safety pharmacology studies evaluate the compound's potential to induce unwanted pharmacodynamic effects on a wide variety of physiological functions. For biologicals, we have noted that these initial studies often are incorporated into the repeat dose IND-enabling toxicology studies, which is consistent with the ICH S7A guideline for safety pharmacology and usually is sufficient at the IND stage *(12)*. Like small molecules, single- and multiple-dose toxicology studies usually are performed with cytokine therapeutics. The duration, route of exposure, test article formulation, and frequency should mimic those planned for the clinic. The potential for cytokine biologicals to interact with deoxyribonucleic acid and cause cellular damage is low and, therefore, genetic toxicity testing is not appropriate. Furthermore, the presence of amino acids such as histidine in the formulation can complicate interpretation of bacterial mutagenesis tests. In the unlikely event that there is an organic contaminant as a manufacturing by product, genetic toxicity may be an issue.

As far as effects on reproduction are concerned, developmental toxicity studies are conducted with biologicals, including cytokines. The ICH S6 regulatory document dealing with preclinical safety evaluation of biotechnology-derived therapeutics provides general guidance on reproduction and fertility testing by emphasizing the need to consider the clinical indication, the targeted patient population and the availability of a physiologically relevant animal species *(13)*. This latter issue is particularly important in light of the species-specific responses of these molecules as the nonhuman primate model may be the only relevant model in which to conduct these studies.

Carcinogenicity testing of endogenous proteins whose hormone-like actions affect cell division, distribution and differentiation must be carefully considered. In a recent review *(14)*, the pros and cons of initiating carcinogenicity testing, using human IL-10 as an example, were discussed. It is important to understand whether the pharmacodynamic action of the compound compromises the generally accepted protective mechanisms of tumorigenicity. Although many cytokines have immunosuppressive activity, the relevance of any effects to the current theories of tumor immunosuveillance must be considered. In addition, if the proposed test species responds to the specific pharmacological action of the compound in the same manner as humans, one needs to consider the evidence that an immune response to the foreign protein potentially would neutralize its activity. For example, although there may be sufficient pharmacodynamic evidence to support the use of rodents, including transgenics, as a relevant test species, exposure may result in formation of antibody with consequential loss of detectable plasma levels and associated pathology. Other conventional species, such as the Syrian hamster, might suffice, but an understanding of the physiological role of the cytokine in this species is needed, as is availability of suitable reagents and standards for immunoanalytical evaluation. Therefore, before moving ahead with testing, these and other key questions must be considered before deciding whether such studies are necessary and whether or not they can be conducted in a way that will yield meaningful results for risk assessment.

Because the general development strategy is the same, there are several critical differences between traditional small MW therapeutics and cytokines that influence the safety assessment strategy *(15)*. They include selecting a relevant animal model, differentiating between toxicity and exaggerated immunopharmacology, and evaluating the impact of antidrug antibody on pharmacokinetics and pharmacodynamics of the compound *(16)*. In addition, the unique pharmacology of these molecules makes them especially challenging to develop as potential therapeutics. The fact that they function as extracellular signaling molecules, and are endogenous to the host makes their actions similar to hormones. Unlike hormones, which act at a distant site, cytokines are active at very low physiological concentrations *(17)*. As a result, their primary area of influence frequently is restricted to the local cellular milieu in which the molecule is produced.

These molecules have other important characteristics that must be considered during development. Paradoxically, many cytokines are pleiotropic in their actions and exhibit different activities across a wide variety of cells. Once produced, many cytokines act in complex feedback loops, inhibiting the production of other molecules or their receptors. A cytokine may also increase the production of another cytokine, or its receptor, leading to a cascade of effects. Much of this is a result of the fact that, to be active, cytokines bind to receptors on target cells, that often have multiple subunits. Furthermore, several act in a redundant fashion, sharing many actions among other, related molecules having known mechanisms. Receptors for several hormones and cytokines, such as IL-2, IL-4, IL-6, the CSF, prolactin and erythropoetin demonstrate common structural features *(18,19)*. Taken as a whole, these unique features make development of cytokine therapeutics a challenge, when compared with traditional small molecules and other proteins, including monoclonal antibodies.

The development of IL-6 illustrates those important differences in responses between animal models and humans do occur and must be accounted for during development. IL-6 has a broad spectrum of biological activities, including stimulation of thrombopoiesis, proliferation of the cells of the mesangium, and induction of acute-phase protein responses in the liver (20,21). The physiologic effects of IL-6 are mediated through membrane receptors expressed on monocytes, nonlymphoid, and lymphoid cells. Safety evaluation studies performed in nonhuman primates and rodents demonstrated that recombinant human IL-6 was well tolerated. As expected, an increase in thrombocyte count levels of acute phase proteins and immune stimulation was observed in the absence of significant target organ toxicity. Patients receiving IL-6 presented with a fever, anemia, and general malaise consistent with an acute phase response. More significantly, the uncontrolled lymphoproliferation and mesangioproliferative nephritis seen in IL-6 transgenic mice was not seen in humans (22). In other mechanistic studies, Ruffel et al. (23) demonstrated that the administration of IL-6 to autoimmune prone (NZB × NZW) F1 female mice enhanced glomerluonephritis after 12 wk of treatment. Moreover, mononuclear cell infiltrates and tubular epithelium expressed high levels of major histocompatability class II antigen. These observations are consistent with the multifunctional immunostimulatory properties of this cytokine and emphasize that animal safety data must be extrapolated with care because of potential species differences in pharmacological and immunological responses.

Most cytokines demonstrate limited species specificity with respect to pharmacologic action. Moreover, the ability of animal models to consistently predict immunotoxicity has been, at time, of limited value. Species-specific immunopharmacology has been demonstrated in studies with GM-CSF and IFN-γ (human vs rodent), recombinant IL-1 (granulopoiesis and neutropenia in the dog vs mouse models), and IL-6 (lymphoproliferation and nephritis in the mouse vs human). When examined in the appropriate species, the pleiotropic actions of many of these molecules make differentiation between toxicity and exaggerated immunopharmacology difficult. For example, production of numerous cytokines, including TNF-α, IL-1, IL-15, and IL-18 are important in the pathogenesis of cytokine-induced shock in humans. Mechanistic studies have suggested that natural killer (NK) cells are important effector cells responsible for this human response (24).

The administration of therapeutic doses of recombinant cytokines to patients with malignant disease can produce systemic complications, which ultimately may present as a systemic inflammatory response. For example, the combination of IL-18 and IL-12 has synergistic antitumor activity in vivo and has been associated with significant toxicity. When examined in a murine model, the coadministration of IL-18 and IL-12 resulted in systemic inflammation and 100% mortality within a week, depending on the strain employed. When the response was investigated further, mice treated with IL-18 plus IL-12 exhibited unique pathological findings as well as elevated serum levels of proinflammatory cytokines and acute-phase reactants. The data suggest that actions of tumor necrosis factor-alpha did not contribute to the observed toxicity, nor did those of T- or B-cells. However, toxicity and death from treatment with IL-18 plus IL-12 could be completely abrogated by elimination of NK cells or macrophages. Further studies in genetically altered mice revealed that gamma interferon produced by NK cells mediated the fatal toxicity via signal transduction pathways *(25)*.

Because of the unique properties of cytokines and other immunoregulatory molecules, traditional dose–response relationships seen with traditional small molecule NCEs often do not apply. Talmadge *(26)* suggested that this is due to many factors, including biodistribution of the molecules in the body, circadian processes, specific receptor-mediated events and indirect, downstream effects from the site(s) of action. Bell shaped or biphasic dose response curves have been seen in studies with IFN-γ (murine tumor metastasis models) as well as TGF-β (wound healing models), among others *(27)*.

Probably the most important factor in safety assessment of cytokines, like other biologicals, is the potential of antibody production to the therapeutic or to endogenous product in the test species. The nature and duration of the antibody response is dependent on many factors. These include homology (or lack thereof) with the endogenous molecule and protein form (conjugation to toxins, drugs, glycol ethers, or creation of new antigenic determinants through fusion molecules). Product-related impurities such as yeast byproducts or endotoxin formed during manufacture can be immunogenic or have adjuvant properties. Other factors, including dose, frequency, and route of exposure also impact antigenicity. From a practical standpoint, subcutaneous and intradermal routes of injection are most likely to favor an immunogenic response because of the proximity of draining lymph nodes, dendritic antigen-presenting cells, and B-cells, followed by intraperitoneal, intramuscular, intravenous, and topical exposure. In repeated-dose preclinical toxicology studies, an antibody response typically is detected within 2 wk, making long-term monitoring of the response important to safety evaluation. In addition to measuring antidrug antibody titers, characterizing the particular isotype produced may explain any subsequent immunopathological, pharmacodynamic or pharmacokinetic changes that may occur *(28,29)*.

Of concern to the toxicologist is appearance of antibody that could complicate nonclinical safety assessment. Documented reactions include: (1) crossreactivity with endogenous proteins; (2) type 2 immune complex formation leading to deposition of antibody-antigen complexes; (3) localized arthus type reactions at injection sites; (4) systemic infusion reactions, including anaphylaxis; and (5) alterations in the pharmacodynamics, half-life and distribution of the cytokine. Despite these concerns, experience has shown that detection of antidrug antibody in animal studies is not necessarily predictive of the same effect in humans. Factors inherent in the host play an important role in predicting immunogenicity of a protein. These include differences in major histocompatability class haplotype, diversity of T- and B-cell repertoires, and immunoglobulin and T-cell receptor gene rearrangements between species. In addition, preexisting autoimmune conditions and host environmental status, including exposure to pathogens, also play a role. Strategies to modify proteins in an attempt to alter immunogenicity include glycosylation and, more recently, modifications of immunodominant peptide epitopes *(30)*. Despite this, there is no clear association between antibody status, degree of glycosylation, MW, and homology to the endogenous protein. Nevertheless, immunogenicity data obtained during preclinical development are most valuable in the design and conduct of subsequent clinical studies.

Aside from antibody effects, direct or indirect immunotoxicity after treatment with therapeutic cytokines is thought to be the cause of many potential adverse events seen during clinical treatment. These range from interferon or TNF-induced flu-like symptoms, vascular leak syndrome (rhIL-12, IL-2), proinflammatory cytokine release syndrome, to more serious and less well understood effects, including thyroid disorders, systemic lupus erythematosis, and diabetes *(31)*. The challenge to the practicing toxicologist is to differentiate adverse immunotoxicity from exaggerated (and sometimes expected) immunopharmacological effects because of treatment. Understanding the role these molecules directly and indirectly play in immune regulation provides important mechanistic information that can aid in identification of specific biomarkers of exposure or of efficacy. These indicators also can serve to bridge the preclinical findings with potential pharmacodynamic effects in humans. When the decision to incorporate measurement of cytokine levels is made, many practical and important assay-related factors must be considered, including the biological source of reagents, the effect of sample processing on activity and, whether or not to measure function or simply presence of the molecule(s) in question *(32)*.

3. REGULATORY CONSIDERATIONS

From 1993 to 2002, therapeutic proteins were regulated by the Office of Therapeutics Research and Review within the Food and Drug Administration (FDA) Centers for Biologics Evaluation and Research. In October 2003, the lead organization responsible for review of safety data supporting development of therapeutic proteins (including cytokines and monoclonal antibodies) was the Office of Drug Evaluation 6 (ODE 6) within the FDA Center for Drug Evaluation and Research. From a regulatory perspective, there are no guidances that specifically regulate development of therapeutic cytokines because they are grouped together with other proteins, including monoclonal antibodies. Those guidances that provide a roadmap for safety assessments have been reviewed elsewhere *(13,33)*. Perhaps the most important of these is the guidance for preclinical safety of biotechnology-derived pharmaceuticals, made public in 1997 under the auspices of the ICH *(34)*. This document forms the basis for developing a preclinical plan for biologicals, discusses which approaches are most appropriate and, more importantly, identifies tests that are not generally required for this class of compounds. This document stresses the importance of choosing the most relevant species for safety evaluation because many biologicals, including cytokines, are highly species specific in their pharmacodynamic action. In contrast to traditional drugs, standard immunotoxicity tier tests are deemed not relevant for initial safety evaluation. However, as discussed previously, measurement and characterization of an antidrug antibody response as it effects either the host or the pharmacodynamics of the test compound in the host is emphasized.

In 2002, the US FDA published its long-awaited guidance document on immunotoxicology evaluation for Investigational New Drugs *(35)*. According to the FDA, all investigational new drugs should be evaluated for immunosuppression. How this is accomplished is subject to individual review and discussion between the drug sponsor and the Agency. It has been our experience that IND-enabling repeat-dose toxicity studies typically do not include specific measures of immune function. Unless there is a specific reason to do otherwise, sufficient information about potential immunotoxicity of the test compound is obtained through the use of standard clinical and anatomic pathology measures rather than immune function tests. With respect to immunogenicity, while acknowledging that evaluation of allergic potential is difficult using the currently accepted animal models nonclinical toxicology, the FDA comments that specific models and methods have been developed. In addition to discussing specific immune function endpoints, the FDA provides guidance on the scope of nonclinical immunotoxicity safety testing as it applies to the disease and targeted patient population(s).

As far as the European Union is concerned, the EMEA Committee for Proprietary Medicinal Products provides more specific recommendations for immunotoxicity and suggests that it be included it in at least one repeat dose toxicity study *(36)*. When it comes to biotechnology products, the EMEA guidance refers to the ICH S6 document *(35)*. Recently, after considerable discussion and debate around harmonization of testing guidelines for immunotoxicology, the ICH Immunotoxicology Working Group (S8) is proceeding to Step 2 of the harmonization process *(37)*.

4. REPRESENTATIVE THERAPEUTICS CURRENTLY IN DEVELOPMENT

Progress to develop cytokines, cytokine receptors, or receptor antagonists as drug therapies has been hampered somewhat by a number of factors, including an incomplete understanding of mechanism(s) of action, lack of suitable and predictive animal models, the potential for immunogenicity, and unintended immunotoxicity. Nevertheless, several of these molecules are continuing in development, particularly for control of treatment of acute and chronic inflammation and the diseases associated with it.

In brief, inflammation is a complex process that encompasses the host response to exogenous stimuli, not the least of which make up the defense mechanisms against infectious agents. Viral and bacterial infection often lead to the release of cytokines, including IL-6, TNF-γ, IL-1 and IFN-γ. Of these molecules, TNF-α plays a major role in the control of inflammation. As a proinflammatory cytokine, TNF-α upregulates other cytokines including GM-CSF, IL-6, IL-1, the prostaglandins and chemokines *(38)*, which leads to the activation of both neutrophils and macrophages, enhancing protease release, induction of leukocyte and vascular adhesion molecule expression and respiratory burst *(39,40)*.

Therapeutic strategies for control and treatment of the inflammatory processes associated with rheumatoid arthritis have lead to the development of various anti-TNF antibodies and soluble receptor antagonists, including etanercept, infliximab, and lenercept, among others. The clinical success realized by these agents has lead to their investigation in other inflammatory conditions, such as etanercept for treatment of psoriatic arthritis *(41)*.

Interleukin-4 is a pleiotropic cytokine that influences development of naïve CD4 helper T-cells into Th2-type cells. This cytokine influences immunoglobulin class switch towards IgE and it is a contributing factor in the proliferation of B- and mast cells. IL-4 mediates important proinflammatory functions in asthma, including induction of the IgE isotype switch, increased expression of vascular cell adhesion molecule 1 and promotion of eosinophil transmigration across the endothelium, stimulation of mucus production, and Th2 lymphocyte differentiation. As an initiator of a Th2 response, it plays a critical role in the pathology of this disease. Blockage of IL-4 action by IL-4 receptor antagonists is being evaluated as a therapeutic strategy for treatment of asthma. Clinical trials using soluble IL-4 receptors to block activity have shown that these molecules are safe and effective in treatment of moderate persistent asthma *(42)*. More recently, gene therapy approaches are being explored in animal models to enable continued therapy with IL-4 antagonists, leading to a reduction in asthma associated cytokines and airway hyperresponsiveness *(43)*.

As previously discussed, IL-6 is a pleiotropic cytokine having central roles in immune regulation, inflammation, and hematopoiesis. Deregulation of IL-6 production is implicated in the pathology of several disease processes. Increased levels of IL-6 and its receptor have been demonstrated in both serum and digestive tissues of patients with Crohn's disease and in patients with rheumatoid arthritis. In animal studies of intestinal inflammation, antibody to the IL-6 receptor prevented inflammation and wasting disease by suppressing adhesion molecule expression by the vascular endothelium. Humanized anti-IL-6 receptor antibody is under development as a therapeutic agent for Crohn's disease *(44)* and rheumatoid arthritis *(45)*, and benefits from treatment have been realized.

IL-10 is a Th2-derived immunoregulatory cytokine with a broad spectrum of biological activities including immunosuppression and modulation of inflammation. The latter effects are caused, in part, by downregulation of proinflammatory cytokines, including TNF-α, IL-1, and IL-6, downregulation of cytokine receptor expression and upregulation of cytokine inhibitors (e.g., soluble TNF-receptor, IL-1 receptor antagonist and TNF-α) *(46)*. In addition, IL-10 has potential in the treatment of allergy as it suppresses mast cell, eosinophil and T-cell specific antibody responses. The inhibitory functions of IL-10 can be exploited clinically as its activity in inhibiting the functions of antigen presenting cells and Th1 cytokine synthesis suggests a possible use as a nonspecific immunosuppressive factor. Many development strategies for IL-10 as a treatment for acute and chronic inflammatory disease, autoimmunity, allograft survival and disorders of the intestinal tract have been described *(47)*.

Like IL-10, IL-18 is a cytokine that, because of its biological activities, is being considered as a immunotherapeutic for cancer and infectious disease *(48,49)*. As a member of the IL-1 superfamily, this cytokine exhibits a broad range of immunoenhancing properties, including induction of IFN-γ, enhancement of NK cell activity, upregulation of functional Fas ligand expression on immune cells, and antiangiogenic activity *(49–51)*. Nonclinical safety studies suggest that the compound is well tolerated in animals at pharmacologically active doses *(52)*.

IL-5 is produced by a number of cell types and is responsible for the maturation and release of eosinophils in the bone marrow. In humans, IL-5 is a very selective cytokine as a result of the restricted expression of the IL-5 receptor on eosinophils and basophils. There are monoclonal antibodies in development that antagonize IL-5 production by preventing receptor binding. One example is mepolizumab (SB-240563), which is a humanized monoclonal antibody specific for human IL-5 that is in development for treatment of Hypereosinophilic Syndrome. Because human and cynomolgus monkey IL-5 differs by only two amino acids, this antibody exhibits comparable inhibition of activity in both human and monkey systems. In efficacy studies, SB-240563 reduced basal levels of eosinpohils in the peripheral blood as well as numbers of eosinophils in bronchoalevolar lavage fluid from *Ascaris suum*-infected monkeys. Nonclinical safety studies with this compound have demonstrated that long-term suppression, via the IL-5 pathway of circulating and fixed tissue eosinophils, is well tolerated and this cytokine antagonist has the potential to be beneficial for chronic inflammatory respiratory diseases *(53)*.

5. CONCLUSION

The safety and efficacy of cytokine-based therapies to treat chronic inflammatory disease states, in particular for rheumatoid arthritis, Crohn's disease, and multiple sclerosis, have shown great progress in recent years in large part because of an increased understanding of the mechanism(s) of action of these molecules in regulating the immune response and autoimmunity. Furthermore, although early in development, targeted delivery by gene therapy of cytokines or cytokine antagonists shows promise and appears to be less a less-toxic alternative treatment for these conditions. As our understanding increases concerning the role cytokines play in modulating the immune response and in regulating cell growth and differentiation, our ability to develop and deliver safer and more effective therapies will improve. In addition to mechanistic knowledge gained in recent years, our ability to successfully engineer creative scientific and regulatory strategies for advancing these molecules forward into development and clinical trials has also improved significantly.

REFERENCES

1. Bloom B, Bennett B. Mechanism of a reaction in vitro associated with delayed-type hypersensitivity. Science 1966;153:80–82.

2. Cohen S, Bigazzi P, Yoshida T. Similarities of T-cell function in cell-mediated immunity and antibody production. Cell Immunol 1974;12:150–159.

3. Mantovani A. Cytokines: A World Apart. In: Pharmacology of Cytokines. Oxford University Press New York, 2000, pp. 1–18.

4. Oppmann B, Lesley R, Blom B, et al. Novel p19 protein engages IL-12p40 to form a cytokine, IL-23, with biological activities similar as well as distinct from IL-12. Immunity 2000;13:715–725.

5. Nicola NA. Cytokine pleiotropy and redundancy: a view from the receptor. Stem Cells 1994;12:(Suppl 1)3–12.

6. Vilcek J. The cytokines: an overview. In: Thompson A, ed. The Cytokine Handbook. 3rd ed. San Diego: Academic Press; 1998;1–20.

7. Homey B. Chemokines and chemokine receptors as targets in the therapy of psoriasis. Curr Drug Targets 2004;3:169–174.

8. Hendrick J, Zlotnick A. Chemokines and lymphocyte biology. Curr Opin Immunol 1996;8:343–347.

9. Baggiolini M, Dewald B, Moser B. Human chemokines: an update. Annu Rev Immunol 1997;15:675–705.

10. Romagnani S. New therapeutic strategies in allergic diseases. Drugs Today 2003;39:849–865.

11. Schooltink H, Rose-John S. Cytokines as therapeutic drugs. J. Interferon and Cytokine Research 2002;22:505–516.

12. International Conference on Harmonisation. Guidance S7A. Safety Pharmacology Studies for Human Pharmaceuticals; July 2001.

13. Weir A. Preclinical safety assessment of therapeutic proteins and monoclonal antibodies. In: Mathieu, M, ed. Biologics Development: A Regulatory Overview. 3rd ed. Waltham MA: Parexel; 2004;17–24.

14. Rosenblum I, Dayan, A. Carcinogenicity testing of IL-10: principles and practicalities. Human Exp Toxicol 2002;21:347–358.

15. Ryan A, Terrell T. Biotechnology and its products. In: Haschek WM, Rousseaux CG, Wallig M, eds. Handbook of Toxicologic Pathology, 2nd ed., Vol 1. San Diego: Academic Press; 2002, pp. 479–500.

16. Thomas PT. Nonclinical evaluation of therapeutic cytokines: immunotoxicologic Issues. Toxicology 2002;174:27–35.

17. Vilcek J. The cytokines: an overview. In: Thompson A, ed. The Cytokine Handbook. 3rd ed. San Diego: Academic Press; 1998, pp. 1–20.

18. Bazan J. A novel family of growth factor receptors: a common binding domain in the growth hormone prolactin, erythropoietin and IL-6 receptors and the p75 IL-2 receptor B-chain. Biochem Biophys Research Commun 1989;164:788–795.

19. Ihle J. Cytokine receptor signaling. Nature 1995;377:591–594.

20. Hirano T, Akira S, Taga T, Hishimoto T. Biological and clinical aspects of interleukin-6. Immunol Today 1990;11:443–449.

21. Van Snick J. Interleukin-6: an overview. Ann Rev Immunol 1990;8:253–278.

22. Kammuler M, Ryffel B. Extrapolation of experimental safety data to humans: the interleukin-6 case. Clin Immunol Immunopathol 1997;83:5–17.

23. Ryffel B, Car B, Gunn H, Roman D, Hiestand P, Mihatsch M. Interleukin-6 exacerbates glomerulonephritis in (NZBxNZW) F1 mice. Am J Pathol 1994; 144:927–937.

24. Carson W, Yu H, Dierksheide J, et al. A fatal cytokine-induced systemic inflammatory response reveals a critical role for NK cells. J Immunol 1999;162: 4943–4951.

25. Carson WE, Dierksheide JE, Jabbour S, et al. Coadministration of interleukin-18 and interleukin-12 induces a fatal inflammatory response in mice: critical role of natural killer cell interferon-gamma production and STAT-mediated signal transduction. Blood 2000;96:1465–1473.

26. Talmadge J. Pharmacodynamic aspects of peptide administration of biological response modifiers. Adv Drug Delivery Rev 1998;33:241–252.

27. Ammaan A, Beck S, DeGutzman L, et al. Transforming growth factor beta. Effect on soft tissue repair. Ann NY Acad Sci 1990;593:124–134.

28. Schellekens H. Immunogenicity of therapeutic proteins: clinical implications and future prospects. Clin Ther 2002;24:1720–1740.

29. Herzyk D. The immunogenicity of therapeutic cytokines. Curr Opinion Mol Ther 2003;5:167–171.

30. Yeung V, Chang J, Miller J, Barnett C, Stickler M, Harding F. Elimination of an immunodominant CD4+ T-cell epitope in human IFN-β does not result in an in vivo response directed at the subdominant epitope. J Immunol 2004;172: 6658–6665.

31. Vial T, Descotes J. Immune-mediated side effects of cytokines in humans. Toxicology 1995;105:31–57.

32. House R. Theory and practice of cytokine assessment in immunotoxicology. Methods 1999;19:17–27.

33. Griffiths SA, Lumley CE. Non-clinical safety studies for biotechnologically-derived pharmaceuticals: Conclusions from an international workshop. Hum Exp Toxicol 1998;17:63–83.

34. International Conference on Harmonization. Guidance S6. Preclinical safety evaluation of biotechnology-derived pharmaceuticals; July 1997.

35. U.S. Food and Drug Administration. CDER Guidance for Industry. Immunotoxicology Evaluation of Investigational New Drugs; October 2002.

36. EMEA Committee for Proprietary Medicinal Products. Note for Guidance on Repeat Dose Toxicity. Appendix B. Guidance on Immunotoxicity. CPMP/SWP/1042/99. July 2000.

37. International Conference on Harmonization Steering Committee Meeting. ICH 2004 Press Release June 9–10, 2004.

38. Old L. Tumor necrosis factor (TNF). Science 1985;230:630–632.

39. Gamble JR, Harlan JM, Klebanoff SJ, Vadas MA. Stimulation of the adherence of neutrophils to umbilical vein endothelium by human recombinant tumor necrosis factor. Proc Natl Acad Sci USA 1985;82:8667–8671.

40. Vilcek J, Lee TH. Tumor necrosis factor. New insights into the molecular mechanisms of its multiple actions. J Biol Chem 1991;266:7313–7316.

41. Mease PJ, Kivitz AJ, Burch FX, et al. Etanercept treatment of psoriatic arthritis: safety, efficacy, and effect on disease progression. Arthritis Rheum 2004;50:2264–2272.

42. Borish LC, Nelson HS, Corren J, Bensch G, Busse WW, Whitmore JB, Agosti JM, for the IL-4R Asthma Study Group. Efficacy of soluble IL-4 receptor for the treatment of adults with asthma. J Allergy Clin Immunol 2001;107:963–970.

43. Zavorotinskaya T, Tomkinson A, Murphy JE. Treatment of experimental asthma by long-term gene therapy directed against IL-4 and IL-13. Mol Ther 2003;7:155–162.

44. Ito H. Il-6 and Crohn's disease. Curr Drug Targets Inflamm Allergy 2003;2:125–130.

45. Nishimoto N, Kishimoto T. Inhibition of IL-6 for the treatment of inflammatory diseases. Curr Opin Pharmacol 2004;4:386–391.

46. Moore K, de Waal Malefyt R, Coffmann R, O'Garra A. Interleukin-10 and the interleukin-10 receptor. Ann Rev Immunol 2001;19:683–765.

47. Li M, He S. IL-10 and its related cytokines for treatment of inflammatory bowel disease. World J Gastroenterol 2004;10:620–625.

48. Herzyk D, Soos J, Maier C, et al. Immunopharmacology of recombinant human interleukin-18 in nonhuman primates. Cytokine 2002;20:38–48.

49. Braddock M, Quinn A, Canvin J. Therapeutic potential of targeting IL-1 and IL-18 in inflammation. Expert Opin Biol Ther 2004;4:847–860.

50. Okamura H, Tsutsui H. Kashiwamura S, Yoshimoto T, Nakanish K. Interleukin-18: A novel cytokine that augments both innate and acquired immunity. Adv Immunol 1998;70:281–312.

51. Hashimoto W, Osaki T, Okamura H, et al. Differential antitumor effects of administration of recombinant interleukin-18 (rIL-18) or rIL-12 are mediated by Fas-Fas ligand and perforin-induced tumor apoptosis. J Immunol 1999;163:583–589.

52. Herzyk D, Bugelski P, Hart T, Wier P. Preclinical safety of recombinant human interleukin-18. Toxicol Pathol 2003;31:554–561.

53. Hart TK, Cook RM, Zia-Amirhosseini P, et al. Preclinical efficacy and safety of mepolizumab (SB-240563), a humanized monoclonal antibody to IL-5 in cynomolgus monkeys. J Allerg Clin Immunol 2001;108:250–257.

10
Flu-Like Syndrome and Cytokines

Jacques Descotes and Thierry Vial

SUMMARY

Flu-like reactions have been described, long before the introduction of therapeutic cytokines, in the clinical setting to treat a variety of pathological conditions. Indeed, flu-like reactions are commonly associated with vaccination as well as a number of infectious diseases unrelated to the influenza virus. Flu-like symptoms have also been described after the early use of supposedly immunostimulating drugs. When the first interferon formulations began to be used to treat cancerous patients, flu-like symptoms with some variation according to the type of interferon, route of administration, schedule, and dose, were observed in most patients. Since then, the flu-like syndrome emerged as a common-if-not universal complication of therapeutic cytokines.

Key Words : Chemical respiratory allergy; skin sensitization; hazard identification; cytokine fingerprinting; cytokines; interleukins.

1. INTRODUCTION

Flu-like reactions have been described, long before the introduction of therapeutic cytokines, in the clinical setting to treat a variety of pathological conditions. Indeed, flu-like reactions are commonly associated with vaccination *(1)* as well as a number of infectious diseases unrelated to the influenza virus. Flu-like symptoms have also been described after the early use of supposedly immunostimulating drugs *(2)*. When the first interferon formulations began to be used to treat cancerous patients, flu-like symptoms with some variation according to the type of interferon (IFN), route of administration, schedule, and dose, were observed in most patients *(3)*. Since then, the flu-like syndrome (FLS) emerged as a common, if not universal, complication of therapeutic cytokines *(4)*.

From: *Methods in Pharmacology and Toxicology: Cytokines in Human Health:*
Immunotoxicology, Pathology, and Therapeutic Applications
Edited by: R. V. House and J. Descotes © Humana Press Inc., Totowa, NJ

2. CLINICAL PRESENTATION

FLS typically consists of fever, chills, fatigue, myalgia, headache, and nausea. Fever is the commonest finding in patients with FLS. It can be of variable magnitude, from a moderate increase in body temperature (38–39°C) to marked hyperpyrexia exceeding 40°C.

Chills are commonly observed in patients with FLS. They can be unrelated to an increase in body temperature, thus suggesting a central mechanism possibly involving cytokines *(5)*. Myalgias are associated with FLS in as many as one half of patients. In rats, interleukin (IL)-1 was shown to cause muscle proteolysis together with a dramatic increase in prostaglandin E2 (PGE2), which promotes muscle protein breakdown *(6)*. Headache is also a frequent occurrence, and the role of cytokines has been suspected in this *(7)*. Malaise and fatigue, which are common features of FLS, may also create clinical consequences of the central effects of cytokines. Changes in blood pressure, from mild hypotension to collapse, rarely are described, but when severe, they are often the result of an abrupt release of tumor necrosis factor (TNF)-α and/or IL-1. Bronchospasm has occasionally been reported *(8)* but probably reflects a hypersensitivity reaction whatever the mechanism involved, immune or nonimmune-mediated.

3. FLS AND TREATMENT
WITH THERAPEUTIC CYTOKINES

3.1. Interferons

FLS is experienced in nearly all patients treated with IFN-α *(9)*. It occurs from 2 to 4 h after IFN-α administration whatever the treatment indication, and usually lasts 4 to 8 h *(10,11)*. The severity of FLS in IFN-α-treated patients is clearly dependent on dose. With low doses, the symptoms are usually mild, with fever exceeding 40°C and/or severe myalgias observed in only 10% of patients. Severe symptoms can be seen in as many as 40% to 60% of patients when high doses are used. They can be treatment-limiting and require either dose reduction or discontinuation of treatment in 5% to 15% of patients. However, FLS generally is well tolerated, can be prevented by paracetamol (acetaminophen), and resolves within the 15 first days without reduction or suspension of treatment because tachyphylaxis usually develops after 7 to 10 d. No differences were seemingly noted depending on the type of IFN-α used *(12,13)*. Fever occurred in 60 to 100% of patients treated with IFN-β for chronic hepatitis C *(14)* or multiple sclerosis *(15)*.

Other common symptoms included fatigue (16–74%), malaise (50%), myalgias (21–42%), and nausea and vomiting (20–26%). No marked differences were noticed between the two recombinant forms IFN-β-1a and IFN-β-1b. The incidence and severity of FLS associated with IFN-β does not seem to be dependent on dose. Thus, the self-administration of six to seven prefilled syringes with IFN-β1a in a suicidal attempt resulted in a modest increase in body temperature *(16)*. Tachyphylaxis usually develops when continuing therapy. Similarly, the most common adverse effects of IFN-γ are fever, chills, dizziness and headache, with decreasing intensity over time *(17)*.

3.2. Interleukins

Fever and flu-like symptoms are universal in patients treated with IL-1α and IL-1β *(18)*. They can be severe and associated with dose-dependent hypotension. FLS is extremely frequent in patients treated with IL-2 whatever the dose, the route, or schedule of administration *(19,20)*, but usually is mild, except when high doses are used *(21)*. The subcutaneous route has been suggested to induce less frequent and severe flu-like symptoms than the intravenous route *(22)*. Although the clinical experience with rIL-3 is still limited, the available data indicate that dose-dependent FLS is the commonest adverse effect usually receding when continuing treatment because of tachyphylaxis *(23)*. A mild flu-like syndrome was seen almost in all phase-I trial patients treated with rhuIL-4. It was more frequent and with increasing severity with higher dose levels and resolved completely on discontinuing therapy *(24)*.

Nearly all patients treated with rIL-6 developed dose-limiting fever, chills, nausea, vomiting and fatigue *(25,26)*. In contrast, flu-like symptoms were usually mild and observed only at high doses in patients treated with rIL-10 *(27)*. Similarly, only mild-to-moderate flu-like symptoms were noted in patients treated with rIl-11 *(28)*. rIL-12 was shown to induce FLS in nearly all treated patients. Symptoms were more severe at high dose and could lead to treatment discontinuation *(29)*.

3.3. TNF-α

Marked FLS is a universal and often dose-limiting complication of treatment with rTNF-α *(30)*. Typical flu-like symptoms are often associated with marked hypotension, general malaise, rigor, and watery diarrhea. rHuTNF when locally applied to 26 patients with diverse advanced tumors and malignant pleural effusions resulted in flu-like symptoms in 41% of the patients *(31)*.

3.4. Growth Factors

As with most cytokines, FLS is commonly observed in patients treated with hemopoietic growth factors, such as granulocyte-macrophage colony-stimulating factor (GM-CSF) or granulocyte colony-stimulating factor (G-CSF). Flu-like symptoms, however, are more frequent with G-CSF than GM-CSF and usually are mild with macrophage colony-stimulating factor *(32)*. It is noteworthy that flu-like symptoms can also be observed following administration of erythropoietin *(33)*.

4. FLS AND OTHER IMMUNOTHERAPEUTICS

FLS is a complication of treatment with many immunotherapeutic agents, including monoclonal antibodies and various immunomodulating drugs.

4.1. Monoclonal Antibodies

Marked FLS was first described with muromonab, a murine anti-CD3 monoclonal antibody (MAb *[34]*). In approximately 50% of patients, the first doses caused hyperpyrexia, chills, tremor, nausea, vomiting and diarrhea, joint pains, and hypotension possibly leading to cardiac ischemia. FLS were associated with a sharp increase in TNF-α and IFN-γ concentrations, suggesting the involvement of endogenous cytokine release. Reduction in the dose and speed of administration and pretreatment with acetaminophen or indomethacin decreased the incidence and severity of these complications. Most adverse effects associated with MAbs are the result of antigen-antibody interactions on specific cells and tissues, and patients commonly experience FLS *(35)* with the first infusion of the anti-CD20 MAb rituximab *(36)*, the humanized anti-Her-2 MAb trastuzumab *(37)*, or the antiTNF-α MAb infliximab *(38)*. Depending on the MAb being used, flu-like symptoms are mild to moderate, even though frequent, as with infliximab, or severe and associated with cardiac toxicity as with trastuzumab. Severe, dose-limiting infusion reactions including have been observed with the use of antibody-targeted immunotoxins *(39)*.

4.2. Immunomodulating Drugs

Clinical manifestations of FLS have long been reported in patients treated with a variety of immunomodulating or immunostimulatory drugs *(2)*. Initially, these biological response modifiers as they used to be called were primarily tested as tentative treatment of human cancer and their mechanism of action was largely speculative. Flu-like symptoms of variable severity were consistently described in human subjects treated with a wide variety of compounds including *Corynebacterium parvum (40)*, mismatched double-stranded RNA *(41)*, and poly (I:C) *(42)*. FLS is also a well-established, although rare and usually mild to moderate complication of treatment with levamisole *(43)* or the interferon-inducer imiquimod *(44)*.

5. DIAGNOSIS

The diagnosis of FLS associated with therapeutic cytokines is usually obvious because flu-like symptoms generally develop within 2 to 4 h after administration. However, the time course of events may be less straightforward because FLS can indeed develop after several days or weeks of treatment. Therefore, other causes may have to be ruled out.

It is important to bear in mind that flu-like symptoms can develop in the context of overlooked infectious diseases unrelated to the influenza virus, such as Q fever *(45)* or psittacosis *(46)*. Flu-like symptoms are also typical clinical manifestations of the sick building syndrome in which latent fungal infections have been suspected to play a critical role *(47)*. The chronic fatigue syndrome, the existence of which is still heavily debated, consists of a variety of nonspecific symptoms, including flu-like symptoms *(48)*. Interestingly, the role of IL-6 has been suggested.

Flu-like symptoms also have been reported in patients treated with a variety of pharmaceutical drugs, although the mechanism involved is usually unknown. Thus, FLS is a reported adverse effect of the lipid-lowering statins *(49)*, the anticancer drug gemcitabine *(50)*, and the anti-bone resorption drugs biphosphonates *(51)*. Because biphosphonates have been suggested to inhibit IL-1, IL-6, and TNF-α release both in vitro and in vivo, a typical cytokine releasing mechanism seems unlikely *(52)*. Flu-like symptoms have been reported with several antidepressant drugs, such as zimeldine *(53)* and fluoxetine *(54)*. Although symptoms generally developed after days or weeks of continuing treatment, they were at least once described after suicidal overdose *(55)*. Another intriguing finding is the development of flu-like symptoms in patients who abruptly withdrawn from long-term antidepressant treatment *(56–58)*. Although similar findings have been also observed following abrupt opiate withdrawal *(59)*, an antiviral activity of antidepressant drugs has been suggested as a possible mechanism *(60)*.

Flu-like symptoms may have toxic causes. In the occupational setting, inhalation of fumes from zinc oxide is the most common cause of metal fume fever presenting as fatigue, chills, fever, myalgias, cough, dyspnea, leukocytosis, thirst, metallic taste, and salivation. Purified zinc oxide fume inhalation was shown to cause an exposure-dependent increase in proinflammatory cytokines and PMNLs in the lung supporting a role for cytokines in metal fume fever *(61)*. Carbon monoxide poisonings are extremely frequent and often overlooked. Flu-like symptoms have repeatedly been reported as the sole clinical manifestation of an overlooked intoxication *(62–64)*.

6. MANAGEMENT

The management of patients with FLS primarily consists of symptomatic measures to relieve fever, muscle pains, and gastrointestinal disturbances. Dose reduction or treatment discontinuation is sometimes required depending on the severity of clinical manifestations. No specific treatment is available.

In an attempt to reduce the incidence and severity of FLS and thus avoid dose limitation or treatment discontinuation for enhance efficacy, various preventive measures have been investigated. In general, pretreatment with minor antipyretic drugs, such as acetaminophen or ibuprofen, are considered to be an effective preventive measure. The following studies in multiple sclerosis patients support this view. Flu-like symptoms at the initiation of IFN-β-1b therapy were only minimal in patients with relapsing-remitting multiple sclerosis who received low-dose prednisone plus paracetamol as compared to paracetamol only during the first 15 days of treatment. At 3 mo, however, both groups showed a similar frequency of flu-like symptoms *(65)*. Eighty-four patients with relapsing-remitting multiple sclerosis treated with intramuscular IFN-β-1a were randomized to compare the efficacy of paracetamol, ibuprofen, and prednisone in the treatment of FLS in a multicenter, randomized, double-blind, controlled trial. 28 patients were given 500 mg of paracetamol or 400 mg of ibuprofen before and 6 and 12 hours after each IFN-β-1a injection, or 60 mg of prednisone daily for 1 wk. No prophylactic treatment for FLS assessed on the severity of fever, myalgia, chills, headache, and asthenia for 27 d was found to be superior to another. However, ibuprofen conferred better control of symptoms immediately following IFN-β-1a injection *(66)*. The percentage of patients with FLS was comparable with paracetamol versus ibuprofen administered 48 h within IFN-β-1a injection to patients in the first weeks of therapy for relapsing-remitting multiple sclerosis *(67)*.

7. MECHANISM

Because fever occurs whatever the nature of the eliciting illness, the role of an endogenous substance has long been suspected *(68)*. The "endogen pyrogen," as it was initially called, released by white blood cells was later shown to be IL-1. Other endogenous proteins were subsequently identified, and nowadays there is evidence that IL-1β, TNF-α, IFN-β, IFN-γ, IL-6, IL-8, and macrophage inflammatory protein-1 act independently as endogenous pyrogens *(69)*. Because of the blood–brain barrier, endogenous pyrogens were thought to be unable to act on the brain in sufficient quantity to induce fever. However, peripheral cytokines can pass the blood–brain barrier by active and saturable specific transport systems *(70)*. Another possibility is that areas of the brain, such as the circumventricular organs, lack a tight blood–brain barrier. Circulating cytokines including those with endogenous pyrogen activity can indeed enter the circumventricular organs through fenestrated capillaries where they induce the production of prostaglandins, such as PGE2, by neurons, microglia and astrocytes. Finally, receptors on the surface of endothelial cells in brain vasculature are potential targets for circulating cytokines *(71)*. The signal for the production of fever is then relayed to cell groups in the hypothalamus and brain stem that coordinate the febrile response. A large number of neurons located in the rostral hypothalamus are thermosensitive. Not only cytokines, but also other endogenous mediators that are involved in fever are produced in the brain. PGE2 is traditionally regarded as a centrally acting mediator of fever since the the seminal findings of Milton and Wendlandt *(72)*. The formation of PGE2 depends on the activity of cyclooxygenase (COX). The induction of COX-2 in response to peripheral injection of a fever-inducing dose of lipopolysaccharide (LPS) was demonstrated in brain endothelial cells, perivascular microglia and meningeal macrophages *(73)*. The central injection of prostaglandins evokes fever *(74)*. LPS appearing in the blood induces circulating cytokines. Then, circulating cytokines and LPS induce cytokine release as well as COX-2 within the brain. Centrally produced cytokines are further triggers for COX-2 induction and thereby for prolonged formation of PGE2 within the preoptic area and the hypothalamus.

Evidence also exists for the recruitment of final brain-derived pyrogenic mediators of fever that are produced and released in response to stimulation of afferent fibers of the vagus nerve. LPS-induced fever is prevented in vagotomized rats *(75)*. Prostaglandins seem to be critical final mediators in the vagally activated fever pathway because bradykinin-induced fever is blocked by indomethacin, a COX-1 and COX-2 inhibitor.

Because of the short latency in fever induction, the role of central mediators has recently been debated and the role of the complement anaphylatoxins suggested *(76)*. Indeed, the intravenous administration of LPS triggers within 2 min the complement cascade via the alternative pathway, resulting in the production in blood of C4a, C3a and C5a. Production of PGE2 could ensue via the hydrolysis of membrane-associated phosphoinositide by phosphoinositide -specific phospholipase C, which is activated by the complement cascade. The anaphylatoxin C5a has been identified as the critical mediator. Thus,

PGE$_2$ released by complement activation could be the immediate factor that stimulates vagal afferents that convey the signals to the preoptic anterior hypothalamus. Interestingly, depletion in complement by cobra venom factor reduced the rise in temperature following intravenous LPS and the usual fever-associated increase in PGE2 in the preoptic anterior hypothalamus *(77)*.

A peripheral instead of a central mechanism in FLS is an attractive hypothesis because it can conciliate findings with therapeutic cytokines, immunodulating drugs as well as Mabs. Indeed, immunomodulating drugs are generally thought to act either via direct cytokine-releasing properties or activation of monocytes/macrophages *(2)*, and complement activation has been demonstrated to play a pivotal role in infusion reactions associated with rituximab *(78)*.

REFERENCES

1. Zhou W, Pool V, Iskander JK, et al. Surveillance for safety after immunization: Vaccine Adverse Event Reporting System (VAERS) - United States, 1991–2001. MMWR Surveill Summ 2003;52:1–24.
2. Descotes J. Adverse consequences of chemical immunomodulation. Clin Res Pract Drug Regul Affairs 1985;3:45–52.
3. Quesada JR, Talpaz M, Rios A, Kurzrock R, Gutterman JU. Clinical toxicity of interferons in cancer patients: a review. J Clin Oncol 1986;4:234–243.
4. Vial T, Descotes J. Immune-mediated side-effects of cytokines in humans. Toxicology 1995;105:31–57.
5. Guieu JD, Hellon RF The chill sensation in fever. Pflugers Arch 1980;384: 103–104.
6. Baracos V, Rodemann HP, Dinarello CA, Goldberg AL. Stimulation of muscle protein degradation and prostaglandin E2 release by leukocytic pyrogen (interleukin-1). A mechanism for the increased degradation of muscle proteins during fever. N Engl J Med 1983;308:553–558.
7. Smith RS The cytokine theory of headache. Med Hypotheses 1992;39:168–174.
8. Gordon MS, Battiato LA, Gonin R, Harrison-Mann BC, Loehrer PJ. A phase II trial of subcutaneously administered recombinant human interleukin-2 in patients with relapsed/refractory thymoma. J. Immunother. Emphasis Tumor Immunol. 1995;18:179–184.
9. Sleijfer S, Bannink M, Van Gool AR, Kruit WH, Stoter G. Side effects of interferon-alpha therapy. Pharm World Sci 2005;27:423–431.
10. Vial T, Descotes J. Clinical toxicity of the interferons. Drug Saf 1994;10:115–159.
11. Pardo M, Marriott E, Moliner MC, Quiroga JA, Carreno V. Risks and benefits of interferon-alpha in the treatment of hepatitis. Drug Saf 1995;13:304–316.
12. Ascione A, De Luca M, Di Costanzo GG, et al. Incidence of side effects during therapy with different types of alpha interferon: a randomised controlled trial

comparing recombinant alpha 2b versus leukocyte interferon in the therapy of naive patients with chronic hepatitis C. Curr Pharm Des 2002;8:977–980.

13. Laguno M, Murillas J, Blanco JL, et al. Peginterferon alfa-2b plus ribavirin compared with interferon alfa-2b plus ribavirin for treatment of HIV/HCV co-infected patients. AIDS 2004;18:F27–36

14. Festi D, Sandri L, Mazzella G, et al. Colecchia A Safety of interferon beta treatment for chronic HCV hepatitis. World J Gastroenterol 2004;10:12–16.

15. Weinstock-Guttman B, Rudick RA. Prescribing recommendations for interferon-beta in multiple sclerosis. CNS Drugs 1997;8:102–112.

16. Falcone NP, Nappo A, Neuteboom B. Interferon beta-1a overdose in a multiple sclerosis patient. Ann Pharmacother 2005;39:1950–1952.

17. Vlachoyiannopoulos PG, Tsifetaki N, Dimitriou I, Galaris D, Papiris SA, Moutsopoulos HM. Safety and efficacy of recombinant gamma interferon in the treatment of systemic sclerosis. Ann Rheum Dis 1996;55:761–768.

18. Veltri S, Smith JW. Interleukin 1 trials in cancer patients: a review of the toxicity, antitumor and hematopoietic effects. Stem Cells 1996;14:164–176.

19. Vial T, Descotes J Clinical toxicity of interleukin-2. Drug Saf. 1992;7:417–433.

20. Atkins MB. Interleukin-2: clinical applications. Semin Oncol 2002;29:S12–17.

21. Kohler PC, Hank JA, Moore KH, Storer B, Bechhofer R, Sondel PM. Phase 1 clinical evaluation of recombinant interleukin-2. Prog Clin Biol Res 1987;244: 161–172.

22. Lopez-Jimenez J, Perez-Oteyza J, Munoz A, et al. Subcutaneous versus intravenous low-dose IL-2 therapy after autologous transplantation: results of a prospective, non-randomized study. Bone Marrow Transplant 1997;19:429–434.

23. De Vries EG, Van Gameren MM, Willemse PH. Recombinant human interleukin 3 in clinical oncology. Stem Cells 1993;11:72–80.

24. Majhail NS, Hussein M, Olencki TE, et al. Phase I trial of continuous infusion recombinant human interleukin-4 in patients with cancer. Invest New Drugs 2004;22:421–426.

25. Weber J, Yang JC, Topalian SL, et al. Phase I trial of subcutaneous interleukin-6 in patients with advanced malignancies. J Clin Oncol 1993;11:499–506.

26. Sosman JA, Aronson FR, Sznol M, et al. Concurrent phase I trials of intravenous interleukin 6 in solid tumor patients: reversible dose-limiting neurological toxicity. Clin Cancer Res 1997;3:39–46.

27. Fedorak RN, Gangl A, Elson CO, et al. Recombinant human interleukin 10 in the treatment of patients with mild to moderately active Crohn's disease. The Interleukin 10 Inflammatory Bowel Disease Cooperative Study Group. Gastroenterology 2000;119:1473–1482.

28. Smith JW. Tolerability and side-effect profile of rhIL-11. Oncology 2000;14, S41–S47.

29. Zeuzem S, Hopf U, Carreno V, et al. A phase I/II study of recombinant human interleukin-12 in patients with chronic hepatitis C. Hepatology 1999;29:1280–1287.

30. Spriggs DR, Sherman ML, Frei E, Kufe DW. Clinical studies with tumour necrosis factor. Ciba Found Symp 1987;131:206–227.
31. Rauthe G, Sistermanns J. Recombinant tumour necrosis factor in the local therapy of malignant pleural effusion. Eur J Cancer 1997;33:226–231.
32. Vial T, Descotes J. Clinical toxicity of cytokines used as haemopoietic growth factors. Drug Saf 1995;13:371–406.
33. Voravud N, Sriuranpong V. Clinical benefits of epoetin alfa (Eprex) 10,000 units subcutaneously thrice weekly in Thai cancer patients with anemia receiving chemotherapy. J Med Assoc Thai 2005;88:607–612.
34. Sgro C Side-effects of a monoclonal antibody, muromonab CD3/orthoclone OKT3: bibliographic review. Toxicology 1995;105:23–29.
35. Dillman RO. Infusion reactions associated with the therapeutic use of monoclonal antibodies in the treatment of malignancy. Cancer Metastasis Rev 1999; 18:465–471.
36. Onrust SV, Lamb HM, Balfour JA. Rituximab. Drugs 1999;58:79–88.
37. McKeage K, Perry CM. Trastuzumab: a review of its use in the treatment of metastatic breast cancer overexpressing HER2. Drugs 2002;62:209–243.
38. Hanauer SB Safety of infliximab in clinical trials. Aliment Pharmacol Ther 1999;13:S16–S22.
39. Thrush GR, Lark LR, Clinchy BC, Vitetta ES. Immunotoxins: an update. Annu Rev Immunol 1996;14:49–71.
40. Fisher B, Rubin H, Sartiano G, Ennis L, Wolmark N. Observations following *Corynebacterium parvum* administration to patients with advanced malignancy. a phase I study. Cancer 1976;38:119–130.
41. Brodsky I, Strayer DR, Krueger LJ, Carter WA. Clinical studies with ampligen (mismatched double-stranded RNA). J. Biol. Response Mod 1985;4:669–675.
42. Giantonio BJ, Hochster H, Blum R, et al. Toxicity and response evaluation of the interferon inducer poly ICLC administered at low dose in advanced renal carcinoma and relapsed or refractory lymphoma: a report of two clinical trials of the Eastern Cooperative Oncology Group. Invest. New Drugs 2001;19:89–92.
43. Scheinfeld N, Rosenberg JD, Weinberg JM Levamisole in dermatology: a review. Am J Clin Dermatol 2004;5:97–104.
44. Gupta AK, Browne M, Bluhm R. Imiquimod: a review. J Cutan Med Surg 2002;6:554–560.
45. Kazar J. *Coxiella burnetii* Infection. Ann NY Acad Sci 2005;1063:105–114.
46. Williams J, Tallis G, Dalton C, et al. Community outbreak of psittacosis in a rural Australian town. Lancet 1998;351:1697–1699.
47. Straus DC, Cooley JD, Wong WC, Jumper CA. Studies on the role of fungi in Sick Building Syndrome. Arch Environ Health 2003;58:475–478.
48. Prins JB, Van der Meer JW, Bleijenberg G. Chronic fatigue syndrome. Lancet 2006;367:346–355.
49. Sinzinger H Flu-like response on statins. Med Sci Monit 2004;8:384–388.
50. Green MR. Gemcitabine safety overview. Semin Oncol 1996;23:S32–S35.

51. Body JJ, Diel I, Bell R. Profiling the safety and tolerability of bisphosphonates. Semin Oncol 2004;31:S73–S78.
52. Santini D, Fratto ME, Vincenzi B, La Cesa A, Dianzani C, Tonini G. Bisphosphonate effects in cancer and inflammatory diseases: in vitro and in vivo modulation of cytokine activities. BioDrugs 2004;18:269–278.
53. Bengtsson BO, Wiholm BE, Myrhed M, Walinder J. Adverse experiences during treatment with zimeldine on special licence in Sweden. Int Clin Psychopharmacol 1994;9:55–61.
54. Wernicke JF The side effect profile and safety of fluoxetine. J Clin Psychiatry 1985;46:59–67.
55. Kim SW, Pentel PR. Flu-like symptoms associated with fluoxetine overdose: a case report. J Toxicol Clin Toxicol 1989;27:389–393.
56. Shrivastava RK, Itil TM. Flu-like illness after discontinuance of imipramine. Biol Psychiatry 1985;20:792–794.
57. Lejoyeux M, Rodiere-Rein C, Ades J. Withdrawal syndrome from antidepressive drugs. Report of 5 cases. Encephale 1992;18:251–255.
58. Rosenbaum JF, Zajecka J. Clinical management of antidepressant discontinuation. J. Clin Psychiatry 1997;58:S37–S40.
59. Farrel M Opiate withdrawal. Addiction 1994;89:1471–1475.
60. Amsterdam JD, Garcia-Espana F, Rybakowski J. Rates of flu-like infection in patients with affective illness. J Affect Disord 1998;47:177–182.
61. Kuschner WG, D'Alessandro A, Wong H, Blanc PD. Early pulmonary cytokine responses to zinc oxide fume inhalation. Environ Res 1997;75:7–11.
62. Grace TW, Platt FW. Subacute carbon monoxide poisoning. Another great imitator. JAMA 1981;246:1698–1700.
63. Dolan MC, Haltom TL, Barrows GH, Short CS, Ferriell KM. Carboxyhemoglobin levels in patients with flu-like symptoms. Ann Emerg Med 1987;16:782–786.
64. Leikin JB, Heckerling P, Maturen A, Perkins JT, Hryhorczuk DO. Carboxyhemoglobin levels in patients with flu-like symptoms. Ann Emerg Med 1988;17:383–384.
65. Rio J, Nos C, Marzo ME, Tintore M, Montalban X. Low-dose steroids reduce flu-like symptoms at the initiation of IFNbeta-1b in relapsing-remitting MS. Neurology 1998;50:1910–1912.
66. Rio J, Nos C, Bonaventura I, et al. Corticosteroids, ibuprofen, and acetaminophen for IFNbeta-1a flu symptoms in MS: a randomized trial. Neurology 2004;63:525–528.
67. Rees J, Haas J, Gabriel K, Fuhlrott A, Fiola M. Both paracetamol and ibuprofen are equally effective in managing flu-like symptoms in relapsing-remitting multiple sclerosis patients during interferon beta-1a (AVONEX) therapy. Mult Scler 2002;8:15–18.
68. Saper CB, Breder CD. The neurologic basis of fever. N Engl J Med 1994;330:1880–1886.

69. Conti B, Tabarean I, Andrei C, Bartfai T. Cytokines and fever. Front Biosci 2004;9:1433–1449.
70. Banks WA, Kastin A. Blood to brain transport of interleukin links the immune and central nervous systems. Life Sci 1991;48: L117–L121.
71. Dinarello CA, Gatti S, Bartfai T. Fever: links with an ancient receptor. Curr Biol, 1999;9:R147–R150.
72. Milton AS, Wendlandt S. Effect on body temperature of prostaglandins of the A, E, and F series on injection into the third ventricle of unanesthetized cats and rabbits J Physiol 1971;218:325–336.
73. Cao C, Matsumura K, Yamagata K, Watanabe Y. Induction by lipopolysaccharide of cyclooxygenase-2 mRNA in rat brain; its possible role in the febrile response. Brain Res 1995;697:187–196.
74. Zeisberger E. From humoral fever to neuroimmunological control of fever. J Thermal Biol 1999;24:287–326.
75. Romanovsky AA, Simons CT, Szekely M, Kulchitsky VA. The vagus nerve in the thermoregulatory response to systemic inflammation. Am J Physiol 1997;273:R407–R413.
76. Blatteis CM, Li S, Li Z, Feleder C, Perlik V. Cytokines, PGE2 and endotoxic fever: a re-assessment. Prostag Other Lipid Mediat 2005;76:1–18.
77. Sehic E, Li S, Ungar AL, Blatteis CM. Complement reduction impairs the febrile response of guinea pigs to endotoxin. Am J Physiol 1998;274: R1594–R1603.
78. Van der Kolk LE, Grillo-Lopez AJ, Baars JW, et al. Complement activation plays a key role in the side-effects of rituximab treatment. Br J Haematol 2001; 115:807–811.

11

Cytokine-Induced Vascular Leak Syndrome

Roxana G. Baluna

SUMMARY

The vascular leak syndrome (VLS) is a major dose-limiting toxicity of cytokine therapy. VLS is characterized by an increase in vascular permeability resulting in tissue edema and, ultimately, multiple organ failure. The most frequent clinical manifestations of cytokine-induced VLS include weight gain, edema, oliguria, hypotension, and dyspnea. Respiratory insufficiency requiring mechanical ventilation and hypotension requiring pressor support have been described as the most severe manifestations of VLS. The pathogenesis of vascular damage is complex and can involve activation of endothelial cells and leukocytes, release of cytokine and inflammatory mediators, and alterations in cell–cell and cell–matrix adhesion with disturbance of vascular integrity. A better understanding of these mechanisms may lead to the development of interventions that will improve the therapeutic efficacy of cytokines. This chapter discusses the clinical manifestation, possible mechanisms, and therapeutic modalities for VLS induced by cytokine therapy.

Key Words: Vascular leak; cytokine; IL-2; endothelial cells; toxicity; cancer therapy.

1. INTRODUCTION

The therapeutic efficacy of interleukin-2 (IL-2) and of other cytokines has been limited by vascular leak syndrome (VLS). Most studies of VLS have focused on IL-2 therapy because it currently is undergoing extensive clinical testing. VLS is characterized by an increase in vascular permeability resulting in tissue edema and, ultimately, multiple organ failure. VLS has been observed in various pathological conditions. VLS occurs after administration of cytokines, including IL-2 *(1–11)*, IL-1 *(12)*, IL-3 *(13)*, IL-4 *(14,15)*, interferon (IFN)-α *(16)*, and IFN-β1b *(17)*. VLS has been reported

From: *Methods in Pharmacology and Toxicology: Cytokines in Human Health:*
Immunotoxicology, Pathology, and Therapeutic Applications
Edited by: R. V. House and J. Descotes © Humana Press Inc., Totowa, NJ

when IL-2 is administered either alone or in combination with lymphokine-activated killer (LAK) cells, tumor infiltrating lymphocytes, other cytokines, monoclonal antibodies, or chemotherapy *(18–24)*. VLS also is induced by granulocyte–macrophage colony stimulating factors *(25,26)*, antiganglioside antibodies *(27)*, cyclosporine *(28)*, cyclophosphamide *(29)*, mitomycin C, FK973, FK317 *(30)*, gemcitabine *(31,32)*, docetaxel *(33)*, monocrotaline pyrrole, cytosine arabinoside *(34)*, and acitretin *(35)*. VLS is a major toxic effect in cancer patients treated with immunotoxins *(10,36–38)*. VLS is also a complication of bone marrow transplantation *(39,40)* and has been observed in patients with T-cell lymphoma *(41)*, non-Hodgkin's lymphoma *(42)*, sepsis, trauma, surgery, burns, pancreatitis, and other diseases *(10,43–55)*.

2. CLINICAL MANIFESTATIONS OF VLS

VLS is characterized by an increase in vascular permeability that leads to increased leakage of fluids, proteins, and electrolytes into interstitial spaces; this leakage results in tissue edema and hypoxia and, ultimately, multiple organ failure (Fig. 1 *[10,18,50,56]*). Interstitial edema in the lungs is manifested by different grades of pulmonary insufficiency. Intravascular hypovolemia caused by fluid leak is responsible for cardiovascular manifestations. Respiratory insufficiency requiring mechanical ventilation and hypotension requiring pressor support have been described as the most severe manifestations of VLS *(6,11)*. Proteinuria and oliguria are early signs of renal failure caused by decreases in renal perfusion *(50)*. Systemic manifestations of vascular leak such as hypoalbuminemia, weight gain, and edema frequently are reported. The relationship between some symptoms and VLS is less clear. For example, anorexia and nausea could be caused by gastric edema, and cerebral edema could be responsible for aphasia *(57)*. Likewise, myalgia and rhabdomyolysis might be related to muscular edema *(58)*. Various other clinical manifestations have been attributed to VLS including low PaO_2, decrease of ventilation/perfusion ratios, decrease in sodium excretion, decrease in creatinine clearance, decrease in plasma oncotic pressure, and fever *(3,50)*. The most frequent clinical manifestations of cytokine-induced VLS include weight gain, edema, oliguria, hypotension, and dyspnea (Fig. 1 *[4,6,10,20,21,25,50,59–66]*). Radiographically, VLS has been associated with pulmonary edema, pleural effusions, and pericardial effusion *(67,68)*. Initial clinical testing identified malaise and weight gain as the dose-limiting toxicities of systemic administration of IL-2 *(1)*, whereas further clinical experience revealed a significant incidence of pulmonary edema in patients treated with high-dose of IL-2 *(69)*.

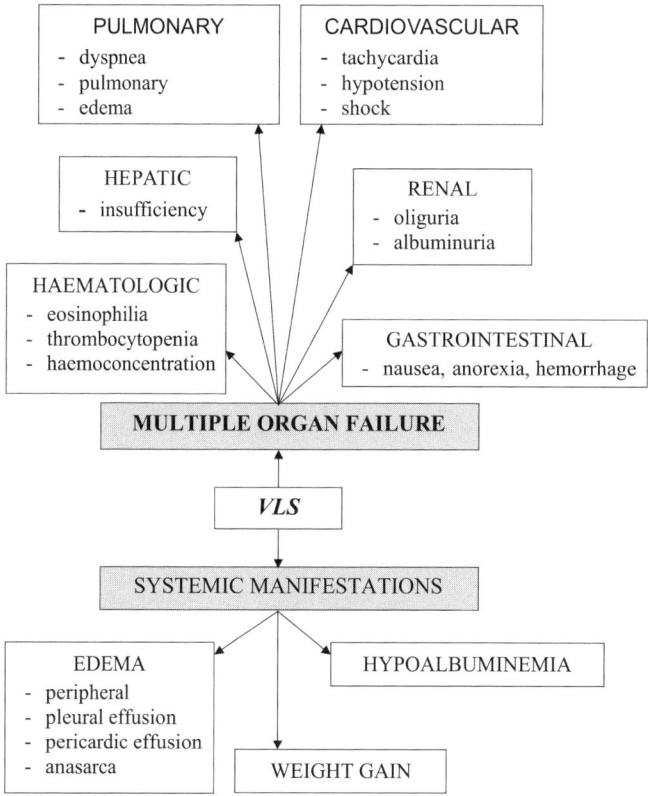

Fig. 1. Clinical manifestation of vascular leak syndrome (VLS): (1) systemic manifestations include: hypoalbuminemia, edema, and weight gain; (2) multiple organ failure include: pulmonary, cardiovascular, renal, gastrointestinal, hepatic, and central nervous system manifestation.

The IL-2-induced pulmonary toxicity is resolved within a few days upon treatment discontinuation *(70,71)*. The incidence rate as recorded in the first large cohorts of patients ranged from 10 to 20%, with a considerable number of treated patients requiring intubation *(68,69,72)*. Of interest, newer clinical data show a decline in IL-2-related pulmonary toxicity as the result of an improvement in patient eligibility screening and optimization of therapeutic conditions *(11)*. Kammula et al. reviewed safety data of high-dose bolus recombinant IL-2 administered in 1241 cancer patients during a 12-yr period and found a clear improvement in IL-2 safety profile, with a decrease from 12 to 3% in the intubation frequency *(11,70,73)*. Approx 75% of patients undergoing intravenous recombinant IL-2 (rIL-2) therapy will demonstrate radiological signs of pulmonary edema *(68,74)*. By contrast, only 25% of patients will develop clinical signs and symptoms of pulmonary disease. Radiological signs are usually found on chest radiography 1 to 5 d after the start of cytokine therapy and include bilateral, symmetric interstitial edema with thickened septal lines. Peribronchial cuffing is observed in 75% of cases. To provide improved standards for measuring edema and for the definition of VLS, subcutaneous-thoracic ratio was calculated and was found to be a useful tool. The ratio can measure objectively the edema and the vascular leak *(75)*.

The earliest clinical manifestations of VLS are hypotension and tachycardia, which can be seen 2 h after the first dose of high-dose IL-2 *(71)*. Intravenous fluids are the initial therapy for hypotension. Therapy with high doses of IL-2 induces hemodynamic changes consistent with a high-output and low-resistance state similar to changes noted during the early phase of septic shock. Patients showed a significant decrease in mean arterial and systemic vascular resistance but an increase in heart rate and cardiac index. No significant change was noted in pulmonary capillary wedge pressure. Although blood pressure normalized in 24 h, the systemic vascular resistance remained below baseline levels 6 d after IL-2 administration had been stopped *(76)*. Within the first 8 h of administering IL-2, decreased urine output is frequent and is a consequence of hypotension and decreased intravascular volume. Renal dysfunction during IL-2 has been described as transient and without evidence of intrinsic renal damage *(71,77)*.

VLS usually starts 3 to 4 d after the initiation of cytokine therapy with albumin decreases and weight gain. VLS becomes dose-limiting within 5 to 10 d of high-dose IL-2 therapy. Although most symptoms disappear within 2 wk, some appear late and take longer to resolve. The criteria for defining different grades of VLS have been developed based on observations in patients treated with a ricin toxin A (RTA)-containing immunotoxin (IT), including Grade I, minimal ankle pitting edema; Grade II, ankle-pitting edema and weigh gain of less than 10 lb; Grade III, peripheral edema with weight gain greater than 10 lb or pleural effusion with no pulmonary function deficit; Grade IV, anasarca, pleural effusion or ascites with pulmonary function deficit or pulmonary edema; Grade V, respiratory failure requiring mechanical ventilation or hypotension requiring pressor support *(10,37)*.

3. MECHANISMS OF VLS

The mechanisms underlying VLS during cytokine therapy are only partially understood. The pathogenesis of VLS is complex and can involve direct or indirect damage of vascular endothelium with activation of endothelial cells (ECs) and leukocytes, release of secondary cytokines and inflammatory

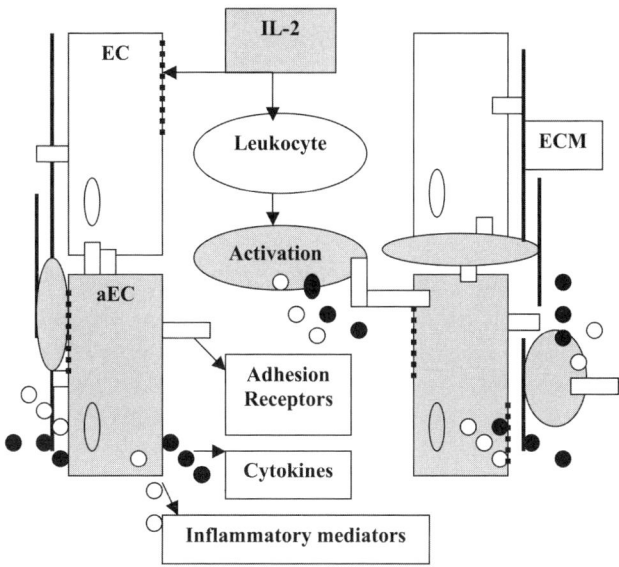

Fig. 2. Mechanisms for endothelial damage in vascular leak syndrome (VLS). Interleukin (IL)-2 may induce toxic effect on the endothelium by interfering directly or indirectly with endothelial monolayer integrity mediated by cell–cell and cell–extracellular matrix interactions. The activation of endothelial cells (EC) or leukocytes by IL-2 results in the expression of adhesion receptors, secretion of secondary cytokines and of inflammatory mediators. Activated leukocytes bind to activated EC (aEC) and damage the endothelium. Secondary cytokines and inflammatory mediators amplify the endothelial damage, which results in increase of vascular permeability and the development of vascular leak.

mediators, and alterations in cell–cell and cell–matrix interaction with increases in vascular permeability, vascular leak, edema, and multiple organ failure *(10,50,56,59 64,79–82)*.

3.1. Vascular Damage Induced by Cytokine Therapy

3.1.1. Direct Effect on ECs

The possible direct effect of IL-2 on cultured EC monolayers has been suggested (Fig. 2). Indeed, it has been shown that IL-2 directly increases the permeability of the vascular endothelium to albumin in vitro. The effects of IL-2 on its target cells are mediated by specific cell surface receptors. This effect is inhibited by anti-IL-2 receptor antibodies *(83)*. Arguing against this, it has been reported by others that IL-2 is not toxic to cultured human umbilical vein endothelial cells (HUVECs) and, in fact, that neither Nude nor irradiated mice treated with IL-2 develop VLS *(78)*. It has been suggested that secondary cytokines may also play a role in the development of VLS through a direct effect on ECs. For example, IFN-γ and IL-1 damage the EC monolayers, in vitro, *(84–86)*. The direct effect of IL-2 on ECs is supported by the identification of a specific sequence in the IL-2 responsible for EC damage *(87)*. It has been shown that the peptides containing the LDL motif in IL-2, specifically damaged HUVECs in vitro *(87)*. Baluna et al. suggested that a $(X)D(Y)$-conserved motif (where x = L,I,G or V and y = V,L,or S) in the IL-2, plant toxin ricin A chain, disintegrins, and other VLS-inducing proteins may be responsible for binding to ECs and initiating VLS. This motif is located in α-helix A of IL-2, centered on Asp-20, and has been reported to damage ECs by caspase-3-mediated apoptosis *(87,88)*. A mutated peptide (p1–30Lys-20) abrogating this motif was tested and was found to retain antitumor activity suggesting the possible production of mutated IL-2 peptides of therapeutic interest *(89)*. Epstein et al. described a fragment in IL-2 molecule, consisting of amino acids 22–58 which retained the vasopermeability activity of IL-2 *(90)*. The vasopermeability activity of IL-2 can be substantially decreased by single point mutations such as Arg38Trp without grossly affecting the immune function of the cytokine *(91,92)*.

3.1.2. Indirect Effect on Vascular Integrity

3.1.2.1. Leukocyte-Mediated Vascular Damage

It has been proposed that VLS, which develops during IL-2 therapy, is a result of the interactions of leukocytes with ECs (Fig. 2 *[93–95]*). The IL-2-activated human lymphocytes exhibit enhanced adhesion to normal vascular ECs and cause their lysis *(81,82,96)*. It has also been suggested that neutrophils play a critical role in VLS by adhering to ECs and inducing damage via reactive oxygen intermediates and proteases *(79,97–100)*. Indeed, the depletion of circulating neutrophils in animals with vascular leak prevents acute pulmonary edema *(101)*. VLS in LAK therapy could be explained by a direct effect of LAKs on ECs. In this regard, LAK cells, but not IL-2 itself, are more cytotoxic to cultured ECs than are stimulated neutrophils *(102,103)*. Direct evidence for the involvement of LAKs, particularly natural-killer (NK) cells, also has been obtained by the demonstration that anti-asialo GM-1 and anti-NK-1.1 antibodies protect against IL-2-induced VLS in mice *(64,104)*. The addition of dexamethasone to IL-2-treated LAKs abolishes their antitumor cytolytic effect but only partially inhibits their ability to induce increased endothelial permeability, suggesting the existence of a noncytolytic mechanism by which activated lymphocytes can increase endothelial permeability *(105)*. Furthermore, IL-2 also upregulates perforin and FasL, which might be responsible for the damage of ECs, leading to extravasation of intravascular fluid. It has been suggested that both NK and polymorphonuclear (PMN) cells play a central role in the late events of IL-2-induced VLS *(106)*. An increase in the numbers of eosinophils in patients treated with IL-2 has been reported and suggests a possible role of extravascular eosinophil degranulation in the pathogenesis of VLS. The eosinophil activation that accompanies the therapy with IL-2 can result in direct toxicity to the lung and a localized VLS *(63,67)*.In addition, T-helper cells may have a role in VLS, which develops in patients with bone marrow transplantation *(107)*. Adherent and extravasating leukocytes produce additional vasoactive agents and proteases, which further aggravate the leakage of macromolecules though the endothelium (Fig. 2 *[108]*). The interactions of activated leukocyte with extracellular matrix (ECM) components have been reported and have been suggested to play roles in VLS. Kaslovsky et al. *(109)* hypothesized that the adhesion of neutrophils to the ECM–protein, fibronectin (Fn) mediates the release of neutrophil products, oxidants and proteases, causing EC injury and an increase in EC permeability. The alteration of the ECM by different enzymes released by activated cells may also be involved in VLS. Thus, a possible role of LAK-derived proteoglycan-degrading enzymes in IL-2-induced VLS has been hypothesized *(96)*. In addition, it has been demonstrated that CD44 knockout mice exhibit marked decrease in IL-2-induced VLS, thereby suggesting a role for CD44 adhesion molecule in VLS *(110)*. The cytotoxic lymphocytes use CD44 in mediating endothelial cell injury. Blocking CD44 in vivo may offer a novel therapeutic approach to prevent endothelial cell injury by cytotoxic lymphocytes *(111)*. A high level of soluble intercellular cell adhesion molecule 1 and vascular cell adhesion molecule 1, has been shown to correlate with endothelial activation in VLS *(112)* whereas the decreases in the level of serum adhesion molecule-Fn was associated with the severity of VLS *(113)*. These data suggest that the leukocytes play a major role in vascular damage induced by cytokine therapy by interfering with vascular integrity and inducing cell-to-cell and cell-matrix interaction disturbance.

3.1.2.2. SECONDARY CYTOKINE-MEDIATED VASCULAR DAMAGE

The role for secondary cytokines in the development of cytokine induced-VLS has been demonstrated. The activation of ECs and leukocytes by IL-2 therapy, results in a cascade of events including the release of secondary cytokines which increase vascular permeability by various mechanisms (Fig. 2 *[93,114]*). For example, plasma tumor necrosis factor (TNF-α) levels increase within 2 h of IL-2 administration *(115)*. Furthermore, a strong correlation between serum levels of TNF-α and weight gain has been reported *(116)*. It has been demonstrated that TNF-α is an important mediator of fluid extravasation in IL-2-induced VLS *(117–119)*. A direct toxic effect on ECs has been demonstrated for TNF-α because treatment of EC monolayers with TNF-α results in increased permeability to proteins *(120)*. The direct cytotoxic effects of TNF-α include G protein-coupled activation of phospholipase, generation of reactive oxygen radicals and damage to nuclear deoxyribonucleic acid by endonucleases *(121)*. It has been shown that TNF-α induces endothelial cell apoptosis *(122)*. In addition, the release of TNF-α may subsequently lead to an activation of the classical pathway of the complement, resulting in vascular leak *(123)*. Interestingly, Puri and Rosenberg found that neither TNF-α nor INF-γ induced vascular leakage in the lungs of mice *(124)*. However, the messenger ribonucleic acid for TNF-α increases in macrophage after treatment with IL-2, and intravenous administration of a soluble TNF-α receptor diminishes IL-2-induced pulmonary VLS, supporting the role for TNF-α in the development of VLS *(116,117)*. In addition, TNF-α binds to ECM proteins as Fn, which has a role in the maintenance of vascular integrity *(125)*, and exogenous Fn prevents the increase in vascular permeability mediated by TNF-α *(120)*. Furthermore, there is evidence to support an active role for cytoskeleton in the TNF-α-mediated vascular barrier dysfunction *(126)*.

Significant changes in the levels of proinflammatory cytokines IL-6 and IL-8 were observed in patients receiving IL-2 *(127)*. In contrast to the anti-inflammatory cytokine IL-10, which did not increase significantly, the serum concentrations of the soluble TNF-α receptors rose continuously and significantly. In parallel, a significant rise in nitrate plasma levels was observed *(127)*. The inflammatory cytokine response may directly alter the cytoskeleton of the endothelium and increase permeability, independently of neutrophils *(128)*. It has been shown that IL-1 induces VLS when administrated experimentally *(84)*. On the other hand, the coadministration of IL-1 decreases IL-2-+INFα-mediated VLS *(129,130)*, and IL-1-receptor antagonists augment IL-2-induced VLS in mice *(131)*. Studies of ECs in vitro have shown that IL-1 can directly antagonize the TNF-α-induced activation of ECs *(132)*. The basis for the IL-1-mediated abrogation of VLS remains unclear. The IL-5 is another cytokine with a possible role in VLS. It has been shown that IL-5 is implicated in eosinophilia, which is associated with vascular damage induced by IL-2 treatment *(133)*. A protective effect against the vascular damage induced by activated neutrophils has been suggested for IL-8 *(134–139)*. It has been shown that IL-8 secreted by activated ECs induces alterations in the molecular conformation and redistribution of actin microfilaments in neutrophils, resulting in the inhibition of adhesion *(140)*. The administration of IL-8 to mice treated with IL-2 resulted in the suppression of IL-2-induced multiple organ edema as well as decreases in antitumor efficacy *(141)*. Furthermore, it has been reported that the administration of IL-10 to mice inhibited IL-2-induced increases in serum TNF-α but was ineffective at reducing IL-2-mediated pleural effusions *(142)*. Standiford et al. *(143)* showed that the neutralization of IL-10 increased lethality in endotoxemia and suggested that IL-10 has a protective effect against the vascular damage. The IL-12 and IL-15 may be implicated in the recruitment of NK and PMN cells, respectively, during IL-2-induced VLS *(106)*. In addition, IL-18 in synergy with IL-2 induces lethal lung injury in mice *(144)*.

Cytokine-induced changes in both the cytoskeleton and ECM of ECs also have been shown. For example, TNF-α and IFN-γ caused human umbilical vein endothelial cells to lose their Fn matrix and to rearrange actin filaments *(145)*. TNF-α can change the interaction of lung EC monolayers with their ECM in association with an increase in endothelial monolayer permeability *(146)*. TNF-α induces cytoskeleton disassembly leading to changes in cell shape, the formation of gaps between cells, and increases in endothelial permeability *(147)*. Modulation of the expression of vitronectin receptors on ECs by TNF-α and INF-γ has been described, and it has been suggested that cytokines can modify the interaction of ECs with the ECM by selectively altering the expression of specific cell surface integrins *(10)*. In conclusion, various secondary cytokines may have different effects on vascular integrity and the development of VLS during cytokine therapy, and may be useful targets for therapeutic manipulation to decrease VLS.

3.1.2.3. INFLAMMATORY MEDIATOR-MEDIATED VASCULAR DAMAGE

Immunotherapy with IL-2 promotes a proinflammatory state (Fig. 2). Inflammatory mediators that may contribute to vascular injury include complement activation products, platelet-activating factor, endothelin, thromboxane A, prostaglandins, leukotrienes, vascular permeability factor, mast cell degranulation products, neutrophil-derived matrix metalloproteinases, elastase, and oxygen radicals *(148–152)*. It has been shown that C3a and C4a complement levels increase during IL-2 therapy and that this increase correlates with symptoms of VLS such as weight gain and hypoalbuminemia *(153)*. In addition, it has been shown that the activated state of the complement system is accompanied by the reduced activity of C1 inhibitors *(154,155)* and that the administration of C1 inhibitor reduces vascular toxicity in patients treated with IL-2 *(156)*. The role of histamine, serotonin, and bradykinin in VLS is controversial *(157)*; however, it has been shown that mast cell degranulation before IL-2 therapy prevents protein leakage in animals with VLS , and it has been suggested that IL-2 leads to complement activation, which induces mast cell degranulation, resulting in the release of vasoactive mediators and an increase in microvascular permeability *(158)*. The formation of gaps between adjacent ECs in the vasculature after exposure to histamine has been described, and histamine induced macromolecular leakage is associated with changes in the EC-actin cytoskeleton *(159–162)*. In addition, it has been found that bradykinin, serotonin, and C5a and C3a complement increase vascular permeability although these agents fail to induce gap formation in EC monolayers in vitro *(93)*. Increases in plasma level of thromboxane have been reported after the administration of IL-2 *(151,163)*. The activation of neutrophils by thromboxane and/or the direct effect of thromboxane on the EC cytoskeletal stress fibers have been proposed as possible mechanisms for the vascular damage *(101,151)*. The disassembly of microfilaments in ECs also has been suggested as a mechanism for leukotrienes-induced increases in vascular permeability *(164)*. Klausner et al. *(152)* have shown that IL-2 therapy leads to increases in LTB4 and have suggested that LTB4 mediates IL-2-induced lung injury. A possible role in VLS induced by IL-2 therapy has been associated with the activation of coagulation. For example, it has been reported that IL-2 activates coagulation and fibrinolysis and suggested that TNF-α might be a factor that mediates these effects *(165)*. In addition, it has been shown by Hack et al. *(166)* that factor XII and prekallikreen decrease in patients treated with IL-2 and that the decrease correlates with both weight gain and decreases in albumin. The contact system may be involved in the increase in vascular permeability either directly by effects on vessels or indirectly by effects on neutrophils. A significant increase of vasoconstrictor peptide endothelin-1 was observed in patients treated with IL-2, indicating activation of endothelial cells. The simultaneous increase of tissue-plasminogen activator and plasminogen activator inhibitor type-1 also was described *(172)*. The activation of ECs by cytokines results in the release of large amounts of nitric oxide (NO *[167]*). Infiltrating phagocytic neutrophils and monocytes cells are also a source of NO *(99)*. NO may induce oxidative injuries to ECs, increase the expression of adhesion receptors, which enhance the adhesion of neutrophils and endothelial damage and/or may mediate smooth muscle relaxation, and increase vascular permeability *(160,168)*. The NO synthesis inhibitors prevent the development of hypotension, suggesting a role for NO in the development of IL-2-induced hypothension *(169)*. The inhibition of NO synthesis by different agents has been reported to decrease IL-2-induced vascular toxicity in mice *(170)*. Others reported no decreases of vascular leak in mice by administration of NO inhibitors *(171)*. In conclusion, there is evidence that inflammatory media-

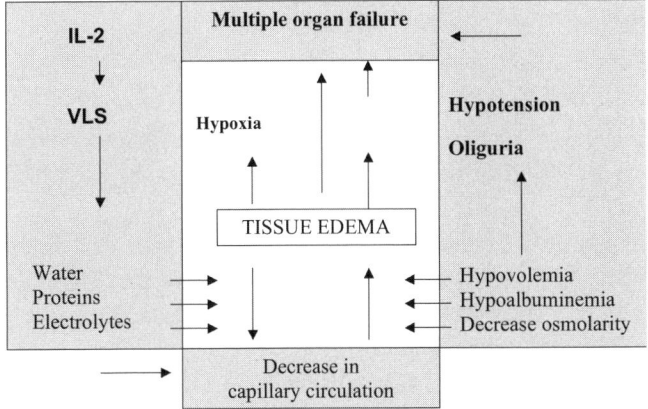

Fig. 3. Mechanisms for multiple organ failure in vascular leak syndrome (VLS). Endothelial damage results in increased vascular permeability, and extravasation of water, proteins, and electrolytes, manifested by hypovolemia, hypoalbuminemia, decreases in plasma osmolarity, hypotension, and oliguria. Tissue edema caused by vascular leak results in weight gain, peripheral edema, anasarca, or pulmonary edema. Tissue edema is accompanied by elevated tissue pressure, which induces decreases in the capillary circulation, hypoxia, and multiple organ failure. IL, interleukin.

tors released during cytokine therapy may contribute to VLS by various mechanisms, and anti-inflammatory therapy may decrease or prevent VLS.

3.2. Multiple Organ Failure Induced by Cytokine Therapy

The direct or indirect toxic effects of IL-2 on vascular endothelium induce a cascade of events terminated with multiple organ failure. The vascular damage results in increased vascular permeability, extravasation of water, proteins, and electrolytes that is manifested by decreases in both serum albumin levels and plasma osmolarity and by hypovolemia and hypotension. Tissue edema caused by vascular leak results in weight gain, peripheral edema, anasarca, or pulmonary edema. On the other hand, tissue edema is accompanied by increased tissue pressure, which induces increases in the venous pressure and decreases in the capillary circulation, resulting in decreased microcirculatory perfusion and hypoxia. Numerous other factors contribute to reduce capillary perfusion including microvascular plugging by leukocytes, hypovolemia, hypotension, and arteriole vasoconstriction. Hypoxia induces multiple organ failure including lung, heart, vessels, liver, gastrointestinal tract, coagulation system, and central nervous system (Fig. 3 *[10,18,50,56]*).

4. MANAGEMENT OF VLS

For IL-2 to become more effective in the treatment of cancer, the inhibition of VLS is highly desirable. However, the optimal methods for treating VLS are not known and most patients receive treatment to decrease symptoms. For example, diuretics are administered to reduce excessive weight gain, edema, and oliguria. Intermittent exogenous oxygen or intubation with mechanical ventilation support is administered in cases of pulmonary insufficiency. Fluid resuscitation and vassopressors are used to maintain renal perfusion and blood pressure. The ideal fluid for resuscitation is still unclear *(173–177)*. Blood pressure is supported with dopamine or phenylephrine *(76)*. Oliguria is treated with fluid boluses and dopamine at renal perfusion doses *(77)*. The plasmapheresis is used in the management of VLS in sepsis *(178)*.

The value of anti-inflammatory, anticoagulant and vasoactive agents in VLS is extensively studied in IL-2 therapy, IT therapy, sepsis, trauma, and surgery. Various agents are tested in vitro, in vivo, or in patients, including corticosteroids, nonsteroidal anti-inflammatory drugs, 5-lipoxygenase inhibitors, leukotriene antagonists, N-acetylcysteine, procystein, pentoxifilline, lisofylline, ketoconazole, prostaglandin E1, IL-10, IL-8, antiadhesion molecules, matrix metalloproteinase and elastase inhibitors, histamine receptor blockers, superoxide dismutase, catalase, dimethylthiourea, dimethyl sulfoxide, heparin, hirudin, antithrombin, almitrine, prostacyclin, prostacyclin, dobutamine, dopexamine, endotheline blockers, sodium nitroprusside, cGMP phosphodiesterase inhibitor, and so on *(148)*.

Several investigators have attempted to develop models of IL-2-mediated VLS in animals to identify potential inhibitors that reduce toxicity while preserving antitumor efficacy *(130)*. It has been suggested that TNF-α is the primary mediator of IL-2-induced VLS. Therefore, treatment strategies aimed at inhibiting the production or the effect of TNF-α may prevent or ameliorate VLS. In support of this, passive immunization against TNF inhibited the IL-2-mediated vascular toxicity (macromolecular leakage and hypotension) but decreased the IL-2-mediated antitumor effect in rats *(179)*. Pentoxifylline, an inhibitor of TNF-α production, has been shown to inhibit IL-2-induced multiple organ edema in mice and its use has been proposed for VLS therapy *(5,97,119)*. Another effect of pentoxifylline is the inhibition of integrin-mediated adherence of IL-2 activated leukocytes *(180)*. The administration of pentoxifylline and ciprofloxan to cancer patients treated with IL-2 and LAKs has a protective effect on VLS without apparent loss of therapeutic efficacy *(181)*. Kemeny et al. *(182)* demonstrated a protective effect by CNI-1493 treatment, an inhibitor of macrophage activation, including the synthesis of TNF-α and other cytokines in the animal. In addition, taurine has been shown to reduce the IL-2-induced acute lung injury. These data suggest that taurine prevents IL-2-induced tissue injury in part by decreasing neutrophil-endothelial interactions *(183)*. Recent research has evaluated the immunomodulatory properties of thalidomide. Thalidomide appears to inhibit TNF-α production in mononuclear cells and may therefore explain the clinical benefit observed in a patient with systemic capillary leak syndrome *(184)*. It has been suggested that the adrenergic agonists may play a role in the production of inflammatory cytokines. In animal model, for instance, dobutamine and dopexamine ameliorated lung injury by intratracheal endotoxin installation and by decreasing proinflammatory cytokine release and neutrophils entrapped in the injured lungs *(185)*. In addition, anti-inflammatory and immunosuppressive therapies have been used to reduce the toxic effects of both IL-2. For example, dexamethasone and cyclophosphamide are effective in inhibiting fluid accumulation in the lungs of mice treated with IL-2, whereas cyclosporin A and azathioprine are not *(181,186)*. The decrease of IL-2-induced VLS has been achieved in mice using oral methotrexate *(187)*. Corticosteroid therapy can suppress some side effects of exogenous IL-2, including fever, chills, confusion, and dyspnea without consistently influencing weight gain *(186)*. Unfortunately, an immunosuppressive regimen may affect the immune status of cancer patients, with unpredictable consequences for long-term prognosis *(174,188)*. Nonsteroidal anti-inflammatory drugs also were effective in limiting VLS in IL-2-treated animals *(186)*. Other proposed therapies for VLS include the administration of C1-inhibitor to decrease complement activation *(189,190)*, therapy with antibiotics for associated infection to decrease cytokine production, and anti-adhesion therapy to prevent the binding of activated leukocytes to the ECs *(39,191,192)*. Thus, it has been suggested that the molecular targeting of CD44 may serve as a useful tool to selectively alter the LAK activity and to prevent EC injury induced by IL-2 *(110)*.

Changes in therapeutic regimens can decrease the toxic effect of IL-2. For example, subcutaneous administration of both IL-2 and INF-α does not lead to VLS, although VLS is dose-limiting when intravenous regimens are used *(193)*. The feasibility and the safety of long-term administration of subcutaneous rIL-2 at conventional doses of 4.5 million IU/d, three times weekly, have been well established *(194)* whereas novel locoregional administration strategies, such as the inhalation of nebulized rIL-2, are being investigated with the aim of improving its therapeutic index *(195)*. The more favorable toxicity profile of subcutaneous, compared with intravenous, bolus administration of rIL-2, is possibly attributed to a lower systemic absorption, and a better pharmacokinetic profile associated with this route *(196)*. High-dose IL-2 is associated with significant morbidity; however, the incidence and severity of toxicities have decreased as clinicians have gained experience with this agent and implemented toxicity prevention and management strategies *(197)*. Practical guidelines for the safe administration of high-dose IL-2 have been recommended *(71)*. In addition, it has been developed a novel use of targeted IL-2, which takes advantage of its vasopermeability activity to induce vascular leakage within the tumor vasculature. Toward this end, it has been demonstrated that pretreatment with antibody/IL-2 chemical conjugates or fusion proteins enhances specific tumor uptake of therapeutic molecules, including radiolabeled monoclonal antibodies and chemotherapeutic drugs, without affecting normal tissue uptake *(91,198–200)*.

At present, there is no optimal treatment to reduce vascular toxicity while preserving antitumor activity. The mechanisms underlying VLS are poorly understood. There are evidences that an inflammatory process amplifies the initial pathologic event. In this regard, prophylactic anti-inflammatory therapy might be useful in decreasing vascular toxicity. A better understanding of the mechanisms underlying VLS induced by cytokine therapy is mandatory in order to find modalities for prevention and/or decrease the vascular toxicity. The identification of a structural motif in the IL-2 molecule responsible for initiation of VLS, suggests that deletions or mutations in this sequence or the use of blocking peptides may increase the therapeutic index of IL-2.

REFERENCES

1. Lotze MT, Raynor AA, Ettinghausen SE, Vetto JT, Seipp CA, Rosenberg SA. Clinical effects and toxicity of interleukin-2 in patients with cancer. Cancer 1986;58: 2764–2772.
2. Rosenberg SA, Lotze MT, Muul, LM, et al. A progress report on the treatment of 157 patients with advanced cancer using lymphokine-activated killer cells and interleukin-2 or high-dose interleukin-2 alone. N Engl J Med 1987;316: 889–897.
3. Glauser FL, DeBlois G, Bechard D, Fowler AA, Merchant R, Fairman RP. Cardiopulmonary toxicity of adoptive immunotherapy. Am J Med Sci 1988; 296:406–412.
4. Chang AE, Rosenberg SA. Overview of interleukin-2 as an immunotherapeutic agent. Semin Surg Oncol 1989;5:385–390.
5. West WH. Clinical application of continuous infusion of recombinant interleukin-2. Eur J Cancer Clin Oncol 1989;25:S11–S15.
6. Vial T, Descotes J. Clinical toxicity of interleukin-2. Drug Safety 1992;7:417–433.
7. Tartour E, Mathiot C, Fridman WH. Current status of interleukin-2 therapy in cancer. Biomed Pharmacother 1992;46:473–484.
8. Vial T, Descotes J. Clinical toxicity of cytokines used as haemopoietic growth factors. Drug Safety 1995:13:371–406.

9. Vial T, Descotes J, Immune-mediated side-effects of cytokines in humans. Toxicology 1995;105:31-57.

10. Baluna R,Viteta ES. Vascular leak syndrome: a side effect of immunotherapy. Immunopharmacology 1997;37:117–132.

11. Rosenberg SA. Interleukin-2 and the development of immunotherapy for the treatment of patients with cancer. Cancer J Sci Am 2000;6(Suppl 1):S2–S7.

12. Worth LL, Jaffe N, Benjamin RS, et al. Phase II study of recombinant interleukin1alpha and etoposide in patients with relapsed osteosarcoma. Clin Cancer Res 1997;3:1721–1729.

13. Hurwitz N, Probst A, Zufferey G, et al. Fatal vascular leak syndrome with extensive hemorrhage, peripheral neuropathy and reactive erythrophagocytosis: an unusual complication of recombinant IL-3 therapy. Leuk Lymphoma 1996;20;337–340.

14. Sosman JA, Fisher SG, Kefer C, Fisher RI, Ellis TM. A phase I trial of continuous infusion interleukin-4 (IL-4) alone and following interleukin-2 (IL-2) in cancer patients. Ann Oncol 1994;5:447–452.

15. Atkins MB, Vachino G, Tilg HJ, et al. Phase I evaluation of thrice-daily intravenous bolus interleukin-4 in patients with refractory malignancy. J Clin Oncol 1992;10:1802–1809.

16. Yamamoto K, Mizuno M, Tsuji T, Amano T. Capillary leak syndrome after interferon treatment for chronic hepatitis C. Arch Intern Med 2002;162:481–482.

17. Schmidt S, Hertfelder HJ, von Spiegel T. Lethal capillary leak syndrome after a single administration of interferon beta-1b. Neurology 1999;53:220–222.

18. Sculier JP, Bron D, Verboven N, Klastersky J. Multiple organ failure during interleukin-2 administration and LAK cells infusion. Intensive Care Med 1988;14:666–667.

19. Margolin KA, Rayner AA, Hawkins MJ, et al. Interleukin-2 and lymphokine-activated killer cell therapy of solid tumors: Analysis of toxicity and management guidelines. J Clin Oncol 1989;7:486–498.

20. Dutcher JP, Gaynor ER, Boldt DH, et al. A phase II study of high-dose continuous infusion interleukin-2 with lymphokine-activated killer cells in patients with metastatic melanoma. J Clin Oncol 1991;9:641–648.

21. PhilipT, Mercatello A, Negrier S, et al. Interleukin-2 with and without LAK cells in metastatic renal cell carcinoma: the Lyon first-year experience in 20 patients. Cancer Treat Rev 1989;16:91–104.

22. Lindenman A, Hoffken K, Schmidt RE, et al. A phase-II study of low-dose cyclophosphamide and recombinant human interleukin-2 in metastatic renal cell carcinoma and malignant melanoma. Cancer Immunol Immunother 1989;28:275–281.

23. Wersall JP, Masucci G, Hjelm AL, et al. Low dose cyclophosphamide, alpha-interferon and continuous infusions of interleukin-2 in advanced renal cell carcinoma. Med Oncol Tumor Pharmacother 1993;10:103–111.

24. Hamblin TJ, Sadullah S, Williamson P, et al.A phase-III study of recombinant interleukin 2 and 5-fluorouracil chemotherapy in patients with metastatic colorectal cancer. Br J Cancer 1993;68:1186–1189.

25. Emminger W, Emminger-Schmidmeier W, Peters C, et al. Capillary leak syndrome during low dose granulocyte-macrophage colony-stimulating factor (rh GM-CSF) treatment of a patient in a continuous febrile state. Blutalkohol 1990;61:219–221.
26. Rechner I, Brito-Babapulle F, Fielden J. Systemic capillary leak syndrome after granulocyte colony-stimulating factor (G-CSF). Hematol J 2003;4:54–56.
27. Bajorin DF, Chapman PB, Wong G, et al. Phase I evaluation of a combination of monoclonal antibody R24 and interleukin 2 in patients with metastatic melanoma. Cancer Res 1990;50:7490–7495.
28. Mackie FE, Umetsu D, Salvatierra O, Sarwal MM. Pulmonary capillary leak syndrome with intravenous cyclosporin A in pediatric renal transplantation. Pediatr Transplant 2000;4:35–38.
29. Carlson K, Smedmyr B, Hagberg H, Oberg G, Simonsson B. Haemolytic uraemic syndrome and renal dysfunction following BEAC (BCNU, etoposide, ara-C, cyclophosphamide)±TBI and autologous BMT for malignant lymphomas. Bone Marrow Transplant 1993;11:205–208.
30. Beckerbauer L, Tepe JJ, Eastman RA, Mixter PF, Williams RM, Reeves R. Differential effects of FR900482 and FK317 on apoptosis, IL-2 gene expression, and induction of vascular leak syndrome. ChemBiol 2002;9:427–441.
31. De Pas T, Curigliano G, Franceschelli L, Catania C, Spaggiari L, de Braud F. Gemcitabine-induced systemic capillary leak syndrome. Ann Oncol 2001;12: 1651–1652.
32. Pulkkanen K, Kataja V, Johansson R. Systemic capillary leak syndrome resulting from gemcitabine treatment in renal cell carcinoma: a case report. J Chemotherapy 2003;15:287–289.
33. Semb KA. Aamdal S. Oian P. Capillary protein leak syndrome appears to explain fluid retention in cancer patients who receive docetaxel treatment. J Clin Oncol 1998;16:3426–3432.
34. Woods WG, Ramsay NK, Weisdorf DJ, et al. Bone marrow transplantation for acute lymphocytic leukemia utilizing total body irradiation followed by high doses of cytosine arabinoside: Lack of superiority over cyclophosphamide-containing conditioning regimens. Bone Marrow Transplant 1990;6:9–16.
35. Estival JL, Dupin M, Kanitakis J, Combemale P. Capillary leak syndrome induced by acitretin. Br J Dermatol 2004;150:150–152.
36. Vitetta ES, Thorpe PE, Uhr JW. Immunotoxins: Magic bullets or misguided missiles. Immunol. Today 1993;14:252–259.
37. Sausville EA, Headlee D, Stetler-Stevenson M, et al. Continuous infusion of the anti-CD22 immunotoxin, IgG-RFB4-SMPT-dgA in patients with B cell lymphoma: a phase I study. Blood 1995;85:3457–3465.
38. Frankel AE, Tagge EP, Willingham MC. Clinical trials of targeted toxins. Semin Cancer Biol 1995;6:307–317.
39. Funke I, Prummer O, Schrezenmeier H, et al. Capillary leak syndrome associated with elevated IL-2 serum levels after allogeneic bone marrow transplantation. Ann Hematol 1994;68:49–52.

40. Schots R, Kaufman L, Van Riet I, et al. Proinflammatory cytokines and their role in the development of major transplant-related complications in the early phase after allogeneic bone marrow transplantation. Leukemia 2003;17:1150–1156.

41. Dereure O, Portales P, Clot J, Guilhou JJ. Biclonal Sezary syndrome with capillary leak syndrome. Dermatology 1994;188:152–156.

42. Jillella AP, Day DS, Severson K, Kallab AM, Burgess R. Non-Hodgkin's lymphoma presenting as anasarca: probably mediated by tumor necrosis factor alpha (TNF-alpha). Leuk Lymphoma 2000;38:419–422.

43. Margarson MP, Soni NC. Changes in serum albumin concentration and volume expanding effects following a bolus of albumin 20% in septic patients. Br J Anaesth 2004;92:821–826.

44. Teelucksingh S, Padfield PL, Edwards CR. Systemic capillary leak syndrome. Q J Med 1990;75:515–524.

45. Nelson BK. Snake envenomation. Incidence, clinical presentation and management. Med Toxicol Adverse Drug Exp1989;4:17–31.

46. Benedetti TJ, Kates R, Williams V. Hemodynamic observations in severe preeclampsia complicated by pulmonary edema. Am J Obstet Gynecol 1985; 152:330–334.

47. Atkinson JP, Waldmann TA, Stein SF, et al. Systemic capillary leak syndrome and monoclonal IgG gammapathy: studies in a sixth patient and a review of the literature. Medicine 1977;56:225–239.

48. Barnadas MA, Cistero A, Sitjas D, Pascual E, Puig X, de Moragas JM. Systemic capillary leak syndrome. J Am Acad Dermatol 1995;32:364–366.

49. Zhang W, Ewan PW, Lachmann PJ. The paraproteins in systemic capillary leak syndrome. Clin Exp Immunol 1993;93:424–429.

50. Zikria BA. Mechanisms of multiple system organ failure. In: Zikria BA, Oz MO, Carlson RW, eds. Reperfusion Injuries and Clinical Capillary Leak Syndrome. Armonk, NY: Futura Publishing Co.; 1994, pp. 443–489.

51. Abramov Y, Galun E, Granat M, et al. Postpartum systemic capillary leak syndrome: A possible etiology. Acta Obstet Gynecol Scand 1995;74:395–398.

52. McGregor JM, Barker JN, MacDonald DM. Pulmonary capillary leak syndrome complicating generalized pustular psoriasis: possible role of cytokines. Br J Dermatol 1991;125:472–474.

53. Karne S. Gorelick FS. Etiopathogenesis of acute pancreatitis. Surg Clin North Am 1999;79:699–710.

54. Tassani P, Schad H, Winkler C, et al. Capillary leak syndrome after cardiopulmonary bypass in elective, uncomplicated coronary artery bypass grafting operations: does it exist? J Thorac Cardiovasc Surg 2002;123:735–741.

55. Fishel RS, Are C, Barbul A. Vessel injury and capillary leak. Crit Care Med 2003;31(8 Suppl):S502–S511.

56. Ettinghausen SE, Puri RK, Rosenberg SA. Increased vascular permeability in organs mediated by the systemic administration of lymphokine-activated killer cells and recombinant interleukin-2 in mice. J Natl Cancer Inst 1988;80: 177–188.

57. Vitetta ES, Stone M, Amlot P, et al. A phase I immunotoxin trial in patients with B cell lymphoma. Cancer Res 1991;51:4052–4058.

58. Dolberg-Stolik OC, Putterman C, Rubinow A, Rivkind AI, Sprung CL. Idiopathic capillary leak syndrome complicated by massive rhabdomyolysis. Chest 1993;104:123–126.

59. Rosenstein M, Ettinghausen SE, Rosenberg SA. Extravasation of vascular fluid mediated by the systemic administration of recombinant interleukin-2. J Immunol 1986;137:1735–1742.

60. Pichert G, Jost LM, Fierz W, Stahel RA. Clinical and immune modulatory effects of alternative weekly IL-2 and interferon -2a in patients with advanced renal cell carcinoma and melanoma. Br J Cancer 1991;63:287–292.

61. Textor SC, Margolin K, Blayney D, Carlson J, Doroshow J. Renal, volume and hormonal changes during therapeutic administration of recombinant interleukin-2 in man. Am J Med 1987;83:1055–1061.

62. Stahel RA, Sculier JP, Jost LM, et al. Tolerance and effectiveness of recombinant interleukin-2 (r-met Hu IL-2) [ala] 125]) and lymphokine-activated killer cells in patients with metastatic solid tumors. Eur J Cancer Clin Oncol 1989;25:965–972.

63. van Haelst Pisani C, Kovach JS, Kita H, et al. Administration of interleukin-2 (IL-2) results in increased plasma concentrations of IL-5 and eosinophilia in patients with cancer. Blood 1991;78:1538–1544.

64. Gately MK, Anderson TD, Hayes TJ. Role of asialo-GM1-positive lymphoid cells in mediating the toxic effects of recombinant IL-2 in mice. J Immunol 1988;141:189–200.

65. Droder RM, Kyle RA, Greipp PR. Control of systemic capillary leak syndrome with aminophylline and terbutaline. Am J Med. 1992;92:523–526.

66. Vogelzang PJ, Droder RM, Kyle RA, Greipp PR. Chest roentgenographic abnormalities in IL-2 recipients. Incidence and correlation with clinical parameters. Chest 1992;101:746–752.

67. O'Hearn DJ, Leiferman KM, Askin F, Georas SN. Pulmonary infiltrates after cytokine therapy for stem cell transplantation. Massive deposition of eosinophil major basic protein detected by immunohistochemistry. Ame J Respi Crit Care Med 1999;160:1361–1365.

68. Mann H, Ward JH, Samlowski. Vascular leak syndrome associated with interleukin-2: Chest radiographic manifestations. Radiology 1990;176:191–194.

69. Conant EF, Fox KR, Miller WT. Pulmonary edema as a complication of interleukin-2 therapy. Am J Roentgenol 1989;152:749–752.

70. Briasoulis E, Pavlidis N. Noncardiogenic Pulmonary Edema: An Unusual and Serious Complication of Anticancer Therapy. Oncologist 2001;6:153–161.

71. Schwartzentruber DJ. Guidelines for the safe administration of high-dose interleukin-2. J Immunother 2001;24:287–293.

72. Lee RE, Lotze MT, Skibber JM. Cardiorespiratory effects of immunotherapy with interleukin-2. J Clin Oncol 1989;7:7–20.

73. Kammula US, White DE, Rosenberg SA. Trends in the safety of high dose bolus interleukin-2 administration in patients with metastatic cancer. Cancer 1998;83:797–805.

74. Ketai LH, Goodwin JD. A new view of pulmonary edema and acute respiratory distress syndrome: state of the art. J Thorac Imaging 1998;13:147–171.

75. Sonntag J, Grunert U, Stover B, Obladen M. [The clinical relevance of subcutaneous-thoracic ratio in preterm newborns as a possibility for quantification of capillary leak syndrome]. Zeitschrift fur Geburtshilfe und Neonatologie 2003;207:208–212.

76. Gaynor ER, Vitek L, Sticklin L, et al. The hemodynamic effects of treatment with interleukin-2 and lymphokine-activated killer cells. Ann Intern Med 1988; 109:953–958.

77. Guleria AS, Yang JC, Topalian SL. Renal dysfunction associated with the administration of high-dose interleukin-2 in 199 consecutive patients with metastatic melanoma or renal cell carcinoma. J Clin Oncol 1994;12:2714–2722.

78. Bechard DE, Gudas SA, Sholley M, et al. Nonspecific cytotoxicity of recombinant interleukin-2 activated lymphocytes. Am J Med Sci 1989;298:28–33.

79. Edwards MJ, Miller FN, Sims DE, Abney DL, Schuschke DA, Corey TS. Interleukin 2 acutely induces platelet and neutrophil-endothelial adherence and macromolecular leakage. Cancer Res 1992;52:3425–3431.

80. Navarro C, Garcia-Bragado F, Lima J, Fernandez JM. Muscle biopsy findings in systemic capillary leak syndrome. Hum Pathol 1990;21:297–301.

81. Fujita S, Puri RK, Yu ZX, Travis WD, Ferrans VJ. An ultrastructural study of in vivo interactions between lymphocytes and endothelial cells in the pathogenesis of the vascular leak syndrome induced by interleukin-2. Cancer 1991; 68:2169–2174.

82. Queluz TT, Brunda M, Vladutiu AO, Brentjens JR, Andres G. Morphological basis of pulmonary edema in mice with cytokine-induced vascular leak syndrome. Exp Lung Res 1991;17:1095–1108.

83. Downie Ryan US, Hayes BA, Friedman M. Interleukine-2 directly increases albumin permeability of bovine and human vascular endothelium in vitro. Am J Respir Cell Mol Biol 1992;7:58–65.

84. Dinarello CA. The proinflammatory cytokines interleukin-1 and tumor necrosis factor and treatment of the septic shock syndrome. J Infect Dis 1991;163: 1177–1184.

85. Campbell WN, Ding X, Goldblum SE. Interleukin-1α and -β augment pulmonary artery transendothelial albumin flux in vitro. Am J Physiol 1992;263: L128–L136.

86. Maruta MK, Burkart V, Gillis S, Kolb H. IL-1 and IFN- increase vascular permeability. Immunology 1988;64:301–305.

87. Baluna R, Rizo J, Gordon BE, Ghetie V. Vitetta ES. Evidence for a structural motif in toxins and interleukin-2 that may be responsible for binding to endothelial cells and initiating vascular leak syndrome. Proc Natl Acad Sci USA 1999;96:3957–3962.

88. Baluna R, Coleman E, Jones C, Ghetie V, Vitetta ES. The effect of a monoclonal antibody coupled to ricin A chain-derived peptides on endothelial cells in vitro: insights into toxin-mediated vascular damage. Exp Cell Res 2000;258:417–424.

89. Rose TJ, Moreau L, Eckenberg R, Theze J. Structural Analysis and Modeling of a Synthetic Interleukin-2 Mimetic and Its Interleukin-2R{beta}2 Receptor J Biol. Chem 2003;278:22,868–22,876.

90. Epstein AL, Mizokami MM, Li J, Hu P, Khawli LA. Identification of a Protein Fragment of Interleukin 2 Responsible for Vasopermeability. J Natl Cancer Inst 2003;95:741–749.

91. Hu P, Mizokami M, Ruoff G, Khawli LA, Epstein AL, Generation of low-toxicity interleukin-2 fusion proteins devoid of vasopermeability activity. Blood 2003;101:4853–4861.

92. Eckenberg R, Rose T, Moreau JL, et al. The first alpha helix of interleukin (IL)-2 folds as a homotetramer, acts as an agonist of the IL-2 receptor beta chain, and induces lymphokine-activated killer cells. J Exp Med 2000;191:529–540.

93. Cotran RS, Pober JS, Gimbrone Jr MA, et al. Endothelial activation during interleukin 2 immunotherapy. A possible mechanism for the vascular leak syndrome. J Immunol 1988;140:1883–1888.

94. Pober JS. Cytokine-mediated activation of vascular endothelium. Am J Pathol 1988;133:426–433.

95. Lentsch AB, Miller FN, Edwards MJ. Mechanisms of leukocyte-mediated tissue injury induced by interleukin 2. Cancer Immunol Immunother 1999;47: 243–248.

96. Damle NK, Doyle LV. IL-2-activated human killer lymphocytes but not their secreted products mediate increase in albumin flux across cultured endothelial monolayers. Implications for vascular leak syndrome. J Immunol 1989;142: 2660–2669.

97. Edwards MJ, Abney DL, Miller FN. Pentoxifylline inhibits interleukin-2-induced leukocyte-endothelial adherence and reduces systemic toxicity. Surgery 1999;110:199–204

98. Elliott MJ, Finn AH. Interaction between neutrophils and endothelium. Ann Thorac Surg 1993;56:1503–1508.

99. Stevens P, Piazza DE. Interleukin-2 increases the oxidative activity and induces migration of murine polymorphonuclear leukocytes in vivo. Int J Immunopharmacol 1990;12:605–611.

100. Saba TM. Kinetics of plasma fibronectin: Relationship to phagocytic function and lung vascular integrity. In: Mosher DF, ed. Fibronectin. San Diego, CA: Academic Press; 1989, pp. 395–439.

101. Welbourn R, Goldman G, Kobzik L, Valeri CR, Shepro D, Hechtman HB. Involvement of thromboxane and neutrophils in multiple-system organ edema with interleukin-2. Ann Surg 1990;212:728–733.

102. Kotasek D, Vercellotti GM, Ochoa AC, Bach FH, Jacob HS. Lymphokine activated killer (LAK) cell-mediated endothelial injury: A mechanism for

capillary leak syndrome in patients treated with LAK cells and interleukin-2. Trans Assoc Am Phys 1987;100:21–27.

103. Kotasek D, Vercellotti GM, Ochoa AC, Bach FH, White JG, Jacob HS. Mechanism of cultured endothelial injury induced by lymphokine-activated killer cells. Cancer Res 1988;48:5528–5532.

104. Anderson TD, Hayes TJ, Gately MK, Bontempo JM, Stern LL, Truitt GA. Toxicity of human recombinant interleukin-2 in the mouse is mediated by interleukin-activated lymphocytes. Separation of efficacy and toxicity by selective lymphocyte subset depletion. Lab Invest 1988;59:598–612.

105. Bechard DE, Fairman RP, Hinshaw DB, Fowler AA, Glauser FL. In vivo interleukin-2 activated sheep lung lymph lymphocytes increase ovine vascular endothelial permeability by non-lytic mechanisms. Eur J Cancer 1990;26:1074–1078.

106. Assier EV, Jullien J, Lefort J, et al. NK cells and polymorphonuclear neutrophils are both critical for IL-2-induced pulmonary vascular leak syndrome. J Immunol 2004;172:7661–7668.

107. Lehmann PV, Schumm G, Moon D, et al. Acute lethal graft-versus-host reaction induced by major histocompatibility complex class II-reactive T-helper cell clones. J Exp Med 1990;171:1485–1496.

108. van Hinsbergh VW, van Nieuw Amerongen GP. Endothelial hyperpermeability in vascular leakage. Vascul Pharmacol 2002;39:171–172.

109. Kaslovsky RA, Lai L, Parker K, Malik AB. Mediation of endothelial injury following neutrophil adherence to extracellular matrix. Am J Physiol 1993; 264:L401–L405.

110. Mustafa A, McKallip RJ, Fisher M, Duncan R, Nagarkatti PS, Nagarkatti M. Regulation of interleukin-2-induced vascular leak syndrome by targeting CD44 using hyaluronic acid and anti-CD44 antibodies. J Immunother 2002; 25:476–88.

111. Rafi-Janajreh AQ, Chen D, Schmits R, et al. Evidence for the involvement of CD44 in endothelial cell injury and induction of vascular leak syndrome by IL-2. J Immunol 1999;163:1619–1627.

112. Carsuzaa F. Pierre C. Morand JJ. Farnarier C. Marrot F. Kaplanski G. Capillary leak syndrome disclosing Ofuji's papuloerythroderma. Ann Dermatol Venereol 1996;123:559–562.

113. Baluna R, Sausville EA, Stone M, Stetler-Stevenson MA, Uhr J, Vitetta ES. Decreases in levels of serum fibronectin predict the severity of vascular leak syndrome in patients treated with ricin A-chain-containing immunotoxins. Clin Cancer Res 1996;2:1705–1712.

114. Pober JS, Cotran RS. Cytokines and endothelial cell biology. Physiol Rev 1990;70:427–451.

115. Baars JW, De Boer JP, Wagstaff J, et al. Interleukin-2 induces activation of coagulation and fibrinolysis: Resemblance to the changes seen during experimental endotoxaemia. Br J Haematol 1992;88:295–301.

116. Deehan DJ, Heys SD, Simpson W, Herriot R, Broom J, Eremin O. Correlation of serum cytokine and acute phase reactant levels with alterations in

weight and serum albumin in patients receiving immunotherapy with recombinant IL-2. Clin Exp Immunol 1994;95:366–372.

117. Dubinett SM, Huang M, Lichtenstein A, et al. Tumor necrosis factor- plays a central role in interleukin-2-induced pulmonary vascular leak and lymphocyte accumulation. Cell Immunol 1994;157:170–180.

118. Zhang J, Yu ZX, Hilbert SL, Yamaguchi, M, Chadwick DP. Cardiotoxicity of human recombinant interleukin-2 in rats. A morphological study. Circulation 1993;87:1340–1353.

119. Edwards MJ, Abney DL, Heniford BT, Miller FN. Passive immunization against tumor necrosis factor inhibits IL-2-induced microvascular alterations and reduces toxicity. Surgery 1992;112:480–486.

120. Wheatley EM, Vincent PA, McKeown-Longo PJ, Saba TM. Effect of fibronectin on permeability of normal and TNF-treated lung endothelial cell monolayers. Am J Physiol 1993;264:R90–R96.

121. Polunovsky VA, Wendt CH, Ingbar DH, Peterson MS, Bitterman PB. Induction of endothelial cell apoptosis by TNF: modulation by inhibitors of protein synthesis. Exp Cell Res 1994;214:584–594.

122. Karsan A, Yee E, Harlan JM. Endothelial cell death induced by tumor necrosis factor-alpha is inhibited by the Bcl-2 family member, A1. J Biol Chem 1996;271:27,201–27,204.

123. Thijs LG, Hack CE, Strack Van Schijndel RJM, et al. Activation of the complement system during immunotherapy with recombinant IL-2. J Immunol 1990;144:2419–2424.

124. Puri RK, Rosenberg SA. Combined effects of interferon alpha and interleukin 2 on the induction of a vascular leak syndrome in mice. Cancer Immunol Immunother 1989;28:267–274.

125. Alon R, Cahalon L, Hershkoviz R, et al. TNF-α binds to the N-terminal domain of fibronectin and augments the beta$_1$-integrin-mediated adhesion of CD4$^+$ T-lymphocytes to the glycoprotein. J Immunol 1994;152:1309–1313.

126. Goldblum SE, Ding X, Campbell-Washington J. TNF-α induces endothelial cell F-actin depolymerization, new actin synthesis, and barrier dysfunction. Am J Physiol Cell Physiol 1993;264:C894–C905.

127. Locker GJ, Kofler J, Stoiser B, et al. Relation of pro- and anti-inflammatory cytokines and the production of nitric oxide in patients receiving high-dose immunotherapy with interleukin-2. Eur Cytokine Network 2000;11:391–396.

128. Dudek,SM ,Garcia JGN. Cytoskeletal regulation of pulmonary vascular permeability. J Appl Physiol 2001;91:1487–1500.

129. Fujita S, Puri R, Yu ZX, Travis W, Yamaguchi M, Ferrans VJ. Interleukin-1 alpha reduces the severity of the vascular leak syndrome produced by interleukin-2 and interleukin-2 plus interferon-alpha. Toxicol Pathol 1994;22:381–397.

130. Puri RK, Travis WD, Rosenberg SA. Decrease in interleukin 2-induced vascular leakage in the lungs of mice by administration of recombinant interleukin 1 alpha in vivo. Cancer Res 1989;49:969–976.

131. Thom AK, Fraker DL, Norton JA. IL-1 receptor antagonist (IL-1ra) augments IL-2-induced pulmonary vascular leak. J Surg Res 1993;54:336–341.
132. Cavender DE, Edelbaum D. Inhibition by IL-1 of endothelial cell activation induced by tumor necrosis factor or lymphotoxin. J Immunol 1988;141:3111–3116.
133. Yamaguchi Y, Suda T, Shiozaki H, Miura Y, Hitoshi Y, Tominaga A, Takatsu K, Kasahara T. Role of IL-5 in IL-2-induced eosinophilia. J Immunol 1990;145:873–877.
134. Gimbrone MA Jr, Obin MS, Brock AF, et al. Endothelial interleukin-8: A novel inhibitor of leukocyte–endothelial interactions. Science 1989;246:1601–1603.
135. Westlin W, Gimbrone MA. Neutrophil-mediated damage to human vascular endothelium. Am J Pathol 1993;142:117–128.
136. Luscinskas FW, Kiely JM, Ding H, Obin MS, Hebert CA, Baker JB, Gimbrone MA. In vitro inhibitory effect of IL-8 and other chemoattractants on neutrophil–endothelial adhesive interactions. J Immunol 1992;149:2163–2171.
137. Hechman DH, Cybulsky MI, Fuchs HJ, Baker JB, Gimbrone Jr. MA. Intravascular IL-8, Inhibitor of polymorphonuclear leukocyte accumulation at sites of acute inflammation. J Immunol 1991;147:883–892.
138. Finn A, Naik S, Klein N, Levinsky RJ, Strobel S, Elliott M. Interleukin-8 release and neutrophil degranulation after pediatric cardiopulmonary bypass. J Thorac Cardiovasc Surg 1993;105:234–241.
139. Li L, Elliott JF, Mosmann, TR. IL-10 inhibits cytokine production, vascular leakage, and swelling during T helper 1 cell-induced delayed-type hypersensitivity. J Immunol 1994;153:3967–3978.
140. Westin WF, Kiely J, Gimbrone MA. Interleukin-8 induces changes in human neutrophil actin conformation and distribution: Relationship to inhibiton of adhesion to cytokine-activated endothelium. J Leukoc Biol 1992;52:43–51.
141. Lentsch AB, Edwards MJ, Sims DE, Nakagawa K, Wellhausen SR, Miller FN. Interleukin-10 inhibits interleukin-2-induced tumor necrosis factor production but does not reduce toxicity in C3H/HeN mice. J Leukoc Biol 1996;60:51–57.
142. Heniford BT, Edwards MJ, Wilson MA, Klar EA, Doak KW, Miller FN. Interleukin-8 suppresses the toxicity and antitumor effect of interleukin-2. J Surg Res 1994;56:82–88.
143. Standiford TJ, Strieter RM, Lukacs NW, Kunkel SL. Neutralization of IL-10 increases lethality in endotoxemia. Cooperative effects of macrophaage inflammatory protein-2 and tumor necrosis factor. J Immunol 1995;155:2222–2229.
144. Okamoto M, Kato S, Oizumi K, et al. Interleukin 18 (IL-18) in synergy with IL-2 induces lethal lung injury in mice: a potential role for cytokines, chemokines, and natural killer cells in the pathogenesis of interstitial pneumonia. Blood 2002;99:1289–1298.
145. Stolpen bAH, Guinan EC, Fiers W, Pober JS. Recombinant tumor necrosis factor and immune interferon act singly and in combination to reorganize human vascular endothelial cell monolayers. Am J Pathol 1986:123:16–24.

146. Gao B, Curtis T, Blumenstock F, Minnear F, Saba T. Increased recycling of (alpha) 5(beta)1 integrins by lung endothelial cells in response to tumor necrosis factor. J Cell Sci 2000;113:247–257.

147. Ferrero E, Villa A, Ferrero ME, et al. Tumor necrosis factor α-induced vascular leakage involves PECAM1 phosphorylation. Cancer Res 1996;56:3211–3215.

148. Groeneveld JA. Vascular pharmacology of acute lung injury and acute respiratory distress syndrome Vasc Pharmacol 2002;39:247–256.

149. Rondeau E, Sraer J, Bens M, Doleris LM, Lacave R, Sraer JD. Production of 5-lipoxygenase pathway metabolites by peripheral leucocytes in capillary leak syndrome (Clarkson disease). Eur J Clin Invest 1987;17:53–57.

150. Cahill RA, Zhao Y, Murphy R, Sala A, Foegh M, Spitzer T, Deeg HJ. High urinary leukotriene E4 (LTE4) and thromboxane 2 (TXB2) levels are associated with capillary leak syndrome in bone marrow transplant patients. Adv Prostaglandin Thromboxane Leukoc Res 1991;21B:525–528.

151. Welbourn R, Goldman G, Kobzik L, Paterson I, Shepro D, Hechtman HB. Interleukin-2 induces early multisystem organ edema mediated by neutrophils. Ann Surg 1991;214:181–186.

152. Klausner JM, Goldman G, Skornick Y, et al. Interleukin-2-induced lung permeability is mediated by leukotriene B4. Cancer 66;1990:2357–2364.

153. Thijs LG, Hack CE, Strack van Schijndel RJM, et al. Activation of the complement system during immunotherapy with interleukin-2: relation to the development of side effects. J Immunol 1990;144:2419–2424.

154. Nurnberger W, Michelmann I, Petrik K, et al. Activity of C1 esterase inhibitor in patients with vascular leak syndrome after bone marrow transplantation. Ann Hematol 1993;67:17–21.

155. Salat C, Holler E, Schleuning M, et al. Levels of the terminal complement complex, C3a-desArg and C1-inhibitor in adult patients with capillary leak syndrome following bone marrow transplantation. Ann Hematol 1995;71: 271–274.

156. Hack CE, Ogilvie AC, Eisele B, Eerenberg AJ, Wagstaff J, Thijs LG. C1-inhibitor substitution therapy in septic shock and in the vascular leak syndrome induced by high doses of interleukin-2. Intensive Care Med 1993;19: S19–S28.

157. Marath A, Man W, Taylor KM. Histamine release in pediatric cardiopulmonary bypass: a possible role in the capillary leak syndrome. Agents Actions 1987;20:299–302.

158. Edwards MJ, Heniford BT, Miller FN. Mast cell degranulation inhibits IL-2-induced microvascular protein leakage. J Surg Res 1992;52:429–435.

159. DeFouw DO, Brown KL, Feinberg RN. Canine jugular vein endothelial cell monolayers in vitro: vasomediator-activated diffusive albumin pathway. J Vasc Res 1993;30:154–160.

160. Hinshaw DB, Burger JM, Beals TF, Armstrong BC, Hyslop PA. Actin polymerization in cellular oxidant injury. Arch Biochem Biophys 1991;288:311–316.

161. Thurston G, Baldwin AL, Wilson LM. Changes in endothelial actin cytoskeleton at leakage sites in the rat mesenteric microvasculature. Am J Physiol 1995;268: H316–H329.

162. Gottlieb AI, Langille BL, Wong MK, Kim DW. Structure and function of the endothelial cytoskeleton. Lab Invest 1991;65:123–137.

163. Klausner JM, Paterson IS, Goldman G, et al. Interleukin-2-induced lung injury is mediated by oxygen free radicals. Surgery 1991;109:169–175.

164. Goldman G, Welbourn R, Alexander S, et al. Modulation of pulmonary permeability in vivo with agents that affect the cytoskeleton. Surgery 1991;109: 533–538.

165. Baars JW, Wolbink GJ, Hart MH, et al. The release of interleukin-8 during intravenous bolus treatment with interleukin-2. Ann Oncol 1994;5:929–934.

166. Hack CE, Wagstaff J, Strack van Schijndel RJ, et al. Studies on the contact system of coagulation during therapy with high doses of recombinant IL-2: Implications for septic shock. Thromb Haemost 1991;65:497–503.

167. Suschek C, Rothe H, Fehsel K, Enczmann J, Kolb-Bachofen V. Induction of a macrophagea-like nitric oxide synthase in cultured rat aortic endothelial cells. IL-1 beta-mediated induction regulated by tumor necrosis factor-alpha and IFN-gamma. J Immunol 1990;151:3283–3291.

168. Mulvin DW, Grosso MA, Kruse CA, Howard RB, Repine JE, Johnston MR. The role of toxic oxygen metabolites in interleukin-2-induced vascular leak syndrome. Curr Surg 1989;46:396–398.

169. Kilbourn RG, Fonseca GA, Griffith OW, et al. NG-methyl-arginine, an inhibitor of nitric oxide synthase, reverses interleukin-2-induced hypotension. Crit Care Med 1995;23:1018–1024.

170. Orucevic A, Lala PK. N-G-nitro-L-arginine methyl ester, an inhibitor of nitric oxide synthesis, ameliorates interleukin-2-induced capillary leak syndrome in healthy mice. J Immunother 1995;18:210–220.

171. Leder GH, Oppenheim M, Rosenstein M, et al. Inhibition of nitric oxide synthesis does not improve interleukin-2-mediated antitumor effects in vivo. Eur Surg Res 1996;28:167–178.

172. Locker GJ, Kapiotis S, Veitl M, et al. Activation of endothelium by immunotherapy with interleukin-2 in patients with malignant disorders. Br J Haematol 1999;105:912–919.

173. Velanovich V. Crystalloid versus colloid fluid resuscitation: A meta-analysis of mortality. Surgery 1989;105:65–71.

174. Pockaj BA, Yang JC, Lotze MT, et al. A prospective randomized trial evaluating colloid versus crystalloid resuscitation in the treatment of the vascular leak syndrome associated with interleukin-2 therapy. J. Immunother. 1994;15: 22–28.

175. Memoli B, De Nicola L, Libetta C, et al. Interleukin-2-induced renal dysfunction in cancer patients is reversed by low-dose dopamine infusion. Am J Kidney Dis 1995;26:27–33.

176. Hasibeder WR. Fluid resuscitation during capillary leakage: does the type of fluid make a difference. Intensive Care Med 2002;28:532–534.

177. Marx G. Fluid therapy in sepsis with capillary leakage. Eur J Anaesthesiol 2003;20:429–442.

178. Koepp A. Lampert R. Use of plasmapheresis in a 62-year-old patient with severe infection. Infusionstherapie und Transfusionsmedizin. 1996;23:92–96.

179. Edwards MJ, Heniford BT, Klar EA, Doak KW, Miller FN. Pentoxifylline inhibits interleukin-2-induced toxicity in C57BL/6 mice but preserves antitumor efficacy. J Clin Invest 1992;90:637–641.

180. Kovach NL, Lindgren CG, Fefer A, Thompson JA, Yednock T, Harlan JM. Pentoxifylline inhibits integrin-mediated adherence of interleukin-2-activated human peripheral blood lymphocytes to human umbilical vein endothelial cells, matrix components and cultured tumor cells. Blood 1994;84:2234–2242.

181. Thompson JA, Bianco JA, Benyunes MC, Neubauer MA, Slattery JT, Fefer A. Phase Ib trial of pentoxifylline and ciprofloxacin in patients treated with interleukin-2 and lymphokine-activated killer cell therapy for metastatic renal cell carcinoma. Cancer Res 1994;54:3436–3441.

182. Kemeny MM, Botchkina GI, Ochani M, Bianchi M, Urmacher C, Tracey KJ. The tetravalent guanylhydrazone CNI-1493 blocks the toxic effects of interleukin-2 without diminishing antitumor efficacy. Proc Natl Acad Sci USA 1998;95:4561–4566.

183. Abdih H, Kelly CJ, Bouchier-Hayes D, Barry M, Kearns S. Taurine prevents interleukin-2-induced acute lung injury in rats. Eur Surg Res 2000;32:347–352.

184. Staak JO, Glossmann JP, Esser JM, Diehl V, Mietz H, Josting A. Thalidomide for systemic capillary leak syndrome. Am J Med 2003;115:332–334.

185. Dhingra VK, Uusaro A, Holmes CL, Walley KR. Attenuation of lung inflammation by adrenergic agonist in murine acute lung injury. Anesthesiology 2001;95:947–953.

186. Butler LD, Mohler KM, Layman NK, Cain RL, Riedl PE, Puckett LD, Bendele AM. Interleukine-2 induced systemic toxicity: Induction of mediators and immunopharmacologic intervention. Immunopharmacol Immunotoxicol 1989;11:445–487.

187. Mier JW, Aronson FR, Numerof RP, Vachino G, Atkins MB. Toxicity of immunotherapy with interleukin-2 and lymphokine-activated killer cells. Pathol Immunopathol Res 1988;7:459–476.

188. DeJoy SQ, Jeyaseelan,R Sr, Torley LW, Schow SR, Wick MM, Kerwar SS, Attenuation of interleukin 2-induced pulmonary vascular leak syndrome by low doses of oral methotrexate. Cancer Res 1995;55:4929–4935.

189. Hack CE, Ogilvie AC, Eisele B, Jansen PM, Wagstaaff J, Thijs LG. Initial studies on the administration of C1-esterase inhibitor to patients with septic shock or with a vascular leak syndrome induced by interleukin-2 therapy. Prog Clin Biol Res 1994;388:335–357.

190. Caliezi C, Wuillemin WA, Zeerleder S, Redondo M, Eisele B, Hack CE. C1-esterase inhibitor: an anti-inflammatory agent and its potential use in the treatment of diseases other than hereditary angioedema. Pharmacol Rev 2000;52:91–112.

191. Ohkubo C, Bigos D, Jain RK. Interleukin 2 induced leukocyte adhesion to the normal and tumor microvascular endothelium in vivo and its inhibition by dextran sulfate: Implications for vascular leak syndrome. Cancer Res 1991;51: 1561–1563.

192. McKallip RJ, Fisher M, Do Y, et al. Targeted deletion of CD44v7 exon leads to decreased endothelial cell injury but not tumor cell killing mediated by interleukin-2-activated cytolytic lymphocytes. J Biol Chem 2003;278:43,818–43,830.

193. Palmer PA, Atzpodien J, Philip T, et al. A comparison of 2 modes of administration of recombinant interleukin-2: continuous intravenous infusion alone versus subcutaneous administration plus interferon alpha in patients with advanced renal cell carcinoma. Cancer Biother 1993;8:123–136.

194. Guida M, Abbate I, Casamassima A, et al. Long-term subcutaneous recombinant interleukin-2 as maintenance therapy: biological effects and clinical implications. Cancer Biother 1995;10:195–203.

195. Huland E, Heinzer H, Mir TS, Huland H. Inhaled interleukin-2 therapy in pulmonary metastatic renal cell carcinoma: six years of experience. Cancer J Sci Am 1997;3(Suppl 1):S98–S105.

196. Anderson PM. Sorenson MA. Effects of route and formulation on clinical pharmacokinetics of interleukin-2. Clin Pharmacokinet 1994;27:19–31.

197. Schwartz RN, Stover L, Dutcher J. Managing toxicities of high-dose interleukin-2. Oncology (Huntington) 2002;16 (Suppl 13):11–20.

198. Khawli LA, Miller GK, Epstein AL. Effect of seven new vasoactive immunoconjugates on the enhancement of monoclonal antibody uptake in tumors. Cancer 1994;73:824–831.

199. Hornick JL, Khawli LA, Hu P, Sharifi J, Khanna C, Epstein AL. Pretreatment with a monoclonal antibody/interleukin-2 fusion protein directed against DNA enhances the delivery of therapeutic molecules to solid tumors. Clin Cancer Res 1999;5:51–60

200. Carnemolla B, Borsi L, Balza E, et al. Enhancement of the antitumor properties of interleukin-2 by its targeted delivery to the tumor blood vessel extracellular matrix. Blood 2002;99:1659–1665.

12

Cytokines and Autoimmune Diseases

*From the Control of Autoimmune Diseases
With Anti-Cytokine Treatment to the Induction
of Autoimmunity With Cytokine Treatment*

Pierre Miossec

SUMMARY

The identification of the role of cytokines in inflammatory and autoimmune diseases has led to significant progress in treatment. The best example is probably the beneficial effect of blocking TNFα in an increasing number of inflammatory diseases, starting with rheumatoid arthritis. However since not all patients respond and since the treatment is suspensive, other cytokine inhibitors are now ready to be tested. The negative consequences of blocking cytokines have demonstrated their role in immunity as observed by the reactivation of tuberculosis following TNFα inhibition. Similarly, adminsitration of other cytokines such as Interferonα has been associated with the induction of autoimmune manifestations, often in a predisposed context.

Key Words: TNF-α; IL-1; IL-6; Th1/Th2; rheumatoid arthritis; Crohn's disease; lupus.

1. INTRODUCTION

The therapeutic use of inhibitors of tumor necrosis factor α (TNF-α) has shown the critical role of a single key cytokine in the pathogenesis of first rheumatoid arthritis (RA) and Crohn's disease (CD) and then of many other inflammatory diseases *(1)*. Although their inducing mechanisms are far from being clarified, these diseases have been classified as autoimmune. The absence of identification of a causal agent or a specific initial mechanism was considered for a long time as the major limitation for the improvement of treat-

From: *Methods in Pharmacology and Toxicology: Cytokines in Human Health:
Immunotoxicology, Pathology, and Therapeutic Applications*
Edited by: R. V. House and J. Descotes © Humana Press Inc., Totowa, NJ

ments. These new results have clearly demonstrated the role of these nonspecific soluble factors in the clinical presentation of diseases, dominated by inflammation, matrix formation abnormalities, and autoantibody production. The nonspecific effect of these cytokines was shown when the same clinical results observed in RA with TNF-α inhibitors also were observed in CD for which the clinical expression, anatomical distribution, and underlying mechanisms are very different *(2)*.

Based on a classification of cytokines according to their contribution to inflammation, their effect on matrix formation, and antibody production, it is possible to define autoimmune diseases according to these cytokine profiles. Such a classification, although oversimplified, allows a better understanding of clinical manifestations, selection of treatments, and of mechanisms of adverse drug reactions.

Cytokines themselves have been used for treatment starting with interferon-α (IFN-α) for viral hepatitis. Later, the inhibition of cytokines was a key step forward for the control of the most severe inflammatory diseases. At the same time, the administration of cytokines or of their inhibitors has been associated with adverse events. Some of these are indicators of the contribution of cytokines and/or regulatory pathways involving cytokines to a wide number of mechanisms.

In this review, cytokines will be classified according to their regulatory properties to associate common autoimmune diseases with these profiles. This classification will allow the analysis of the clinical results obtained with cytokine inhibitors, mainly against TNF-α. Finally, we will examine the adverse events associated with the administration of cytokine inhibitors and of cytokines, focusing on IFN-α.

2. CYTOKINES WITH REGULATORY PROPERTIES

A classification of T-cells according to their cytokine profile has been proposed for the mouse and then the human situation. Th2 responses are characterized by the production of interleukin-4 (IL-4), IL-5, IL-10, and IL-13, whereas Th1 responses are characterized by that of IL-2 and IFN-γ production *(3)*. Results in animal models have indicated the role of Th1 cells in delayed type hypersensitivity and that of Th2 cells in allergy and some parasitic infections. The same dichotomy was later applied to human T cells *(4)*. However, a more simple classification into type 1/type 2 has been proposed as these cytokines are produced by other cell types (Fig. 1). For instance IL-10 is largely produced by monocytes and B-cells, whereas IL-4 is also produced by mast cells and basophils.

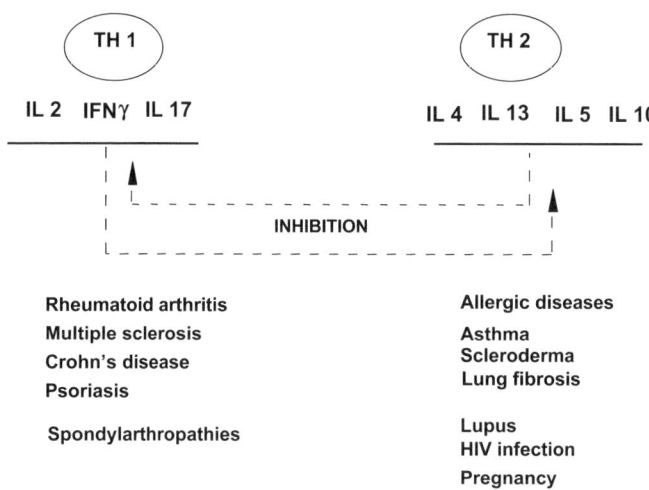

Fig. 1. Association between T-cell-derived cytokines and autoimmune diseases.

An important feature of type 1 and type 2 cells is the ability of one subset to regulate the activities of the other in a dynamic process. Normal T cells initiate a production of type 1 cytokines in response to mycobacterial antigens and in the presence of type 1 cytokines. Conversely, they produce type 2 cytokines in response to allergens and in the presence of type 2 cytokines. Both IL-4 and IL-10 are strong inhibitors of IFN-γ production, whereas IFN-γ inhibits IL-10 production and action *(5)*. Typical Th1-inducing conditions combine the use of IL-12 and of IL-4 inhibition, whereas the Th2-inducing conditions combine the addition of IL-4 and of IL-12 inhibition. Accordingly, IL-12 increases TNF-α and IL-1 production, which is inhibited by IL-4.

In addition to the Th1/Th2 cytokine polarization, T cells have been classified according to their regulatory properties. A Th3 subset was defined with the production of transforming growth factor β (TGF-β) with immunosuppressive properties. Additional regulatory T-cell subsets have been defined in particular by the continuous expression of CD25. These cells are producers of IL-10 and TGF-β. These cells are involved in the inhibition of the immune response and in the induction of tolerance.

More recently, inflammatory and autoimmune diseases have been classified using only two sets of cytokine pairs *(6)*. TNF-α and IFN-α are the members of the first pair, where TNF-α is associated with inflammation such as RA and IFN-α with autoimmunity such as systemic lupus erythematosus (SLE). The second pair is made of IFN-γ and IL-4, which are the prototypic cytokines of the Th1 and the Th2 patterns, respectively.

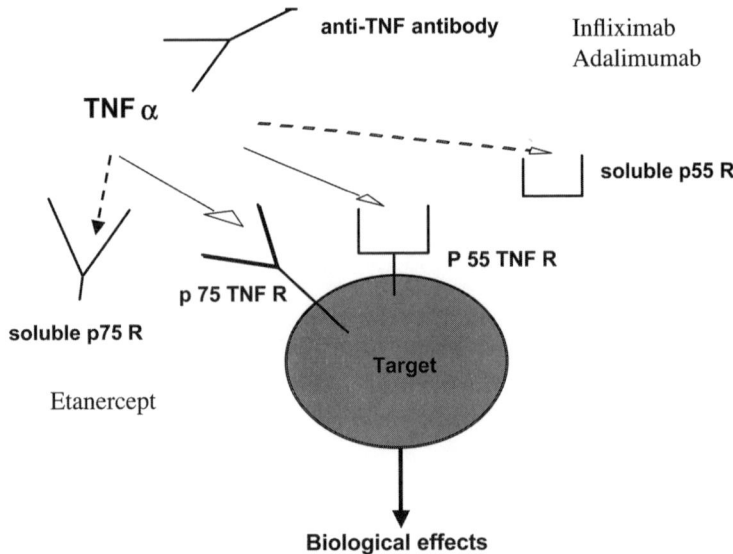

Fig. 2. Interactions between tumor necrosis factor (TNF)-α, its receptors, and current specific inhibitors.

3. PROINFLAMMATORY CYTOKINES

These cytokines have major functions in the protection against any aggressive situation. They share a large number of activities and are also involved in the pathogenesis of diseases with various anatomical expressions.

3.1. Tumor Necrosis Factor-α

TNF-α is the founding father of a growing family, which includes many critical factors of the immune system. TNF-α and lymphotoxin α (LT-α) act as trimers on two receptors: the type I TNF receptor (p55-TNF-R) and the type II TNF receptor (p75-TNF-R, Fig. 2). These two cytokines control inflammation and apoptosis. TNF-α is secreted as a trimer but exists also as a trans-membrane biologically active monomer, which is important for local cell–cell interactions *(7)*. This molecule is released as a soluble form under the effect of a membrane metallo-proteinase, the TNF-converting enzyme (TACE). We will focus below on the consequences of TNF-α inhibition.

The fixation of TNF-α to its receptors leads to an activation of the signal transduction pathways, including mitogen-activated protein kinases and nuclear factor-κB. The p55 but not the p75 TNF receptor has a death domain involved in apoptosis controlled by Fas, after activation by TNF-α. The other

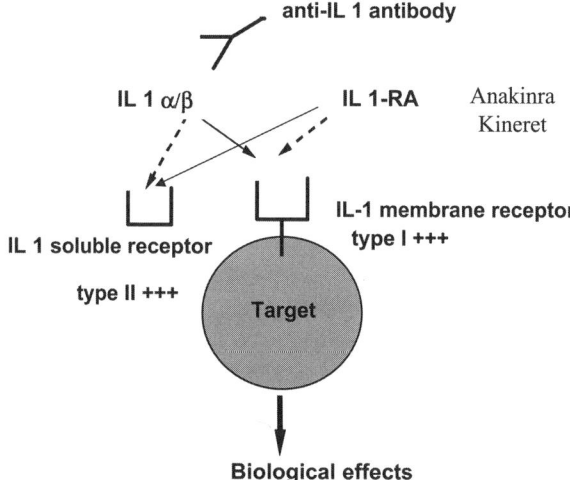

Fig. 3. Interactions between interleukin (IL)-1, its receptors, and current specific inhibitors.

pathway is involved in the inflammatory reaction with the synthesis of cytokines, chemokines, and proteases. For a given activated cell, there is a selective choice between the inflammatory or the apoptotic pathways so that only one of them prevails.

4. OTHER CYTOKINES
WITH INHIBITORS IN DEVELOPMENT

More recently, additional cytokines have been the targets for treatment. At this early stage, it is still difficult to compare the current results with those obtained with TNF-α inhibitors.

4.1. Interleukin-1

IL-1 has been classified as a critical cytokine in chronic inflammation. The mode of action of IL-1 is similar to that of TNF-α but with important differences *(8)*. The two IL-1 (α and β) act through two receptors (Fig. 3). The critical receptor for biological response is the membrane type I IL-1 receptor, which transduces a signal, when combined with an accessory protein to form the fully functional IL-1 receptor. The membrane type II IL-1 receptor does not transduce a signal. It is rather an endogenous regulator, which is released as a soluble form and traps soluble IL-1. In addition to the two IL-1 molecules with an agonistic effect, IL-1 receptor antagonist (IL-1RA) can bind to the membrane type I receptor but does not induce a signal since the accessory chain is not recruited.

IL-1RA (Anakinra) has been used for the treatment of RA. The beneficial effects, although significant enough to get marketing approval from the health authorities for the treatment of RA, appear to be fewer than those of TNF-α inhibition *(9)*. A major limitation with IL-1RA is the need for a continuous high level of the compound, because at least a 1:100 ratio has to be obtained between the IL-1 and IL-1RA levels. At present, other means of blocking IL-1 are in progress, such as anti-IL-1 antibodies and IL-1 soluble receptors used alone or in complex structures in an IL-1 trap.

4.2. Interleukin-6

As for IL-1 and TNF-α, IL-6 is highly present in the context of chronic inflammation. There has been some debate on the classification of IL-6 as a pro- or an anti-inflammatory cytokine. Controlling IL-6 as a therapeutic approach was indeed the best way to clarify this issue. For that purpose, an anti-IL-6 receptor antibody named MRA has been developed and has shown efficacy for the treatment of RA and CD. The molecule is now in phase III trials *(10,11)*.

4.3. Interleukin-15

IL-15 is a cytokine involved in chronic inflammation, especially in the activation of T cells. Recent positive results have been obtained with a monoclonal anti-IL-15 antibody for the treatment of RA *(12)*.

5. OTHER CYTOKINES AS TREATMENT TARGETS

The favorable clinical results obtained with TNF-α and IL-1 inhibitors may suggest that the story of the contribution of cytokines to arthritis is almost over (Fig. 4). However, additional cytokines, which also contribute to destructive inflammation, have been considered as possible targets (Fig. 4).

IL-17 is a T-cell-derived cytokine, which could be classified as a Th1 cytokine. IL-17 often acts in synergy with TNF-α and IL-1 *(13)*. In addition, IL-17 increases IL-1 and TNF-α production by monocytes. In vitro and animal models have strongly indicated the interest in blocking IL-17 for the treatment of chronic inflammation.

IL-18, IL-12, and IL-23 are cytokines that act in association because they favor a Th1 cytokine profile and are produced by monocytes and other antigen-presenting cells *(14)*. They interact through synergistic mechanisms. IL-12 targeting is on going with an anti-IL-12 antibody. IL-18 action is regu-

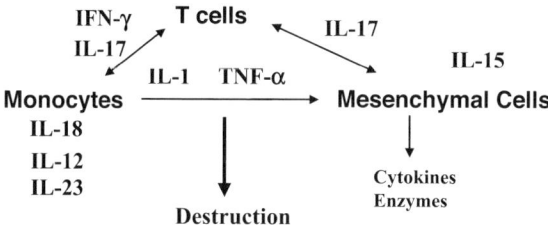

Fig. 4. Interactions between the major proinflammatory cytokines and matrix destruction.

lated by its endogenous inhibitor IL-18 Binding Protein (IL-18 BP). Preclinical results have been obtained to suggest the use of IL-18 BP as treatment *(15)*. IL-23 has been more recently described as a cytokine responsible for some of the proinflammatory effects first associated with IL-12. In addition, IL-23 increases IL-17 production by T-cells *(16)*.

6. CLASSIFICATION OF AUTOIMMUNE DISEASES ACCORDING TO THEIR CYTOKINE PROFILES

Having defined these cytokine profiles, the next step is to classify the major autoimmune diseases according to the presence of inflammation with matrix destruction, of inflammation with matrix deposition, and of autoantibody production. The goal of such classification is to help understanding common features and differences between these diseases.

6.1. Inflammatory Diseases Associated With Matrix Destruction

This group of diseases includes conditions in which a chronic inflammatory reaction at disease sites leads to matrix destruction (Fig. 4). Such local reaction appears very similar between diseases with changes related to the anatomical situation. RA is the prototypic example of such conditions in which chronic inflammation of the synovium membrane leads to the accumulation of blood-derived cells such as monocytes, dendritic cells, and T and B lymphocytes, which interact with resident mesenchymal cells named synoviocytes. These interactions result in an increased production of destructive enzymes, which induce bone and cartilage destruction. CD has similar profile in which the migration of cells inside the intestinal mucosa leads to ileitis and gut matrix destruction (Fig. 5). The final extension of CD is the formation of fistulas between the intestine and the skin. Migration of inflam-

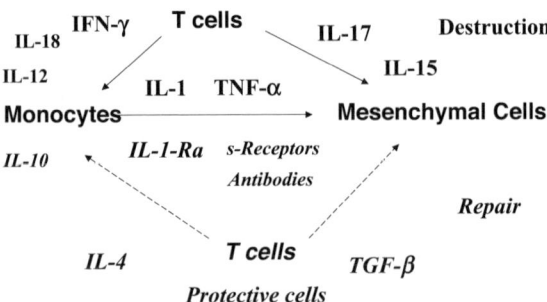

Fig. 5. Interactions between regulatory cytokines, matrix destruction, and repair defects.

matory cells to the skin dermis is the hallmark of psoriasis, which can also be associated with destructive arthritis with common features with RA. Multiple sclerosis is the transfer of this pathway to the central nervous system. The clinical manifestations result from myelin destruction after focal inflammation. Studies of these lesions have indicated the presence of a T-cell infiltrate with interactions with other immune cells such as B-cells, monocytes, and dendritic cells. Characterization of the T-cell subsets in all these diseases has indicated the overexpression of a Th1 profile combined with the high production of monocytes derived cytokines such as TNF-α and IL-1 *(17)*. Synergistic interactions occur between these cytokines under the enhancing effects of signals from T-cell–derived cytokines.

In addition to the increased level of destruction, these diseases have in a major defect in repair activity (Fig. 5). Such defect is responsible common for the rapid destruction after such type of inflammation. As an example, RA-associated inflammation can lead to a joint destruction within 2 yr, whereas 20 yr are needed to come to a stage of joint replacement in osteoarthritis.

Spondyloarthropathies can be classified as an in-between situation. Locally, the same proinflammatory cytokines, such as TNF-α, are highly detected at the site of the sacroiliitis *(18)*. Howeve,r such destruction is followed by a step of repair activity as expressed by syndesmophyte formation.

6.2. Inflammatory Diseases Associated With Matrix Formation

This is the mirror image of the previous situation. After local inflammation, there is an intense stimulation of matrix formation. This leads to collagen deposition in organs with functional consequences, which can be more severe than the destructive pattern described above (Fig. 6). As opposed to

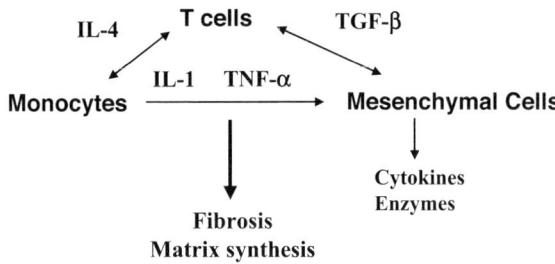

Fig. 6. Interactions between regulatory cytokines and matrix formation.

the Th1 cytokine pattern associated with destruction, here a Th2 pattern is overexpressed *(19)*. Such switch in phenotype induces the contribution of cytokines such as IL-4 and of growth factors such as TGF-β. These factors have a major inducing effect on extracellular matrix synthesis and deposition. Conversely, such deposits in critical organs are the consequences of a defect in matrix degrading enzymes.

Such pattern is observed in conditions such as scleroderma, lung interstitial fibrosis either idiopathic or secondary to exposure to silica, plastic and other fibrotic chemicals (Fig. 6).

6.3. Inflammatory Diseases Associated With Autoantibody Production

Cytokines have a major effect on B-cell activation and differentiation. Enhancement of these pathways during diseases may induce the production of autoantibodies associated with clinical symptoms. Lupus is the prototype of such disease (Fig. 1). Regarding cytokine contribution, IL-10 is probably the cytokine whose contribution to lupus has been the best demonstrated up to a therapeutic modulation *(20)*. Circulating IL-10 levels are increased during lupus and are correlated to various indices of disease activity *(21)*. The blood cell origin of IL-10 includes Th2 T-cells, monocytes, and B-lymphocytes. Production by keratinocytes after exposure to ultraviolet light may explain the flares of disease activity observed after sun exposure. The placenta is also a source of IL-10. Production during pregnancy may contribute to disease flares. Conversely, the anti-inflammatory effect of IL-10 may explain the beneficial effect of pregnancy on RA activity (Fig. 1). This dichotomy indicates differences of regulatory pathways between the two diseases.

In vivo cell interactions with stromal mesenchymal cells in bone marrow and pathological sites of B-cell activation play a critical role in the protection of B cells from elimination by apoptosis. In addition to the effect of IL-10, the chemokine stromal cell-derived factor-1 produced by mesenchymal cells, plays an essential role in these local cell interactions *(22)*.

The therapeutic modulation of the action of IL-10 was evaluated in models of mouse lupus. Administration of an anti-IL-10 antibody in mice induced a protection from renal disease associated with a better survival *(23)*. When applied to human lupus with a mouse anti-IL-10 antibody, an open study showed improvement of skin manifestations *(24)*.

A greater production of IFN-α has been described for a long time in patients with lupus. More recently, work on the differentiation of dendritic cells has confirmed the importance of IFN in human lupus *(25)*. High expression of genes regulated by IFN-α was detected in children with active lupus *(26)*. Such expression was sensitive to the effect of a treatment by steroids. Accordingly, control of the action and the production of IFN-α could have an interest in lupus treatment.

7. INHIBITION OF TNF-α

7.1. Mode of Action of the Specific TNF-α Inhibitors

In reference to the natural control of the action and the production of a cytokine, therapeutic control can be specific of a given cytokine (antibody, soluble receptor), or more global, by acting on a group of proinflammatory cytokines (methotrexate, leflunomide, inhibitors of common intracellular pathways).

When considering the specific inhibitors of TNF-α, inhibition with an anti-TNF-α antibody represents the simplest concept (Fig. 2). There are several types of anti-TNF-α antibodies, either fully human (adalimumab) or chimeric, keeping a more or less important part of the antibody binding site of murine origin (infliximab). These monoclonal antibodies bind specifically to epitopes of the TNF-α trimer, which interact with the membrane receptors. This results in an inhibition of the TNF-α capacity to bind to the membrane receptors, resulting in a functional inhibition. In addition these antibodies will also recognize membrane TNF-α, thus inhibiting cell interactions.

On the contrary, the use of the soluble receptors (p55 or p75) represents the enhancement of a natural regulation because the same molecules already control the action of endogenous TNF-α. Etanercept is a fusion protein with two p75 receptors, connected to an Fc fragment of an IgG1. This Fc fragment improves the pharmacokinetics of the complex.

TNF-α and lymphotoxin (LT)-α use the same two p55 and p75 TNF receptors to control inflammation and apoptosis. Advantages or limita-

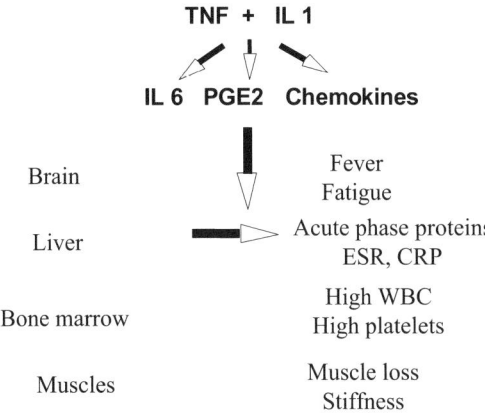

Fig. 7. Contribution of tumor necrosis factor (TNF)-α and interleukin (IL)-1 to the systemic manifestations of chronic inflammation.

tions of the exclusive inhibition of TNF-α (with a specific anti-TNF-α antibody) or associated to that of the LT-α (with a soluble receptor) have not been really clarified.

7.2. Local and Systemic Effects of TNF-α Inhibition

The initial clinical results obtained with the administration of anti-TNF-α antibodies and TNF-α-soluble receptors have justified their large scale development. The major indications are RA, CD, spondylarthropathies, psoriatic arthritis, and chronic juvenile arthritis.

The rapid effect on systemic manifestations and on the levels of acute phase proteins confirms the importance of TNF-α in systemic inflammation (Fig. 7). The feeling of well-being reported rapidly by these patients is the confirmation of the effect of TNF-α on brain, in particular on the hypothalamus.

The anti-inflammatory effect results from local actions on the expression and the composition of the inflammatory reaction (Fig. 8). The migration of inflammatory cells contributes to the initiation and to the chronicity of the inflammatory process, leading to matrix destruction. The formation of any inflammatory reaction relies on new blood vessel formation, which is critical for the migration of these inflammatory cells. TNF-α induces the expression of adhesion molecules on endothelial cells. These effects favor the migration of T cells with a memory phenotype. These cells express preferentially particular chemokine receptors (CCR1, CCR5). These selective mechanisms direct cells towards skin, joint, gut, eye sites resulting in the clinical pictures, each specific of the respective disease *(27)*.

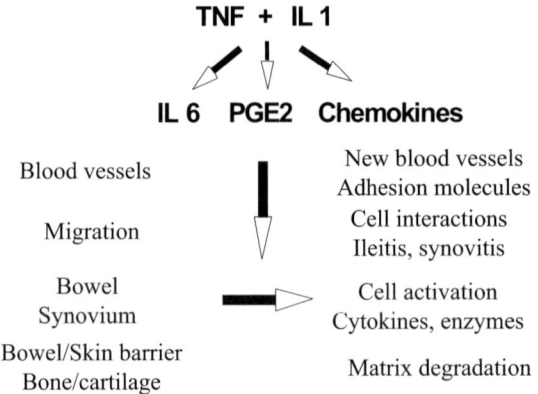

Fig. 8. Contribution of tumor necrosis factor (TNF)-α and interleukin (IL)-1 to the local manifestations of chronic inflammation.

The increase of angiogenesis is an important characteristic of the rheumatoid synovitis. Local concentrations of vascular endothelium growth factor were found to decrease in response to infliximab. The sequential biopsies of synovial membrane of treated patients showed a reduction of the cell infiltrate and of angiogenesis (28). This reduction of the number of inflammatory cells and of their interactions contributes directly to a protective effect by lowering the local production of enzymes involved in destruction.

The rapid reduction of joint swelling is an impressive effect of anti-TNF-α treatment. In vitro, infliximab, but not etanercept, can induce death of cells expressing membrane TNF-α in the presence of complement through a mechanism of antibody-dependent cell-mediated cytoxicity. However, no increase in cell death by apoptosis has been found in synovial biopsies after treatment with infliximab (29). This aspect would have been able to allow a better understanding of the differences in the mode of action of the inhibitors. The absence of induction of apoptosis with etanercept could explain its lack of efficacy in CD, where infliximab is effective (30).

Clinical results showed a reduction of the rate of articular destruction measured by the absence of new radiological joint damage. This effect is critical because of a major depression of repair capacities. In CD, such effect results in the closing of fistulas. An action on proteases involved in the destruction of bone and cartilage, and of gut and skin matrix represents the main mode of action of these inhibitors. This mechanism has been clarified in vitro and in animal models (31). Bone destruction in RA results from an activation of osteoclasts combined with the recruitment of osteoclast precursors, which is amplified by TNF-α. This inhibitory effect can influence the formation of osteoclasts by reducing the recruitment of precursors, which are common to monocytes and dendritic cells. Such effect implies a direct interaction in the RANK/RANK-L pathway. RANK-L activates the receptor RANK on osteoclasts and plays a critical role in bone destruction. RANK-L is expressed by osteoblasts, T lymphocytes and synoviocytes. RA blood concentrations of soluble RANK-L were normalized with infliximab *(32)*.

Induction of repair remains the final achievement to protect joint. In transgenic mice expressing human TNF-α, its neutralization inhibits bone and cartilage degradation *(33)*. An effect on repair is fully obtained by the coadministration of anti-TNF-α and of OPG, blocking at the same time TNF-α-TNF receptor and RANK-RANK-L interactions *(34)*.

7.3. Understanding the Adverse Reactions of TNF-α Inhibitors

The inhibition of a central molecule, such as TNF-α, has been associated with adverse events, which allow a better understanding of the physiological role of TNF-α. In addition, the heterogeneity of response and the risk of disease reactivation when stopping the inhibition indicate that these inflammatory diseases cannot be simplified as diseases of the TNF-α pathway.

TNF-α is a critical molecule in the control of acute and chronic infections. A greater mortality was observed during the treatment of toxic shock with inhibitors of TNF-α. For long-term treatment, the major complication with the use of TNF-α inhibitors has been the onset of opportunistic infections. If all types of opportunistic infections were observed, tuberculosis has been by far the most frequent *(35)*. Its severity associated to an unusual mortality rate quickly drew attention. Epidemiological studies have showed that it was essentially a reactivation of known or undiagnosed tuberculosis.

Such a notion of reactivation is highly suggestive of an acquired defect of cell-mediated immunity. Primary immune defects have been described in association with mutations of genes coding for the receptors for IFN-γ and IL-12 *(36)*. These defects are responsible for severe mycobacterial infections usually secondary to an immunization with BCG.

In the context of chronic inflammation as in RA, there is already a systemic response defect to IL-12 and IL-18, which are key cytokines for the production of IFN-γ *(37)*. There is a synergy between these two cytokines related to the effect of IL-12 on the induction of a functional IL-18 receptor *(38)*. Such defect in RA results in a lower production of IFN-γ by blood cells in response to IL-12 and IL-18. This defect is proportional to the level of systemic inflammation, as measured by CRP levels. This could explain the increased frequency of tuberculosis in RA even in the absence of anti-TNF-α treatment *(39)*. At initiation of a treatment with a TNF-α inhibitor, the disease is usually still very active. The additive effect of the inhibitor explains the rapid onset and the severity of these reactivations. Later, the risk is reduced because the improvement of disease activity has been able to correct the systemic immune defect related to inflammation. This lowers the risk of a reactivation. The specific anti-TNF effect results from an inhibition of cell–cell interactions. Such effect has a positive impact on local inflammation and results in the beneficial clinical effect. Conversely, inhibition of these interactions also results in granuloma disintegration leading to the diffusion of mycobacteria, which were kept under control in these granulomas.

The induction of antinuclear antibodies is common, but rarely associated with clinical manifestations of lupus *(40)*. A positive connection between TNF-α and lupus has been established in lupus mice where the inhibition of TNF-α increases incidence and mortality from renal disease *(41)*. Conversely, administration of TNF-α has a protective effect on mouse lupus. The anti-inflammatory cytokine IL-10 is directly involved in the production of autoantibodies and IgG *(42)*. These properties indicate a mutual inhibition between TNF-α and IL-10 actions. Furthermore, the inhibition of TNF-α favors the production of IL-10, leading to the production of autoantibodies and the orientation of the Th1-Th2 cytokine balance toward Th2 *(17)*.

These notions must be taken into account also to estimate the possible effect of TNF-α inhibitors on the incidence of lymphomas *(43)*. The interpretation of these observations is extremely difficult because of a greater frequency of lymphomas in the general population, further increased in association with RA and even more with Sjögren's syndrome. The contribution of TNF-α to the mechanisms of apoptosis, the immunosuppressive effect of IL-10, which production and action are increased with TNF-α inhibitors are mechanisms to be considered. Conversely, TNF-α is involved in lymph node hypertrophy, a common sign of activity of inflammatory diseases *(44)*.

The beneficial effect obtained in the treatment of RA and CD would suggest that all inflammatory diseases in particular those associated with a Th1 profile could be controlled with the same TNF-α inhibitors. However TNF-α inhibition during multiple sclerosis was associated to an increase in clinical and radiological signs *(45)*. The mechanism has not been clarified, but could be related to the anatomical site and to the contribution of the blood–brain barrier *(46)*. The contribution of TNF-α inhibition could be demonstrated with the reintroduction of infliximab in a case of aseptic meningitis *(47)*.

7.4. Targeting One or More
Than One Cytokine at the Same Time

It remains to be understood how blocking a single cytokine such as TNF-α can still be effective. As of today the list of cytokines, chemokines, and growth factors involved in inflammation is up to 100. It was first thought that TNF-α could be located upstream of a cytokine cascade. However, cytokines such as IL-12, IL-18, and IL-17 have been shown to act on the production of TNF-α, itself.

It should be recognized that the concentrations, which may be expected in vivo, are much lower that those tested in culture systems with isolated cells stimulated with a single cytokine. In addition, interactions between cytokines can be synergistic through common intracellular pathways. These pathways are located downstream of the receptors, which confer specificity. Transcription pathways, such as nuclear factor-κB, among others, are shared for the activation of IL-1, TNF-α, and IL-17. These common pathways are the therapeutic targets of small molecules now in clinical development.

A combination of low concentrations of cytokines was used to dissect these synergistic interactions. As an example, combination of low concentrations of TNF-α, IL-1, and IL-17, with no effect when used alone, was able to induce a level of transcription factor activation higher than the one observed with high concentrations of a single cytokine *(48)*. More importantly, such a combination was able to increase the recruitment of transcription factors, including some not activated even by high concentrations of a single cytokine *(49)*. Such findings reflect the in vivo situation with local interactions between cytokines produced by monocytes such as TNF-α and IL-1 and by T cells such as IL-17.

Similar conclusion was obtained when low doses of soluble receptors for IL-1, TNF, and IL-17 were combined with samples of synovium and of juxta-articular bone *(50)*. As observed in patients, two thirds of the RA synovium samples responded to etanercept. The rate of response was increased up to 90% when the three receptors were combined. As of today, such combinations have not been fully tested in the clinic. One trial tested the combination of IL-1RA and TNF sR in RA. No improvement of efficacy was observed, whereas incidence of adverse events mainly infectious was increased *(51)*.

7.5. Understanding the Heterogeneity of Response to TNF-α Inhibitors

In clinical practice, it appears that about one third of the RA patients do not improve. This heterogeneity could be at least partially explained by the absence of a TNF-α contribution in some patients *(52)*. As indicated previously, the same rate of response is observed when incubating fragments of RA synovial membrane with etanercept *(50)*. Although never directly proven, such differences could be related to cytokine gene polymorphisms directly affecting TNF-α or other cytokines *(26,27,53)*.

A progressive loss of response to inhibitors previously effective is also commonly reported. For compounds such as infliximab, induction of an anti-body response directed against the mouse part of the antibody site has been demonstrated in patients with CD *(54)*. Progressive induction of a TNF-α-independent inflammatory pathway is another mechanism further support-ing the downstream situation of TNF-α in the pathogenesis of these diseases.

Finally, these treatments have a suspensive effect, with reappearance of symptoms after prolonged treatment discontinuation. This reflects an effect on the action and not so much on the production of TNF-α. Stopping the inhibition would be then followed by a rebound effect. Furthermore, TNF-α inhibition decreases the production of soluble TNF-α receptors, thus reduc-ing the endogenous anti-inflammatory regulation. These data also suggest the contribution of other factors and other cell types. The possible role of combined or sequential control of T cells, B cells, and dendritic cells, and of their interactions on the induction of a remission has yet to be demonstrated.

8. AUTOIMMUNE MANIFESTATIONS WITH CYTOKINE ADMINISTRATION

The list of cytokines of possible therapeutic use has been growing quickly. As for any adverse event associated with a new compound, the interpreta-tion of the underlying mechanisms often is difficult. The incidence is obvi-ously related to the properties of the molecule itself, its dose regimen, and route of administration but also to the underlying disease as well as to the size of the exposed population. Among the cytokines already in clinical appli-cation, IFN-α has been the most commonly used, particularly in patients with chronic viral hepatitis. For these reasons, this cytokine administered in this indication has been associated with the largest list of adverse events. Thus, the adverse events observed with this cytokine first will be reviewed first.

8.1. Interferon-α

The major indications are type B and C chronic viral hepatitis in addition to cancer, mostly renal cell carcinoma and melanoma, and hematological malignancies. The large number of treated patients, the largest for any cytokine, allows a good assessment of its safety *(55)*.

Induction of autoimmune events appears to be a frequent feature *(56,57)*. This includes an exhaustive list of manifestations of autoimmunity and as-sociated diseases *(55)*. Therefore it was recommended to exclude patients with concomitant clinically overt autoimmune disease from the use of IFN-α for the treatment of viral hepatitis.

The range goes from the mere presence or induction of autoantibodies with no clinical consequence to the most severe autoimmune disease. The pathogenicity of such autoantibodies is often unclear and far from being always associated with disease manifestations as for the spontaneous auto-immune diseases. The targets of these antibodies include among many others, blood cells (red cells, leukocytes, platelets), coagulation factors (factor VIII, lupus anticoagulant), immunoglobulins (rheumatoid factor with or without cryoglobulin activity), intracellular components (nucleus, enzymes), hor-mones (thyroid, insulin), skin (epidermis [58]).

In particular, exacerbation of hepatitis C-related cryoglobulinemia with possible severe clinical consequences including polyneuritis and even fatal cases has been reported (59). This has to be taken into account since an association with hepatitis C is found in almost 50% of all cases of mixed cryoglobulinemia (60), although treatment with IFN-α is usually helpful, particularly in combination with ribavirin (61,62).

Blood cells and coagulation factors are frequent targets. These manifesta-tions include idiopathic or autoimmune hemolytic anemia and thrombocy-topenia (63). Regarding acquired factor VIII inhibitor, some patients with hemophilia A and hepatitis C developed antibodies to factor VIII (64). Induc-tion of lupus anticoagulant has been implicated in the development of throm-botic events.

Regarding the list of organ specific diseases, thyroid abnormalities appear to be the most common manifestations (65). The exact incidence remains unclear. In a survey including 11,241 patients with hepatitis, 71 developed autoimmune thyroid disease during IFN-α treatment (56). Various thyroid abnormalities have been observed. Half of them had hypothyroidism, 30% hyperthyroidism, and 20% a biphasic (hyperthyroidism followed by hypothy-roidism) pattern. This includes the two ends of the spectrum ranging from patients clinically and biochemically hypothyroid but negative for thyroid autoantibodies to patients remaining euthyroid but with thyroid autoanti-bodies. Idiopathic thyroiditis is a very common autoimmune disease, often associated with Sjögren syndrome resulting in lymphocytic infiltration of the exocrine glands. It is important to keep in mind that such features are commonly found in patients with hepatitis C in the absence of any treatment with cytokines (66).

In the same line, induction of insulin antibodies and onset of insulin depen-dent diabetes with increased antiglutamic acid decarboxylase antibody levels have been observed. This occurred in 10 of the 11,241 patients included in the aforementioned survey (56).

The other manifestations include almost the entire list of autoimmune diseases *(55)*. Regarding arthritis manifestations, IFN-α was responsible for the induction or flare of various types of inflammatory arthritis either associated with RA, psoriasis, lupus, spondylarthropathy or yet unclassified *(67,68)*. Regarding muscular manifestations, cases of dermatomyositis, polymyositis have been described mostly in patients with chronic hepatitis C *(67)*. Similarly, cases of myasthenia gravis with antiacetylcholine receptor antibodies and Guillain-Barré syndrome have been observed *(69)*.

Features of SLE, including severe cases with nephropathy, cerebral Vasculitis, and chorea, have been reported *(57)*. As indicated previously, the role of IFN-α in the pathogenesis of lupus has been demonstrated making IFN-α a target for treatment *(70,71)*. It should be indicated however that induction of lupus remains a rare event.

Interference with graft survival has been a major consequence of such treatment. This included graft versus host disease following therapy by allogeneic bone marrow transplantation or allograft rejection following treatment of hepatitis C after liver as well as renal transplantation *(72)*. Treated recipients may also develop progressive cirrhosis despite achieving a sustained virological response.

9. OTHER CYTOKINES

IL-2 was the first cytokine used as a recombinant product. In the early studies, IL-2 was used for the ex vivo culture of autologous peripheral blood lymphocytes before re-injection. Reduction of metastases was observed in patients with extensive melanoma and renal cell carcinoma *(73)*. Most recent studies have used IL-2 either alone or often combined with other cytokines, mainly IFN-α.

One of the major adverse reactions with IL-2 was the acute accumulation of body fluid related to a capillary leak syndrome *(74)*. In vivo studies have shown the activation of vascular endothelial cells by proinflammatory cytokines, the production of which was stimulated by IL-2. When used on a more chronic basis, such effect may contribute to the migration of inflammatory cells, mostly lymphocytes, to the perivascular site. A number of adverse reactions observed with IFN-α have been described with IL-2 *(75)*. This includes the induction of chronic arthritis, myositis, thyroid manifestations, and induction of various antibodies.

Most probably related to the accumulation of body fluid, carpal tunnel syndrome has been observed in some patients treated with the combination of IL-2 and IFN-α. Skin manifestations included allergic reactions with angioneurotic edema and urticaria in association with activation of neutrophils and complement, local and systemic hypersensitivity reactions.

IFN-β currently is used for the treatment of multiple sclerosis *(76)*. Although the clinical experience is not as large as with IFN-α, autoimmune manifestations do not appear to be a frequent concern, although thyroiditis is here also the most common autoimmune reported event.

IFN-γ is an NK/Th1-derived cytokine. Some of the comments regarding IFN-α apply to IFN-γ. The experience with this latter cytokine although limited, indicates its autoimmune potential. In patients with hepatitis C, thyroid dysfunction was uncommon in contrast to the induction of antinuclear antibodies. In patients with psoriasis arthritis or spondylarthropathy, increased arthritis activity was observed. In patients with rheumatoid arthritis, no benefit was demonstrated but in some cases, induction of antinuclear antibodies was observed *(77)*. In some patients with multiple sclerosis, treatment with IFN-γ led to increased disease activity *(78)*. More recently, however, in patients with cancer or opportunistic infections related to HIV infection, autoimmune manifestations were not observed *(79,80)*.

IL-12 is a cytokine produced by monocytes and antigen-presenting cells with a major effect on cell-mediated immunity in part through the production of IFN-γ. IL-12 has been recently used as protein and gene therapy for the treatment of cancer *(81,82)*. Autoimmune manifestations have not been observed, but further data with prolonged exposure are needed.

The family of colony-stimulating factors (CSF) includes IL-3 or multi-CSF, granulocyte-CSF (G-CSF), and granulocyte-monocyte-CSF (GM-CSF). Such cytokines are used for the stimulation of bone marrow precursors after bone marrow depression either spontaneous or post-chemotherapy.

Most autoimmune manifestations have been observed with G-CSF or GM-CSF. They have been used in patients with RA, particularly those with Felty's syndrome, which is defined as the combination of rheumatoid factor-positive destructive RA, severe neutropenia, and splenomegaly. Improvement of the neutropenia has been observed, sometimes with increased thrombocytopenia and anemia *(83)*. Some patients also showed increased arthritis activity *(84)*. However, such flare-up was not a constant observation.

10. CONCLUSION

The use of TNF-α inhibitors has provided clear evidence for the direct role of cytokines in complex inflammatory diseases. The simplest approach, already in practice, is the specific inhibition of their action. The stimulation of the endogenous production of regulatory mechanisms can represent a more physiological way to restore an adequate balance. The lack of response and the occurrence of adverse reactions observed in some patients exposed to cytokines and their inhibitors imply to take into account the level of production and regulation of the target cytokines. Part of such heterogeneity is genetically determined. Considering these issues will allow a better risk benefit assessment of treatment choice for each patient.

REFERENCES

1. Feldmann M, Maini RN. Anti-TNF alpha therapy of rheumatoid arthritis: what have we learned? Annu Rev Immunol 2001;19:163–196.
2. Van Assche G, Rutgeerts P. Anti-TNF agents in Crohn's disease. Expert Opin Invest Drugs 2000;9:103–111.
3. Mosmann TR, Coffman RL. TH1 and TH2 cells: different patterns of lymphokine secretion lead to different functional properties. Annu Rev Immunol 1989;7:145–173.
4. Romagnani S. Human TH1 and TH2 subsets: doubt no more. Immunol Today 1991;12:256–257.
5. Chomarat P, Rissoan MC, Banchereau J, Miossec P. Interferon gamma inhibits interleukin 10 production by monocytes. J Exp Med 1993;177:523–527.
6. Banchereau J, Pascual V, Palucka AK. Autoimmunity through cytokine-induced dendritic cell activation. Immunity 2004;20:539–550.
7. Burger D, Dayer JM. Cytokines, acute-phase proteins, and hormones: IL-1 and TNF-alpha production in contact-mediated activation of monocytes by T lymphocytes. Ann N Y Acad Sci 2002;966:464–473.
8. Arend WP, Dayer JM. Inhibition of the production and effects of interleukin-1 and tumor necrosis factor alpha in rheumatoid arthritis. Arthritis Rheum 1995;38:151–160.
9. Bresnihan B. Anakinra as a new therapeutic option in rheumatoid arthritis: clinical results and perspectives. Clin Exp Rheumatol 2002;20:S32–S34.
10. Nishimoto N, Yoshizaki K, Miyasaka N, et al. Treatment of rheumatoid arthritis with humanized anti-interleukin-6 receptor antibody: a multicenter, double-blind, placebo-controlled trial. A pilot randomized trial of a human anti-interleukin-6 receptor monoclonal antibody in active Crohn's disease. Arthritis Rheum 2004;50:1761–1769.
11. Ito H, Takazoe M, Fukuda Y, Hibi T, et al. A pilot randomized trial of a human anti-interleukin-6 receptor monoclonal antibody in active Crohn's disease. Gastroenterology 2004;126:989–996.
12. McInnes IB, Gracie JA. Interleukin-15: a new cytokine target for the treatment of inflammatory diseases. Curr Opin Pharmacol 2004;4:392–397.
13. Miossec P. Interleukin-17 in rheumatoid arthritis: if T cells were to contribute to inflammation and destruction through synergy. Arthritis Rheum 2003;48:594–601.
14. Trinchieri G, Pflanz S, Kastelein RA. The IL-12 family of heterodimeric cytokines: new players in the regulation of T cell responses. Immunity 2003;19:641–644.

15. Kawashima M, Novick D, Rubinstein M, Miossec P. Regulation of interleukin-18 binding protein production by blood and synovial cells from patients with rheumatoid arthritis. Arthritis Rheum 2004;50:1800–1805.

16. Aggarwal S, Ghilardi N, Xie MH, de Sauvage FJ, Gurney AL. Interleukin-23 promotes a distinct CD4 T cell activation state characterized by the production of interleukin-17. J Biol Chem 2003;278:1910–1914.

17. Miossec P, van den Berg W. Th1/Th2 cytokine balance in arthritis. Arthritis Rheum 1997;40:2105–2115.

18. Braun J, Bollow M, Neure L, et al. Use of immunohistologic and in situ hybridization techniques in the examination of sacroiliac joint biopsy specimens from patients with ankylosing spondylitis. Arthritis Rheum 1995;38:499–505.

19. Wynn TA. Fibrotic disease and the T(H)1/T(H)2 paradigm. Nat Rev Immunol 2004;4:583–594.

20. Llorente L, Richaud-Patin Y. The role of interleukin-10 in systemic lupus erythematosus. J Autoimmun 2003;20:287–289.

21. Houssiau FA, Lefebvre C, Vanden Berghe M, Lambert M, Devogelaer JP, Renauld JC. Serum interleukin 10 titers in systemic lupus erythematosus reflect disease activity. Lupus 1995;4:393–395.

22. Balabanian K, Couderc J, Bouchet-Delbos L, et al. Role of the chemokine stromal cell-derived factor 1 in autoantibody production and nephritis in murine lupus. J Immunol 2003;170:3392–3400.

23. Ishida H, Muchamuel T, Sakaguchi S, Andrade S, Menon S, Howard M. Continuous administration of anti-interleukin 10 antibodies delays onset of autoimmunity in NZB/W F1 mice. J Exp Med 1994;179:305–310.

24. Llorente L, Richaud-Patin Y, Garcia-Padilla C, et al. Clinical and biologic effects of anti-interleukin-10 monoclonal antibody administration in systemic lupus erythematosus. Arthritis Rheum 2000;43:1790–1800.

25. Pascual V, Banchereau J, Palucka AK, Bennett L, Arce E, Cantrell V, Borvak J. The central role of dendritic cells and interferon-alpha in SLE. Interferon and granulopoiesis signatures in systemic lupus erythematosus blood. Curr Opin Rheumatol 2003;15:548–556.

26. Bennett L, Palucka AK, Arce E, et al. Interferon and granulopoiesis signatures in systemic lupus erythematosus blood. J Exp Med 2003;197:711–723.

27. Hjelmstrom P, Fjell J, Nakagawa T, Sacca R, Cuff CA, Ruddle NH. Lymphoid tissue homing chemokines are expressed in chronic inflammation. Am J Pathol 2000;156:1133–1138.

28. Paleolog E. Target effector role of vascular endothelium in the inflammatory response: insights from the clinical trial of anti-TNF alpha antibody in rheumatoid arthritis. Mol Pathol 1997;50:225–233.

29. Tak PP, Taylor PC, Breedveld FC, et al. Decrease in cellularity and expression of adhesion molecules by anti-tumor necrosis factor alpha monoclonal antibody treatment in patients with rheumatoid arthritis. Arthritis Rheum 1996; 39:1077–1081.

30. Van den Brande JM, Braat H, van den Brink GR, et al. Infliximab but not etanercept induces apoptosis in lamina propria T-lymphocytes from patients with Crohn's disease. Gastroenterology 2003;124:1774–1785.
31. Van den Berg WB. Lessons from animal models of arthritis. Curr Rheumatol Rep 2002;4:232–239.
32. Ziolkowska M, Kurowska M, Radzikowska A, et al. High levels of osteoprotegerin and soluble receptor activator of nuclear factor kappa B ligand in serum of rheumatoid arthritis patients and their normalization after anti-tumor necrosis factor alpha treatment. Arthritis Rheum 2002;46:1744–1753.
33. Shealy DJ, Wooley PH, Emmell E, et al. Anti-TNF-alpha antibody allows healing of joint damage in polyarthritic transgenic mice. Arthritis Res 2002;4:R7.
34. Zwerina J, Hayer S, Tohidast-Akrad M, et al. Single and combined inhibition of tumor necrosis factor, interleukin-1, and RANKL pathways in tumor necrosis factor-induced arthritis: effects on synovial inflammation, bone erosion, and cartilage destruction. Arthritis Rheum 2004;50:277–290.
35. Keane J, Gershon S, Wise RP, et al. Tuberculosis associated with infliximab, a tumor necrosis factor alpha-neutralizing agent. N Engl J Med 2001;345:1098–1104.
36. Casanova JL, Abel L. Genetic dissection of immunity to mycobacteria: the human model. Annu Rev Immunol 2002;20:581–620.
37. Kawashima M, Miossec P. Decreased response to IL-12 and IL-18 of blood cells in rheumatoid arthritis. Arthritis Res Ther 2004;6:R39–R45.
38. Kawashima M, Miossec P. Heterogeneity of response of rheumatoid synovium cell subsets to interleukin-18 in relation to differential interleukin-18 receptor expression. Arthritis Rheum 2003;48:631–637.
39. Carmona L, Hernandez-Garcia C, Vadillo C, et al. Increased risk of tuberculosis in patients with rheumatoid arthritis. J Rheumatol 2003;30:1436–1439.
40. Shakoor N, Michalska M, Harris CA, Block JA. Drug-induced systemic lupus erythematosus associated with etanercept therapy. Lancet 2002;359:579–580.
41. Kontoyiannis D, Kollias G. Accelerated autoimmunity and lupus nephritis in NZB mice with an engineered heterozygous deficiency in tumor necrosis factor. Eur J Immunol 2000;30:2038–2047.
42. Llorente L, Zou W, Levy Y, et al. Role of interleukin 10 in the B lymphocyte hyperactivity and autoantibody production of human systemic lupus erythematosus. J Exp Med 1995;181:839–844.
43. Brown SL, Greene MH, Gershon SK, Edwards ET, Braun MM. Tumor necrosis factor antagonist therapy and lymphoma development: twenty-six cases reported to the Food and Drug Administration. Arthritis Rheum 2002;46:3151–315#8.
44. McLachlan JB, Hart JP, Pizzo SV, et al. Mast cell-derived tumor necrosis factor induces hypertrophy of draining lymph nodes during infection. Nat Immunol 2003;4:1199–1205.
45. Mohan N, Edwards ET, Cupps TR, et al. Demyelination occurring during anti-tumor necrosis factor alpha therapy for inflammatory arthritides. Increased

MRI activity and immune activation in two multiple sclerosis patients treated with the monoclonal anti-tumor necrosis factor antibody cA2. Arthritis Rheum 2001;44:2862–2869.

46. Robinson WH, Genovese MC, Moreland LW. Demyelinating and neurologic events reported in association with tumor necrosis factor alpha antagonism: by what mechanisms could tumor necrosis factor alpha antagonists improve rheumatoid arthritis but exacerbate multiple sclerosis? Arthritis Rheum 2001; 44:1977–1983.

47. Marotte H, Charrin JE, Miossec P. Infliximab-induced aseptic meningitis. Lancet 2001;358:1784.

48. Granet C, Maslinski W, Miossec P. Increased AP-1 and NF-kappaB activation and recruitment with the combination of the proinflammatory cytokines IL-1beta, tumor necrosis factor alpha and IL-17 in rheumatoid synoviocytes. Arthritis Res Ther 2004;6:R190–R198.

49. Granet C, Miossec P. Combination of the pro-inflammatory cytokines IL-1, TNF-alpha and IL-17 leads to enhanced expression and additional recruitment of AP-1 family members, Egr-1 and NF-kappaB in osteoblast-like cells. Cytokine 2004;26:169–177.

50. Chabaud M, Miossec P. The combination of tumor necrosis factor alpha blockade with interleukin-1 and interleukin-17 blockade is more effective for controlling synovial inflammation and bone resorption in an ex vivo model. Arthritis Rheum 2001;44:1293–1303.

51. Genovese MC, Cohen S, Moreland L, et al Combination therapy with etanercept and anakinra in the treatment of patients with rheumatoid arthritis who have been treated unsuccessfully with methotrexate. Arthritis Rheum 2004;50:1412–1419.

52. Ulfgren AK, Andersson U, Engstrom M, Klareskog L, Maini RN, Taylor PC. Systemic anti-tumor necrosis factor alpha therapy in rheumatoid arthritis downregulates synovial tumor necrosis factor alpha synthesis. Arthritis Rheum 2000;43:2391–2396.

53. Mugnier B, Balandraud N, Darque A, Roudier C, Roudier J, Reviron D. Polymorphism at position -308 of the tumor necrosis factor alpha gene influences outcome of infliximab therapy in rheumatoid arthritis. Arthritis Rheum 2003;48:1849–1852.

54. Baert F, Noman M, Vermeire S, et al. Influence of immunogenicity on the long-term efficacy of infliximab in Crohn's disease. N Engl J Med 2003;348: 601–608.

55. Miossec P. Cytokine-induced autoimmune disorders. Drug Saf 1997;17:93–104.

56. Fattovich G, Giustina G, Favarato S, Ruol A. A survey of adverse events in 11,241 patients with chronic viral hepatitis treated with alfa interferon. J Hepatol 1996;24:38–47.

57. Wilson LE, Widman D, Dikman SH, Gorevic PD. Autoimmune disease complicating antiviral therapy for hepatitis C virus infection. Semin Arthritis Rheum 2002;32:163–173.

58. Fattovich G, Betterle C, Brollo L, et al. Autoantibodies during alpha-interferon therapy for chronic hepatitis B. J Med Virol 1991;34:132–135.
59. Batisse D, Karmochkine M, Jacquot C, Kazatchkine MD, Weiss L. Sustained exacerbation of cryoglobulinaemia-related vasculitis following treatment of hepatitis C with peginterferon alfa. Eur J Gastroenterol Hepatol 2004;16:701–703.
60. Agnello V, Chung RT, Kaplan LM. A role for hepatitis C virus infection in type II cryoglobulinemia. N Engl J Med 1992;327:1490–1495.
61. Ferri C, Marzo E, Longombardo G, et al. Interferon-alpha in mixed cryoglobulinemia patients: a randomized, crossover-controlled trial. Blood 1993;81: 1132–1136.
62. Mazzaro C, Zorat F, Comar C, et al. Interferon plus ribavirin in patients with hepatitis C virus positive mixed cryoglobulinemia resistant to interferon. J Rheumatol 2003;30:1775–1781.
63. Murakami CS, Zeller K, Bodenheimer HC Jr., Lee WM. Idiopathic thrombocytopenic purpura during interferon-alpha 2B treatment for chronic hepatitis. Am J Gastroenterol 1994;89:2244–2245.
64. Stricker RB, Barlogie B, Kiprov DD. Acquired factor VIII inhibitor associated with chronic interferon-alpha therapy. J Rheumatol 1994;21:350–352.
65. Lisker-Melman M, Di Bisceglie AM, Usala SJ, Weintraub B, Murray LM, Hoofnagle JH. Development of thyroid disease during therapy of chronic viral hepatitis with interferon alfa. Gastroenterology 1992;102:2155–2160.
66. Haddad J, Deny P, Munz-Gotheil C, et al. Lymphocytic sialadenitis of Sjogren's syndrome associated with chronic hepatitis C virus liver disease. Lancet 1992;339:321–323.
67. Conlon KC, Urba WJ, Smith JW 2nd, Steis RG, Longo DL, Clark JW. Exacerbation of symptoms of autoimmune disease in patients receiving alpha-interferon therapy. Cancer 1990;65:2237–2242.
68. Kiely PD, Bruckner FE. Acute arthritis following interferon-alpha therapy. Br J Rheumatol 1994;33:502–503.
69. Batocchi AP, Evoli A, Servidei S, Palmisani MT, Apollo F, Tonali P. Myasthenia gravis during interferon alfa therapy. Neurology 1995;45:382–383.
70. Pascual V, Banchereau J, Palucka AK. The central role of dendritic cells and interferon-alpha in SLE. Curr Opin Rheumatol 2003;15:548–556.
71. Schmidt KN, Ouyang W. Targeting interferon-alpha: a promising approach for systemic lupus erythematosus therapy. Lupus 2004;13:348–352.
72. Saab S, Kalmaz D, Gajjar NA, et al. Outcomes of acute rejection after interferon therapy in liver transplant recipients. Liver Transpl 2004;10:859–867.
73. Rosenberg SA, Packard BS, Aebersold PM, et al. Use of tumor-infiltrating lymphocytes and interleukin-2 in the immunotherapy of patients with metastatic melanoma. A preliminary report. N Engl J Med 1988;319:1676–1680.
74. Ballmer-Weber BK, Dummer R, Kung E, Burg G, Ballmer PE. Interleukin 2-induced increase of vascular permeability without decrease of the intravascular albumin pool. Br J Cancer 1995;71:78–82.

75. Gaspari AA. Autoimmunity as a complication of interleukin 2 immunotherapy. Many unanswered questions. Arch Dermatol 1994;130:894–898.
76. Francis G. Benefit-risk assessment of interferon-beta therapy for relapsing multiple sclerosis. Expert Opin Drug Saf 2004;3:289–303.
77. Cannon GW, Emkey RD, Denes A, et al. Prospective 5-year followup of recombinant interferon-gamma in rheumatoid arthritis. J Rheumatol 1993;20: 1867–1873.
78. Panitch HS, Hirsch RL, Haley AS, Johnson KP. Exacerbations of multiple sclerosis in patients treated with gamma interferon. Lancet 1987;1:893–895.
79. Stuart K, Levy DE, Anderson T, et al. Phase II study of interferon gamma in malignant carcinoid tumors (E9292): a trial of the Eastern Cooperative Oncology Group. Invest New Drugs 2004;22:75–81.
80. Riddell LA, Pinching AJ, Hill S, et al. A phase III study of recombinant human interferon gamma to prevent opportunistic infections in advanced HIV disease. AIDS Res Hum Retroviruses 2001;17:789–797.
81. Cebon J, Jager E, Shackleton MJ, et al. Two phase I studies of low dose recombinant human IL-12 with Melan-A and influenza peptides in subjects with advanced malignant melanoma. Cancer Immun 2003;3:7.
82. Sangro B, Mazzolini G, Ruiz J, et al. Phase I trial of intratumoral injection of an adenovirus encoding interleukin-12 for advanced digestive tumors. J Clin Oncol 2004;22:1389–1397.
83. Hoshina Y, Moriuchi J, Nakamura Y, Arimori S, Ichikawa Y. CD4+ T cell-mediated leukopenia of Felty's syndrome successfully treated with granulocyte-colony-stimulating factor and methotrexate. Arthritis Rheum 1994;37: 298–299.
84. Yasuda M, Kihara T, Wada T, et al. Granulocyte colony-stimulating factor induction of improved leukocytopenia with inflammatory flare in a Felty's syndrome patient. Arthritis Rheum 1994;37:145–146.

13

Evaluation of Antibodies in Clinical Trials of Cytokines

Steven Swanson

SUMMARY

A key component for the success of any therapeutic cytokine is the degree of immunogenicity mediated by the cytokine. There are many factors that can contribute to a cytokine being immunogenic, such as aggregation. The understanding of cytokine immunogenicity requires a thorough knowledge of the antibody testing that was performed. Some of the methods used to test for antibodies against cytokines are the enzyme-linked immunosorbent assay, radioimmune precipitation assay, electrochemiluminescence, and surface plasmon resonance platforms. To understand whether an anticytokine antibody can neutralize the biological effect of a cytokine, it is necessary to perform a bioassay. The significance of cytokine immunogenicity can only be understood in the context of antibody testing data coupled with clinical data related to effects on the patient.

Key Words: Antibody; immunogenicity; ELISA; RIP; bioassay; surface plasmon resonance; Biacore.

1. INTRODUCTION

An important issue in determining the success of a cytokine therapeutic is whether that cytokine induces an immune response. The immunogenicity of the cytokine could range from nonimmunogenic, where antibodies are not generated, to extremely immunogenic, where most subjects exposed to the cytokine develop specific antibodies in response to that exposure. The clinical relevance of an anticytokine antibody is a result of the quantity of antibody generated, what region of the cytokine the antibody binds to, and whether the cytokine has a redundant mechanism of action. When an anti-

From: *Methods in Pharmacology and Toxicology: Cytokines in Human Health: Immunotoxicology, Pathology, and Therapeutic Applications*
Edited by: R. V. House and J. Descotes © Humana Press Inc., Totowa, NJ

body neutralizes the effect of a cytokine, it is much more significant to the patient if there is not an alternate means of accomplishing the biological action of that cytokine. There are many contributing factors that can lead to immunogenicity in all protein therapeutics, and these factors may work independently or in concert. Some of the factors that can cause a protein to be more immunogenic include sequence differences from the native molecule, impurities, and aggregation.

Before considering the effects of immunogenicity, it is important to have an understanding of the methodology used to detect and monitor immunogenicity in clinical studies. There are two major classes of assays used to explore the immune response: immunoassay and bioassay. Although an immunoassay can detect antibodies capable of binding to a therapeutic protein, only a bioassay is able to determine whether those antibodies are capable of neutralizing the biological effect of the protein. A bioassay is typically a cell-based platform in which the cytokine is added to a culture of growing cells and induces a measurable change in those cells. If that induced change is blocked when antibodies from a subject are added along with the cytokine, the antibodies are defined as being neutralizing. A neutralizing antibody is so named because it blocks or neutralizes the biological effect a protein can exert on a cell. It is neutralizing antibodies that have the greatest potential to impact the efficacy of the therapeutic cytokine. It is important to determine whenever neutralizing antibodies occur; however, tests should be conducted to detect the presence of all antibodies capable of binding to the protein. Even antibodies that are not identified as neutralizing may have important clinical ramifications and therefore must be carefully examined.

2. FACTORS THAT CAN CONTRIBUTE TO IMMUNOGENICITY

Humans have developed tolerance for native cytokines and except in rare exceptions do not produce antibodies capable of binding to these native cytokines. When therapeutic cytokines are administered to subjects, this tolerance is sometimes broken and antibodies are generated that can bind and sometimes neutralize the effect of the cytokine. The presence of antibodies is suspected when a subject that has previously responded well to a therapeutic cytokine no longer has the same response. The lack of response is generally the first evidence that the subject has mounted an immune response to the therapeutic.

The reasons why a given cytokine breaks tolerance and leads to an immune response, sometimes after being treated with this cytokine for many months, are not fully understood, but some of the factors associated with immunogenicity can be categorized. Studies have not yet been conducted that can quantify the risk of immunogenicity based on these risk factors. Immunogenicity to a protein therapeutic can not be reliably predicted using empirical methods or even conducting animal studies. The only test that can determine whether a protein will be immunogenic is to conduct an appropriately powered clinical study. Animal models may provide some insight into immunogenicity in humans; however, the results from such studies should not be viewed as conclusive evidence for or against the likelihood of a particular cytokine to induce an immune response. There are many companies that are currently developing various models to predict immunogenicity; however, we are still many years away from reliable immunogenicity prediction.

When therapeutic cytokines deviate from the native sequence it is more likely that the immune system will be able to identify these as foreign and respond by producing antibodies *(1)*. Certain contaminants and process related impurities could lead to an increase in immunogenicity, either by acting as an adjuvant or by causing oxidation *(2)* or aggregation *(3)* of the cytokine. When proteins are aggregated, it is more likely that the immune system will respond. Aggregated material is more prone to phagocytosis and subsequent antigen processing that can lead to an immune response.

The route chosen for the administration of the cytokine also can influence the rate of immunogenicity. A given cytokine administered subcutaneously would be more likely to generate an immune response than the same cytokine administered intravenously. Because therapeutic cytokines are foreign proteins, it is always a possibility that regardless of the route of administration, an immune response could be generated. There is not a route of administration that could absolutely protect against raising an immune response. Another factor that can influence the rate of immunogenicity is the frequency and duration of treatment. The immune system responds best to foreign proteins when they are administered over a long period of time and when the protein is allowed to clear before subsequent administration. This cycling of exposure, clearance, and re-exposure allows the immune system to respond fully. By contrast, if circulating cytokine levels are maintained, the immune system would more likely become tolerant and no longer recognize the therapeutic cytokine as a foreign protein. Because it does take time for the immune system to mount a robust response, those therapeutic cytokines that are not administered chronically are less likely to induce a strong immune response. Patients that are receiving immunosuppressive therapy conjointly with the therapeutic cytokine are also less likely to mount an immune response to the cytokine.

Additional factors that can influence immunogenicity of therapeutic cytokines are the formulation used and the storage and handling conditions. Proteins can be fragile compounds that must be protected from environmental challenges such as temperature extremes, vigorous shaking, and excessive freeze-thaw cycles. Therapeutic cytokines are formulated by adding reagents such as human serum albumin to help stabilize the protein and inhibit aggregation. The formulation of a protein therapeutic is specific for that cytokine and is designed to protect the integrity of the cytokine. Each formulated cytokine has its own unique stability profile and some cytokines are more fragile than others. When cytokines become unstable one key event that can occur is aggregation.

3. METHODS FOR ANTIBODY DETECTION

There are many different platforms available for testing samples to determine whether anticytokine antibodies are present. None of the available methods are 100% accurate for detecting specific antibodies; rather, each assay has specific strengths and weaknesses. The most popular of these methods are described with advantages and disadvantages indicated. When trying to interpret results from antibody analyses and determine the impact on a clinical program, it is important to understand and evaluate the results from these assays. As an example, an assay may indicate that most subjects that are administered a particular cytokine mount an immune response. If the assay used to detect these antibodies is not specific, the assay could falsely identify samples as positive for antibodies that are in fact negative. If the characteristics of the assay to detect immunogenicity are not known, it is not possible to assess the clinical impact of a cytokine program. To fully understand how to interpret the results from antibody analyses, it is imperative that the assays used be validated. The validation process will allow an understanding of the limitations of the assay and this will help prevent misinterpretation of the data. As an example, if an assay indicates there are no antibodies present, the data must be interpreted with the knowledge of the sensitivity of that assay. If an assay's sensitivity is 10 µg/mL (the assay will detect antibodies in a sample only if the concentration of circulating antibody exceeds the level of 10 µg/mL), a negative result would provide less confidence that there were no antibodies present than if the assay sensitivity was 10 ng/mL. Only through the full understanding of the assay's limits can clinical results of an antibody analysis be correctly interpreted. Every assay for the detection and characterization of an immune response is limited by the sample that is taken. If that sample is taken at a time when there is a large amount of circulating cytokine, it will be difficult to impossible for the assay to identify the presence of the antibody. Circulating drug can have the effect of binding to the antibody in the serum sample and thereby blocking the binding sites on the antibody that prevent the antibody's detection in the assay, leading to underreporting of the immune response.

3.1. Immunoassays

There are many different types of immunoassays for the detection of anticytokine antibodies, but each is designed to detect the presence of antibodies that are capable of binding to the cytokine. Each of the described platforms has distinct advantages as well as limitations that need to be considered when interpreting immunogenicity results. Four of the commonly used assay platforms are enzyme-linked immunosorbent assay (ELISA), radioimmune precipitation assay (RIP), surface plasmon resonance (SPR), and electrochemiluminescence (ECL).

3.1.1. ELISA

The ELISA *(4)* has been a mainstay for the detection of antibodies since its development more than 20 yr ago. This platform (Fig. 1) is based upon first immobilizing the cytokine onto a microtiter plate through passive adsorption, covalent coupling, biotinylation, or other methods. After the cytokine is attached, a blocking step is performed during which an irrelevant protein such as bovine serum albumin is added to the wells of the plate for a period of time to occupy or block any region of the well not coated with the cytokine. This blocking step is critical to prevent nonspecific sticking of subsequent reagents that could be perceived as anticytokine antibodies. After the wells are blocked, a dilution of patient serum is added to the well and antibodies specific for the cytokine are allowed to bind to the immobilized cytokine. After the unbound serum components are washed away, a labeled secondary reagent capable of binding to antibodies that have bound to the immobilized cytokine is added. This secondary reagent could be an antihuman immunoglobulin reagent that has been conjugated to the enzyme phosphatase or peroxidase. After incubation and removal of unbound reagents through a washing step, the substrate for the conjugated enzyme is added and the colorimetric result of the enzymatic reaction is monitored. The color observed is proportional to the amount of anticytokine antibody that has been bound to the immobilized cytokine on the plate. Another version of the ELISA uses labeled cytokine as the detector. This bridging format is a very specific way of detecting antibodies and is less likely to be influenced by nonspecific binding. A disadvantage of the bridging platform is that it tends to be less sensitive than the direct ELISA.

The ELISA platform is preferred by many investigators because it is an easy technique to learn and requires minimal equipment. ELISAs can be

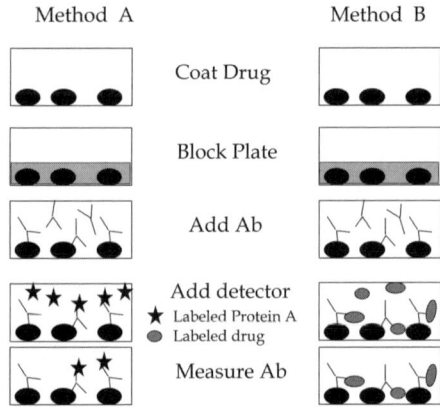

Fig. 1. Two versions of an enzyme-linked immunosorbent assay (ELISA) are shown. In both methods, the drug is coated onto the surface of the wells of a microtiter plate. The wells are then blocked using bovine serum albumin, gelatin, or a powdered milk solution to reduce nonspecific interaction. The serum sample is added and any antibodies capable of binding to the immobilized drug are allowed to bind. After incubation and washing, in Method A, a detecting antibody or labeled protein A is added, and in Method B, labeled drug is added. After incubation and washing, a colorometric solution is added and the change in color intensity is proportional to the amount of antibody present in the serum sample.

very sensitive and robust analytical procedures. Care must be taken when developing an ELISA that the assay is specific and that it only detects antibodies that can bind to the cytokine. If the ELISA plate is improperly blocked or if reagents are chosen that demonstrate cross-reactivity it is possible for the assay to falsely identify samples as positive when that sample does not contain anticytokine antibodies. An additional concern with the ELISA is that antibodies that have low affinity and tend to rapidly dissociate from the cytokine after binding may not be identified as positive. The inability of ELISAs to detect low affinity antibodies is likely related to the multiple washing and incubation steps inherent with this method. These low-affinity antibodies could dissociate from the cytokine and then be rinsed away during a subsequent wash cycle.

3.1.2 Radioimmune Precipitation Assays

Some investigators still prefer the more traditional RIP method for antibody detection (Fig. 2). This method relies on a liquid-phase interaction between the cytokine and the antibody. A predetermined amount of radiolabeled cytokine is added to a dilution of a subject's serum sample. After an

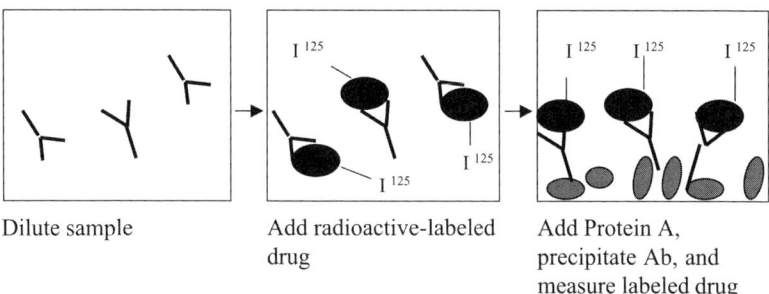

| Dilute sample | Add radioactive-labeled drug | Add Protein A, precipitate Ab, and measure labeled drug |

Fig. 2. The radioimmune precipitation test involves first diluting a serum sample. Radiolabeled drug is then added to the diluted serum sample and any antibodies are allowed to bind. Antibodies present in the serum sample are precipitated by adding protein A or some other precipitating agent. The mixture is then centrifuged, the supernatant decanted, and the amount of radioactivity in the pellet is determined. The amount of antibody able to bind to the drug is proportional to the amount of radioactivity identified in the pellet.

appropriate incubation time, a precipitating reagent, such as protein A or polyethylene glycol, is added to the sample. The precipitating reagent will cause all of the immunoglobulin in a serum sample to precipitate and the amount of specific antibody contained in the original sample will be proportional to the amount of radiolabel that is contained in the precipitate. Advocates of this method point to the simplicity of the method, high sensitivity that can be obtained, relative low cost for the required equipment, and the fluid-based nature of the assay. A fluid-based assay eliminates the need for the immobilization of the cytokine. During immobilization, it is possible that the conformation of the cytokine could be altered or that the cytokine could immobilize in a nonrandom way that could obscure an important epitope or antibody binding region. This could result in false-negative results in an ELISA.

The RIP assays that use protein A as the precipitating reagent are not likely to detect an early immune response to a cytokine because the early immune response is composed primarily of IgM molecules. Although protein A does bind to IgM, it does so only weakly. This weak interaction between protein A and IgM prevents full precipitation of labeled cytokine that has been bound by an IgM molecule and results in under-reporting of antibody positive samples. An additional concern with the RIP platform is that precipitation of all immunoglobulins in a serum sample could result in nonspecific trapping of the radiolabel in the precipitate. The nonspecific trapping of label could result in the over-reporting of specific anticytokine antibodies. Because

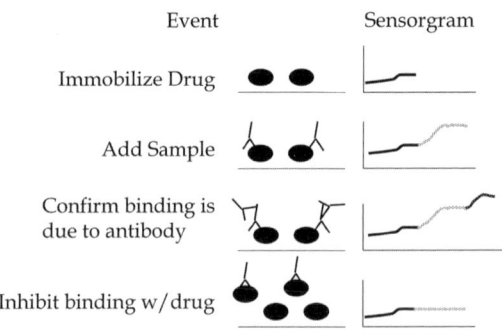

Fig. 3. The surface plasmon resonance test (in this instance using the Biacore 3000) involves first immobilizing the drug onto the surface of a sensorchip, which results in a signal monitored on the sensorgram. When a sample is added that contains antibody able to bind to the immobilized drug, that binding is indicated on the sensorgram. To verify that the binding observed is caused by immunoglobulin, an antihuman antibody is added and a further increase in the sensorgram verifies the initial sample contained antibodies capable of binding to the drug. The sensorgram signal is directly proportional to the amount of mass that binds to the sensor chip. Verification that the antibody is specifically binding to the immobilized drug can be made by first removing the bound antibody (typically with an acid solution) followed by adding the serum sample again in the presence of soluble drug. If the soluble drug prevents binding to the immobilized drug and there is a reduced or absent signal on the sensorgram, the original binding represents a specific antigen-antibody interaction.

this method relies on radioactive materials for detection of antibodies, there are additional costs and required documentation practices that are required for the safe disposal of the spent radioactive material. Additional training is also necessary before initiating the use of radiolabeled materials.

3.1.3. Surface Plasmon Resonance

One example of an instrument that relies on SPR *(5)* is the Biacore (Fig. 3). This instrument has been gaining support throughout the industry over the last several years. This platform relies on first coupling the cytokine to a sensor chip *(6)*. The sensor chip is a glass slide covered with a gold film that has carboxymethyl dextran attached to it. This provides an advantage to coupling the cytokine nonspecifically to a polysorbate plate in that attachment to carboxymethyl dextran is less likely to distort the conformation of the molecule. Having less distortion presents the cytokine to any antibodies in a way that is more like the way the antibody sees the cytokine in vivo. As the cytokine immobilizes onto the surface, the instrument provides a read out in the form of a sensorgram that is proportional to the mass accumulation on the surface of the sensorchip (Fig. 4). This provides data confirming the cytokine has been immobilized.

Once the cytokine has been immobilized onto the sensorchip, the serum sample can be added and, if antibodies bind to the cytokine, the instrument generates an increase in the sensorgram proportional to the amount of antibody that binds. The instrument thereby provides a real-time assessment of binding events as they occur. Unlike more traditional immunoassay platforms such as ELISA and RIP, this instrument does not wait until the conclusion of an assay to provide data. This feature makes the Biacore better suited for the detection of low affinity antibodies or antibodies that dissociate quickly from the immobilized cytokine *(7)*. There have been instances in which low-affinity or rapidly dissociating antibodies have been clinically important, and therefore it is important for any immunoassay platform to be designed in a way to promote the detection of these antibodies.

After the antibody has bound to the immobilized cytokine on the sensorchip, it is possible through the addition of specific isotyping reagents to determine what class of antibody has been generated. The first or primary antibody that is typically produced in response to a foreign protein is IgM. After subsequent exposure to the cytokines, in cases where a robust immune response is generated, IgG is produced. The ability to monitor the class of antibody that is circulating in response to the cytokine can help in the understanding of the progression of the immune response.

This technology offers the advantages of automated sample analysis, real-time detection to better enable detection of low-affinity antibodies, sample immobilization that better preserves the native conformation of the cytokine, and the ability to further characterize the immune response. When second-generation products are investigated, the Biacore allows detection of antibodies to four different entities from the same serum sample. This feature allows analysis of both the original therapeutic and the second generation therapeutic at the same time *(8)*. The Biacore is considerably more expensive than the equipment required for other assay platforms and is somewhat limited by throughput in that a typical assay of 100 samples could take longer than 8 h. A further limitation of this instrument is in assay sensitivity. It is generally possible to achieve greater sensitivity for antibody assays using other platforms, but the advantages offered by this platform in many instances outweigh the deficiencies.

3.1.4. Electrochemiluminescence Assay

ECL assays have many features in common with an ELISA. The method for detection of antibodies is different because ECL assays use detection of

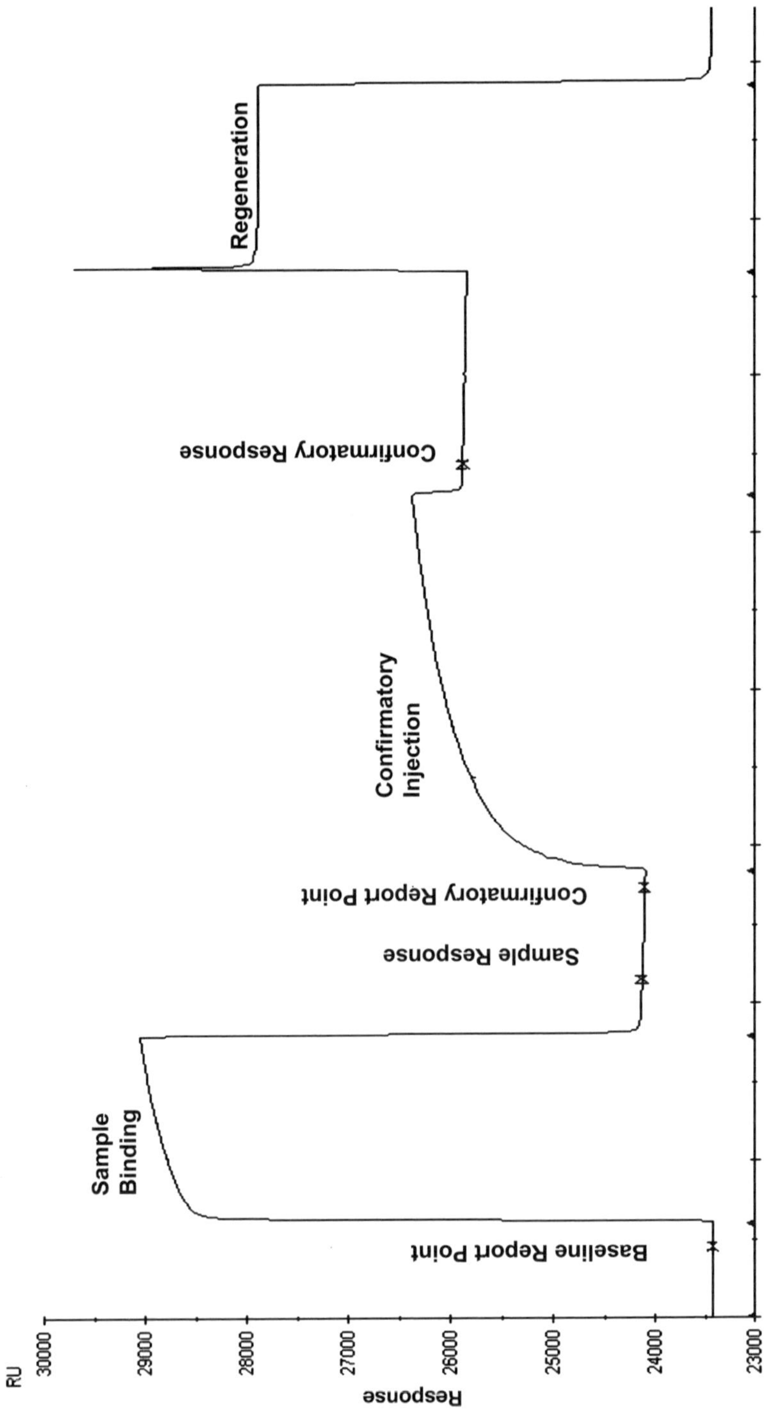

RU
Response
30000
29000
28000
27000
26000
25000
24000
23000
0
100
200
300
400
500
600
700
800
Baseline Report Point
Sample Binding
Sample Response
Confirmatory Report Point
Confirmatory Injection
Confirmatory Response
Regeneration

light that is emitted proportional to the amount of antibody contained in the subject's sample. One version of this assay using the Bioveris system *(9)* relies on immobilization of the cytokine onto magnetic beads. The beads can be blocked with bovine serum albumin or other non-reactive reagents and are then ready for analysis. The subject's serum sample is added to the beads and any antibodies present have the opportunity to bind. Because antibodies are bivalent, there is still one binding site available after the antibody has bound to the immobilized cytokine. This free site can then bind to soluble cytokine that has been coupled to ruthenium. This complex is then pulled toward a magnet and in the presence of the luminescence reagent, light is emitted when an electrical pulse is made. The amount of light emitted is directly proportional to the amount of anti-cytokine antibody present in the original sample.

This assay is very sensitive and typically very specific, in part due to the use of the magnetic beads as a base for the immobilized cytokine. Versions of this assay that rely on microtiter plates for immobilizing the cytokine, such as Meso Scale Discovery, have a higher throughput. The ECL assay in this format is able to detect all classes of antibodies because the antibody in the serum sample creates a bridge between the cytokine immobilized on the magnetic bead and the cytokine that is soluble and conjugated to rhuthenium. A challenge with this assay lies in detection of low-affinity antibodies; however, through careful examination of washing parameters, this assay platform is capable of detecting low affinity antibodies.

3.2. Bioassays

Biological assays differ from immunoassays in that they rely on the interaction of a cell-based system. In the bioassay, a cell line is identified that is capable of responding to the cytokine in a way that can be measured. The types of cellular response that have been used as measurements include cell proliferation, production of a cytokine, apoptosis, and production of specific messenger ribonucleic acid. To perform a bioassay for neutralizing

---

Fig. 4. *(opposite page)* A typical sensorgram is shown. The first report point represents the baseline or sensorgram value prior to sample addition. The difference between the sample response and the baseline report point represents the magnitude of the antibody binding. The difference between the confirmatory response and confirmatory report point represents the magnitude of the confirmatory reagent (typically an antihuman immunoglobulin) and verifies the original binding observed with the sample is a result of an antibody.

antibodies, cells, the cytokine and the subject's serum sample are added together and incubated. If there is an antibody present in the serum sample that is capable of neutralizing the biological effect of the drug, the measured response of the cells is reduced compared with the negative control. The negative control contains the cells, cytokine, and a serum sample known to be negative for the presence of neutralizing antibodies. The positive control for this assay would substitute a known positive sample for the subject's serum.

Because most cells require time to exert the biological response from the cytokine, these assays take on average 2 to 3 d to complete. Because the bioassay is a cell-based system there is more variability associated with this type of assay than is observed with most immunoassay platforms. Understanding the variability of a bioassay is very important to the interpretation of clinical results obtained from that assay. When designing a bioassay, it often is difficult to identify a cell line that has a robust immune response to the therapeutic cytokine. In some of these cases, it is necessary either to engineer a cell line that can respond or occasionally to use primary cells. The use of primary cells is a powerful technique because it better accounts for population variability, however, primary cell assays tend to be less reproducible than assays relying on well-characterized cell lines. When conducting an entire clinical development program for a therapeutic cytokine, it is important to have consistent assays. When a large degree of assay variability occurs, it is possible to misinterpret assay results that lead to incorrect assumptions on a cytokine's immunogenicity.

The bioassay is important because it is the only way to understand if a serum sample contains an antibody that can neutralize the effects of the drug. Neutralizing antibodies typically are of greater concern to the physician because it strongly suggests that the patient has lost (or will soon lose) the biological effect of the drug.

4. EXAMPLES OF IMMUNOGENICITY OF THERAPEUTIC CYTOKINES

There have been many instances of therapeutic proteins that have generated an immune response in patients. Antibodies typically take time to develop and may not appear for several months after initiation of cytokine therapy. In other instances, the immune response is quick and robust. One recent example was reported in Europe when subjects receiving recombinant human erythropoietin (rHuEPO) developed pure red cell aplasia *(10)*. This is a rare condition in which a bone marrow biopsy reveals a marrow that is devoid of red blood cell precursors while maintaining white cell and platelet precursors. It was identified that patients were developing neutralizing antibodies against rHuEPO and that these antibodies were resulting in pure red cell aplasia. The antibodies were capable of not only neutralizing the therapeutic rHuEPO but also neutralizing endogenous erythropoietin. It has not been determined why these patients developed neutralizing antibodies.

The antibodies have been characterized *(11)* as belonging to the IgG class, notably both IgG1 and IgG4. The antibodies are capable of binding to all commercial forms of rHuEPO and have been identified using a variety of assay methodologies. The concentration of these anti-rHuEPO antibodies has been reported between 1 and 69 µg/mL *(11)*.

The antibodies causing pure red cell aplasia are examples of highly clinically relevant antibodies. It is not common for antibodies, even neutralizing antibodies, to exert such a dramatic effect. When a cytokine is involved in a key step in a non-redundant pathway an immune response can be very problematic for the patient. In the case of these antierythropoietin antibodies, the only mechanism for producing red blood cells was shut down when their erythropoietin was neutralized.

Another example of clinically relevant antibodies is the immunogenicity of the recombinant thrombopoietin molecules *(12,13)*. This represents another class of compounds in which neutralizing antibodies crossreact and neutralize the endogenous molecule, in this case, thrombopoietin.

In the case of interferon alpha 2a, antibody formation has been identified as high as 30% to 40% *(14–16)*. This can be contrasted to the lower immunogenicity reported with interferon alpha 2a *(17)*. These products underscore the challenges in trying to compare similar products based solely upon immunogenicity. What seems on the surface to be a straightforward comparator is complicated when different assay methodologies are used for the antibody assessment *(18)*. Although there is great value in comparing immunogenicity between marketed products, it is important to fully understand the limitations of the methods used for antibody detection as well as the details on the dosing schedule for the cytokines being compared.

Interleukin-2 also has had reported immunogenicity *(19,20)* ranging >50% with neutralizing antibodies between 1% and 10%. GM-CSF also has been indicated as immunogenic with antibodies seen in a range between 4% and 55% of patients tested. In some cases, these antibodies are reported as neutralizing and responsible for a diminished clinical effect *(21)*. Another example of different manufacturers of a cytokine producing products with different immunogenicity is interferon beta *(22–25)*. In this case, the version manufactured in *Escherichia coli* has apparently a higher rate of immunogenicity than the Chinese hamster ovary (CHO)-derived product *(22–24)*. The overall immunogenicity for this class of compounds ranges from 10% to 45%. One possible explanation is that the glycosylation associated with the CHO products helps prevent the initiation of an immune response.

These examples demonstrate the strong potential for recombinant cytokines to induce an immune reaction in patients. The importance of antibodies is that they can have a clinical effect. The antibodies have the potential for affecting the pharmacokinetics profile of the cytokine *(26)*. The antibodies could either cause the cytokine to be cleared more quickly or could delay the clearance. Accelerating the clearance of the drug would result in less of the cytokine being available for the patient which could result in a loss of efficacy and force a dose escalation. In this case, the effect of the antibody is to clinically neutralize the drug by making it unavailable. This phenomenon could occur even if the antibodies are not neutralizing as identified by bioassay. When antibodies prolong the exposure of the cytokine it is important to monitor the patient for any signs of overdosing. In addition to altering the pharmacokinetics or biologically neutralizing the effects of a therapeutic cytokine, it is also possible for the antibody to mediate an allergic reaction *(26)*.

5. CONCLUSION

When considering the immunogenicity of therapeutic cytokines, it is important to evaluate all of the data available. The most important data are provided by the analytical assays that identify the presence of the antibody and then characterize those antibodies. By knowing the isotype, concentration, affinity, and ability to neutralize, an informed decision concerning the likely effect on the patient can be made. The most important effect an antibody could exert would manifest itself clinically. The entire scope of the immune response should always be considered, which includes clinical data from the patient in addition to the antibody assay result. Many antibodies that are identified in patients do not exert any clinical effect This lack of clinical effect could be the result of insufficient antibody being produced, the antibody binding to a noncritical portion of the cytokine, or the pathway of the cytokine has sufficient biological redundancy to allow back-up means to achieve the cytokine's biological activity.

There exist many platforms for the detection and characterization of antibodies. In order to interpret results from an analytical procedure there needs to be a clear understanding of the assay parameters and limitations. In some instances it may be necessary to perform a panel of assays to fully understand and describe the immune response to a therapeutic cytokine.

REFERENCES

1. Schellekens H. Bioequivalence and the immunogenicity of biopharmaceuticals. Nat Rev Drug Discov 2002;1:457–462.
2. Hochuli E. Interferon immunogenicity: technical evaluation of interferon-α2A. J Interferon Cytokine Res 1997;17:S15–S21.
3. Moore WV, Leppert P. Role of aggregated human growth hormone (hGH) in development of antibodies to hGH. J Clin Endocrinal Metab 1980;51:691–697.
4. Voller A, Bidwell D, Bartlett A. Enzyme-linked immunosorbent assay. In: Rose NR, Friedman H, eds. Manual of clinical immunology. 2nd ed. Washington, DC: American Society of Microbiology, 1980, pp. 359–371.
5. Knower T, Carlson J, Yarmuch DM, Yarmush ML, Granzow R. The characterization of antibody-antigen interactions by a novel sensor technology: BIACORE™. Biophys J 1991;59:170a.
6. Swanson SJ, Mytych D, Ferbas J. Use of biosensors to monitor the immune response. Dev Biol 2002;109:71–78.
7. Swanson SJ. New technologies for the detection of antibodies to therapeutic proteins. In: Immunogenicity of Therapeutic Biological Products. Dev. Biol, vol 112. Basel: Karger; 2003, pp 127–133.
8. Takacs MA, Jacobs SJ, Bordens RM, Swanson SJ. Detection and characterization of antibodies to PEG-interferon alfa-2b using surface plasmon resonance. J Interferon Cytokine Res 1999;19:781–789.
9. Swanson SJ, Jacobs SJ, Mytych D, Shah C, Indelicato SR, Bordens RW. Applications for the new electrochemiluminescent (ECL) and biosensor technologies. Dev Biol Stand 1999;97:135–147.
10. Casadevall N, Nataf J, Viron B, et al. Pure red cell aplasia and antierythropoietin antibodies in patients treated with recombinant erythropoietin. N Engl J Med 2002;346:469–475.
11. Swanson SJ, Ferbas J, Mayeux P, Casadevall N. Evaluation of methods to detect and characterize antibodies against recombinant human erythropoietin. Nephron Clin Pract 2004;96:c88–95.
12. Li J, Yang C, Xia Y, Bertino A, Glaspy J, Roberts M, Kuter DJ. Thrombocytopenia caused by the development of antibodies to thrombopoietin. Blood 2001; 98:3241–3248.
13. Hardy L, Rogers B, Thomas D, et al. Thrombocytopenia and antigenicity assessment of thrombopoietin treated chimpanzees and rhesus monkeys. Toxicologist 1997;36:277.
14. Hanley JP, Haydon GH. The biology of interferon-alpha and the clinical significance of anti-interferon antibodies. Leuk Lymphoma 1998;29:257–268.
15. Kontsek P, Liptakova H, Kontsekova E. Immunogenicity of interferon-alpha 2 in therapy: structural and physiological aspects. Acta Virol 1999;43:63–70.
16. Ryff, JC. Clinical investigation of the immunogenicity of interferon-α2A. J Interferon Cytokine Res 1997;17:S29–S33.

17. Lindsay KL, Trepo C, Heintges T, et al. Hepatitis interventional therapy group. A randomized, double-blind trial comparing pegylated interferon alfa-2b to interferon alfa-2b as initial treatment for chronic hepatitis C. Hepatology 2001; 34:395–403.
18. Schellekens H, Ryff JC, van der meide PH. Assays for antibodies to human interferon-α: the need for standardisation. J Interferon Cytokine Res 1997;17:5–8.
19. Allegretta M, Atkins MB, Dempsey RA, et al. The development of anti-interleukin-2 antibodies in patients treated with recombinant human interleukin-2 (IL-2). J Clin Immunol 1986;6:481–490.
20. Scharenberg JG, Stam AG, von Blomberg BM, et al. The development of anti-interleukin-2(IL-2) antibodies in cancer patients treated with recombinant IL-2. Eur J Cancer 1994;30:1804–1809.
21. Ullenhag G, Bird C, Ragnhammar P, et al. Incidence of GM-CSF antibodies in cancer patients receiving GM-CSF for immunostimulation. Clinical Immunology 2001;99:65–74.
22. Larocca AP, Leung SC, Marcus SG, Colby CB, Borden EC. Evaluation of neutralizing antibodies in patients treated with recombinant interferon-beta ser. J Interferon Res 1989;9:S51–S60.
23. Perini P, Facchinetti A, Bulian P, et al. Interferon-beta (INF-beta) antibodies in interferon-beta1a- and interferon-beta1b-treated multiple sclerosis patients. Prevalence, kinetics, cross-reactivity, and factors enhancing interferon-beta immunogenicity in vivo. Eur Cytokine Netw 2001;12:56–61.
24. Abdul-Ahad AK, Galazka AR, Revel M, Biffoni M, Borden EC. Incidence of antibodies to interferon-beta in patients treated with recombinant human interferon-beta 1a from mammalian cells. Cytokines Cell Mol Ther 1997;3:27–32.
25. Ross C, Clemmesen KM, Svenson M, et al. Immunogenicity of interferon-beta in multiple sclerosis patients: influence of preparation, dosage, dose frequency, and route of administration. Ann Neurol 2000;48:706–712.
26. Koren E, Zuckerman LA, Mire-Sluis AR. Immune response to therapeutic proteins in humans—clinical significance, assessment and predicition. Curr Pharm Biotechnol 2002;3:349–360.

14

Cytokines and Pharmacokinetic Drug Interactions

Kenneth W. Renton

SUMMARY

The expression of cytochrome P450 and P-glycoprotein is altered during the operation of host defense mechanisms. The basis for this interaction is predominantly through cytokine-mediated pathways. Most of the major cytokines, including interleukin (IL)-1α, IL-1β, IL-2β, IL-6, tumor necrosis factor-α, inteferon-α, inteferon-γ, and transforming growth factor-β, are known to downregulate the major forms of cytochrome P450 and P-glycoprotein. In most cases individual cytochrome P450 forms are downregulated at the level of gene transcription, with a resulting decrease in the corresponding messenger ribonucleic acid, protein, and enzyme activity. The cytokine-mediated loss in drug metabolism is channeled predominantly through the modification of specific transcription factors. Similar pathways appear to alter the expression of P-glycoprotein. In clinical medicine, there are numerous examples of a decreased capacity to handle drugs during infections and disease states that involve an inflammatory component and the production of cytokines. The direct administration of cytokines to humans depresses the levels of several cytochrome P450-mediated pathways. The production of cytokines in humans often results in altered drug responses and increased toxicities, which has major implications in inflammation and infection when the capacity of the liver and other organs to handle drugs is severely compromised. Changes in drug-handling capacity during inflammation/infection will continue to be one of the many factors that complicate therapeutics.

Key Words: Cytochrome p450; P-gycoprotein; cytokines; inflammation; infection; adverse drug response; pharmacokinetics.

From: *Methods in Pharmacology and Toxicology: Cytokines in Human Health:*
Immunotoxicology, Pathology, and Therapeutic Applications
Edited by: R. V. House and J. Descotes © Humana Press Inc., Totowa, NJ

1. INTRODUCTION

The absorption, distribution, and metabolism of a large number of commonly used drugs is dependent on the activity of the family of enzymes commonly known as cytochrome P450 and by a number of drug transporter proteins present in different organs. There is a large degree of interindividual variability in both of these systems that are primary determinants for the pharmacokinetic properties of drugs and chemicals. Although some of the variability in drug handling can be explained on the basis of differences in the expression of these enzymes and proteins at the genetic level, most of the observed variation results from exposure to external factors, such as drugs, chemicals, or hormones, and to changes in development, disease state, or diet. Many of the changes elicited by these nongenetic factors are highly variably between individuals, nonpredictable, and intermittent in nature, resulting in variable and unpredictable patient responses to commonly used drugs.

It is now nearly 30 yr since Renton and Mannering predicted that drug-elimination capacity in the liver may be altered by the interferon that is produced during periods of viral infection and that the cause of this effect was the loss in the activity of cytochrome P450 enzymes (1,2). This was followed by several reports indicating that the levels of theophylline, a drug commonly used at that time for asthma treatment, often was increased during periods of relatively mild infectious disease (3–5). It was predicted that the production of interferon during the virus infection contributed to the loss of drug biotransformation activity and subsequent elevation in drug levels (6). It is now known that a number of cytokines and other components of host defense mechanisms evoke a loss of cytochrome P450 enzymes and that this is a multifactorial and widespread consequence of inflammation and infection (7–10). More recent work also has implicated cytokines in the loss of drug transporter proteins in the membranes of a number of different organs, further complicating the pharmacokinetics of drugs during inflammation and infection (11,12). The fact that both major systems, which deal with the transport and elimination of chemicals or drugs, are affected by cytokines suggests that any disease state that includes an inflammatory component or an infectious condition would have a major influence on the way drugs and chemicals are handled and eliminated by the body. This review will consider the alteration of drug biotransformation and transportation processes and the consequent changes in pharmacokinetics in response to the production of cytokines or similar factors produced during periods of inflammation and infection. Consideration will be given to the mechanism of these responses and to the relevance of interactions in human medicine and toxicology.

1.1. Drug Pharmacokinetics

Pharmacokinetics is the term used to describe the absorption, distribution, and elimination or clearance of drugs and chemicals from the body. Clearance rate of drugs, which is a measure of their elimination from the body, often is dependent on the activity of enzymes, commonly known as cytochromes P450, in the liver. These enzymes play a large role in determining the steady state of drug in plasma, drug dosage, and dosage interval. In addition, cytochromes P450 may play some role in the absorption of certain drugs from the small intestine after oral absorption. Another major determinant involved in drug disposition and distribution is the activity of certain transporter proteins such as P-glycoprotein. These transporter proteins have a number of drug substrates and often play a major role in determining the levels, distribution, and elimination of certain drugs.

1.2. Cytochrome P450

The cytochrome P450 enzymes are a family of heme-containing proteins present in large quantities in the liver, although they do exist in other organs, such as kidneys, lungs, and brain. Although a large number of human cytochrome P450 forms have been identified to date, only a few are responsible for the metabolism of the majority of drugs used in clinical medicine *(13)*. More than 70% of the cytochrome P450 found in the human liver is accounted for by the gene families CYP3A4, CYP2D6, CYP1A2, and CYP2C. Although some drugs are metabolized by one specific cytochrome P450, there are a number of instances in which a drug is metabolized by several different cytochrome P450 forms. The use of drugs in the human population is complicated by the fact that several cytochrome P450 gene families express genetic polymorphism, which leads to a large variation in the ability of individual humans to deal with drugs. Superimposed on this variability is the susceptibility of these enzymes to either induction or inhibition by the large number of external and endogenous factors such as disease, gender, age, hormonal state, or the concomitant use of other drugs. Because the activity of cytochrome P450 largely determines the clearance and elimination of many drugs, the changes to this enzyme system caused by these factors is a major determinant in the large variation in drug pharmacokinetics that is observed in the human population.

1.3. P-Glycoprotein

P-glycoprotein (P-gp) is an ATP-dependent efflux pump with a function to move lipophilic drugs and chemicals across the membrane layers and to pump these compounds out of cells *(14)*. P-gp can confer multidrug resistance to tumor cells by pumping these agents from the cell to the external environment. Although the number of substrates for this transporter is relatively limited, some important drugs, including anticancer agents, cardiac glycosides, cyclosporine, and calcium channel blockers depend on P-gp for absorption and distribution characteristics. P-gp is found in the membranes of a large number of organs, including the blood–brain barrier, the gastrointestinal tract, renal proximal tubules, and hepatocytes. In the blood–brain barrier, for example, P-gp prevents the entry of certain substrates into the central nervous system (CNS) by pumping them back across membranes into the capillary blood. In some cases, this activity protects the brain against toxic injury; however, in other cases, this prevents the entry of useful drugs and into the CNS tissue and thereby prevents therapeutic intervention. The loss of P-gp in blood–brain barrier would be a major disadvantage in that it would increase the susceptibility of the brain toxic injury from drugs and chemicals; however, it would be a major advantage in allowing the entry of useful drugs into brain tissue. The function of P-gp in the gastrointestinal tract is to prevent the absorption of drugs by pumping them back into the lumen of the gut. The loss of P-gp in the small intestine would lead to the increased absorption of any drug that is a substrate for this particular transport protein. Any alteration in P-gp in either a positive or negative direction would lead to major changes in pharmacokinetics of any drug that is a substrate for the protein. Such changes would lead to a major redistribution of drugs in an individual and would lead to major changes in steady-state level, dosage requirements, or dosage interval.

2. MODIFICATION OF CYTOCHROME P450 BY CYTOKINES

The inflammatory response usually is initiated by the activation of inflammatory cells, such as macrophages or neutrophils and is characterized by the release of cytokines, mediators, acute phase proteins, and hormones. It is now well established that during inflammation, the expression of cytochrome P450 and resulting drug biotransformation is downregulated, although there are a few examples of upregulation. The loss of cytochrome P450 during inflammation and infection has largely been attributed to the production of cytokines and subsequent loss of gene expression (8,15). Because inflammation and cytokine production is such a key component of many disease states, the alteration of drug biotransformation during any inflammatory process can produce major changes in pharmacokinetics that directly affect therapeutic success, as shown in Fig. 1.

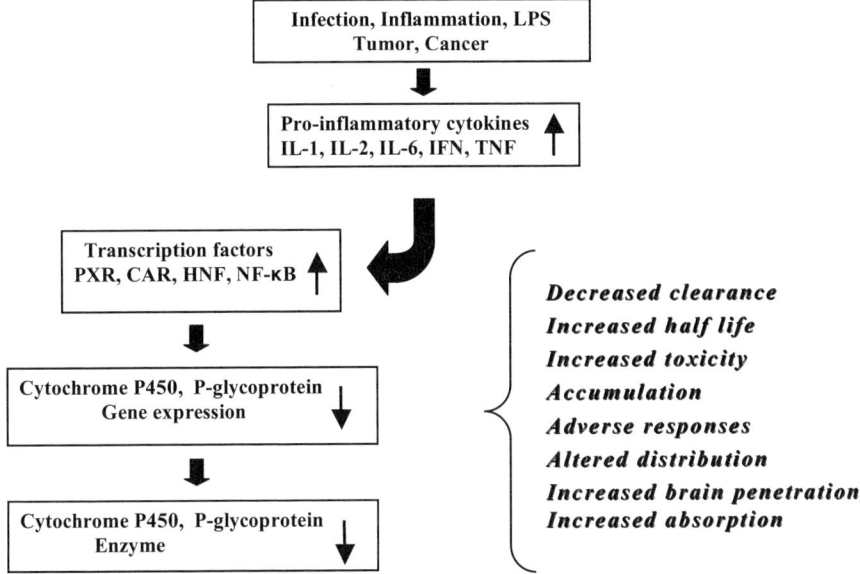

Fig. 1. The alteration of drug biotransformation during any inflammatory process can produce major changes in pharmacokinetics that directly affect therapeutic success.

2.1. Interferons

The loss of cytochrome P450 in response to interferon was first reported in two simultaneously published reports in 1976 *(1,16)*. In response to infections with Newcastle Disease Virus, cytochrome P450 could only be lowered in strains of mice that produced interferon *(17)*. Since then, most forms of the enzyme in the liver of rodents have been shown to be depressed by recombinant interferons of all three major classes *(18–21)*. Constitutive enzymes, including CYP3A2 *(22)*, CYP2C11 *(23)*, CYP2C12 *(24)*, and CYP2E1 *(25)* are downregulated in the rat, and CYP1A2 and CYP2C6 *(26)* are downregulated in the mouse. Inducible forms of the enzyme, including CYP1A1, CYP1A2, and CYP2E1 *(27)*; CYP2B *(28)*; and CYP3A1 *(29)*, also are downregulated. There was a strong correlation in species specificities of interferons in their ability to act as antiviral agents and their ability to lower drug biotransformation *(30)*. In humans, the administration of recombinant α-interferon depresses a number of different cytochrome P450-dependent drug biotransformation pathways *(31–36)*. High doses of IFN-2b given to patients with a melanoma, resulted in a 60% loss in CYP1A2, but CYP2E1 was relatively

Table 1
The Forms of Cytochrome P450 Downregulated by Specific Cytokines

Cytochrome P450 form	Cytokine
CYP1A1	IFN-α, IFN-γ, IFN-2B, IL-1β, IL-6, TNF-α
CYP1A2	IFN-α, IFN-γ, IL-6, TNF-α
CYP2B	TNF-α
CYP2B10	IL-1β, IL-6, TNF-α
CYP2C11	IL-1β, Il-2, TNF-α
CYP2C12	
CYP2C19	IL-6, TNF-α
CYP2D9	TNF-α
CYP2E1	IFN-γ, IL-1β, IL-6, TNF-α
CYP3A2	IL-2, IL-6, TNF-α
CYP3A4	IFN-γ, IL-1β, IL-6, TNF-α
CYP3A6	IFN-γ, IL-1β, IL-6, TNF-α
CYP4A	IL-6

unaffected *(37)*. The magnitude of cytochrome P450 loss appears related to the extent of neurological toxicity and fever often experienced in individuals treated with interferon.

2.2. Interleukins and Other Cytokines

Many of the common cytokines, interleukin (IL)-1α, IL-1β, IL-2β, IL-6, tumor necrosis factor (TNF)-α, and transforming growth factor (TGF)-β depress several cytochrome P450 forms in rodents, and isolated hepatocytes as shown in Table 1 *(38–44)*. In the case of IL-1, the loss of cytochrome P450 can be blocked by an IL-1 receptor antagonists *(45)*. Most of the cytochrome P450 forms that are involved in steroid synthesis pathways are modulated by cytokines *(46)*. Although provoking the loss of cytochrome P450 appears to be a universal property of cytokines, it must be pointed out that different cytokines do not have an equal spectrum of activity toward all enzyme forms. IL-1, which depresses the oxidation of benzphetamine, ethoxycoumarin, and debrisoquine, has no effect on the oxidation of p-nitroanisole *(47)* and only depresses CYP2E1 if the enzyme is induced by previously by dexamethasone *(48)*. TNF-α depresses CYP2C11 and CYP3A2, but CYP2A1 and CYP2C6 are spared *(44)*. When pentoxyfylline is used to block TNF-α production in response to the administration of lipopolysaccharide (LPS), the loss of CYP1A2 and CYP2B are attenuated but CYP2E1 and CYP3A2 are unaffected *(49)*. Of all of the cytokines, IL-6 is the most variable, with widely differentiated response for cytochrome P450 forms

(50,51). LPS-mediated changes in CYP4A and CYP2E1 are dependent on IL-6 but involve another pathway for CYP2D9 *(52)*.

A rather puzzling feature about the response of cytochrome P450s to cytokines is that the loss in specific messenger RNA (mRNA) does not always result in a loss in the corresponding protein. The use of single cytokines in many of the cited studies produces anomalous results. It is well known that in many other responses to inflammation and infection involve the production of several cytokines or cytokine cascades and that the timing and order of appearance of these cytokines often is a critical component of response. The report of Siewart et al. *(53)* supports this contention as the loss of cytochrome P450 is IL-6 dependent in an aseptic inflammation but is dependent on other cytokines in a bacteria-evoked inflammation. Also others have suggested that the variability in response may be caused by the degree of stress that cytokines produce in animal models *(54)*. Despite the fact that some authors have promulgated the idea that a single common cytokine promotes the loss of cytochrome P450, it is much more likely that the nature and timing of the immune activator or activator cascade will determine the extent and magnitude of the response in the intact animal.

2.3. The Effects of Cytokines on Cytochrome P450 in the Brain

Although cytochrome P450 is only present in small amounts in the CNS, it plays a role in the metabolism, activation, and inactivation of a number of drugs and endogenous substances. During CNS inflammation induced by the injection of LPS into the lateral ventricle of rats, CYP1A1/2 is depressed in a number of brain regions, which correlates with the expression of TNF-α, IL-1β, and IL-6 *(55–57)*. The direct administration of these cytokines into the brain also causes a loss in CYP1A in that organ *(57)*. In cultured astrocytes, the addition of LPS, TNF-α, or IL-1β to the cells depressed the levels of CYP1A *(58)*. In the case of LPS, both TNF-α and IL-1β appear in the culture media, and this is accompanied by a concomitant loss in CYP1A. Both the appearance of cytokines and the loss of cytochrome P450 are blocked by the addition of dexamethazone to the media. The production of cytokines within the brain during inflammation is likely to play a role in the alteration of cytochrome P450-based pathways that are critical to the metabolism of drugs and endogenous compounds in the CNS.

2.4. Mechanisms Involved in Cytokine-Mediated Loss in Cytochrome P450

It is now widely accepted that the loss in cytochrome P450 usually results from a decrease in expression of the specific mRNA and subsequent protein synthesis after the administration of cytokines *(8,15)*. The loss in mRNA level precedes the loss in enzyme and activity and is independent of enzyme induction processes. The cytokine-evoked loss in cytochrome P450 has been attributed to the regulation by transcription factors such as nuclear factor (NF)-κB or C/EBP (CCAAT-enhancer binding protein [59,60]). NF-κB is involved in transcription changes of CYP2C11 by IL–1β and NF-1 is involved in the transcription changes to CYP1A1 by TNF-α (59,60). If the CYP2C11 promoter site is altered to prevent NF-αB binding, this enzyme is not altered by IL-1 or LPS (60). IL-6 depresses CYP3A via the induction of C/EBPβ-LAP, a 20kDa C/EBPβ isoform lacking a transactivation domain (61). A loss in CYP2C11 and CYP3A has been attributed to the induction of c-myc by IL-2 (62).

The nuclear hormone receptors CAR and PXR likely play a role in the cytokine mediated loss for CYP2B and CYP3A (63–65). The LPS-mediated decrease in P450 mRNA appears to be associated with the repression of CAR and PXR and LPS prevented the upregulation of CYP3A by the PXR ligand RU486. TNF-α modulation of the phenobarbital evoked induction of CYP2B results from a decrease in nuclear CAR (63). In contrast, it has been suggested that PXR plays a minor role in changes to the expression of constitutive CYP3A (64). Cytokines may alter HNF-1, which is the transcription factor regulating CYP27, the sterol 27-hydroxylase converting cholesterol to 27 hydroxcholesterol (66). There is a parallel loss in the mRNA and protein levels of HNF-1 and CYP27 in response to LPS. The promoter region for CYP27 may be compromised as the binding of HNF-1 to nuclear extracts was diminished. HNF-1α expression and other transcriptional factors that act on the CYP2E1 5'-upstream region may be involved in changes to CYP2E1 (67). HNF-1 is highly dependent on the expression of HNF-4, and both are decreased in response to LPS in the rat (68). HNF-1 pathways are known to depress cytochrome P450 mRNA while suppressing the binding of HNF-4 to deoxyribonucleic acid (DNA [69]). These authors suggest that nitrosylation of the HNF-4-binding domain is a major mechanism to explain the loss of cytochrome P450 in the liver. It is unfortunate that most of the studies on the involvement of transcription factors have involved experiments using LPS rather than cytokines; however, it is likely that these effects of LPS are mediated via certain cytokine pathways. The evidence to date suggests that changes in specific transcription factors in response to cytokines are likely targeted to specific cytochromes P450. It is also very unlikely that a single common transcription factor is involved for all cytochrome P450 forms or for all cytokines.

Studies concerning the effects of inflammation on CYP2E1 have shown that the response of this enzyme to cytokines is very complex and highly dependent on the cytokine involved. This is an important form of the enzyme as it is responsible for a wide variety of drug, solvent, and pro-carcinogen metabolism in addition to a number of key endogenous pathways *(70)*. Although the levels of CYP2E1 are normally controlled at a post transcriptional level, it is down regulated at the transcriptional level by the pro-inflammatory cytokines Il-1β, Il-6, and TNF-α during inflammatory responses *(71,72)*. In contrast the anti-inflammatory cytokine, IL-4, causes an upregulation of human CYP2E1 mRNA and protein level *(67)*. This effect is further complicated by the observation that only the human gene responds in this manner as the rat gene was unresponsive. A recent study proposes that the induction of CYP2E1 by IL-4 in human cells is mediated by a 128-bp DNA sequence which is a target for the binding of several transcription factors *(73)*. In contrast the loss of CYP2E1 in response to IL-1β in the same study was mediated by a separate DNA fragment. The predominant inductive effect of IL-4 therefore could easily mask the smaller depressant effect of IL-1β in an inflammatory response as both cytokines act at different sites. It is interesting to speculate that opposing effects of various cytokines on other cytochrome P450 forms may occur during infection or inflammation and the observed effect on drug disposition is simply dependent on the direction of the predominant action.

Post-transcriptional mechanisms may account for some cytochrome P450 loss in response to cytokines. Transcription block by itself cannot account for the rapid loss in mRNA for CYP2C11 in cultured hepatocytes in response to LPS *(74)*. In mice, the presence of interferon increased the rate of degradation of CYP1A1 mRNA in addition to blocking transcription *(75)*. The loss of CYP2E1 in response to some cytokines likely involves changes to the stability of mRNA in addition to changes in transcription rates *(67)*. Although some losses in cytochrome P450 might occur at a post-transcriptional stage, this mechanism, however, accounts for a small proportion of the losses that occur in response to cytokines.

3. MODIFICATION OF P-GP BY CYTOKINES

During generalized inflammation, a number of studies have shown that P-gp levels are lowered in the liver and in isolated cultured hepatocytes. In the first reported study, Piquette-Miller et al. *(11)* showed that the levels of P-gp and mRNA levels of all three multidrug-resistant isoforms were depressed in the livers of rats treated with LPS or turpentine. In the same experiments, the efflux function of the transporter in hepatocytes obtained from these animals was significantly diminished. Studies by Vos et al. *(76)* confirmed the loss of P-gp protein in response to LPS, but these authors reported that mRNA for mdr1a remained unaffected. In contrast, a loss in the mRNA for mdr1a and mdr1b in both rats and mice with inflammatory responses has been observed *(77,78)*. Although these studies in models of inflammation are not definitive, they suggest that the losses of P-gp are related to the production of cytokines.

3.1. Individual Cytokine Effects on P-gp

The effects of individual cytokines on P-gp has been studied by a number investigators; however, the results obtained are widely divergent. In the rat, the administration of IL-1β resulted in an increase in both mdr1b mRNA and protein expression; however, mRNA levels of mdr1a, mdr2 and spgp were reduced significantly *(78)*. When IL-1β was added to cultured hepatocytes mdr1a and mdr1b, mRNA was unaltered, but the protein level and functional activity using rhodamine 123 as substrate were significantly depressed *(79)*. In the case of IL-6, there is a general agreement that this cytokine causes a loss in mRNA, protein, and functionality in P-gp in both the intact rat and in isolated hepatocytes *(77,78,80)*. Reports studying the effects of interferon on P-gp have been quite divergent. Akazawa *(81)* has shown that INF-γ depresses P-gp function in hepatocytes with no changes in protein level whereas IFN-α had no effect. On the other hand, Kang and Perry *(82)* reported that mdr1a mRNA and protein levels for P-gp were significantly elevated in hamster ovary ChR C5 cells treated with IFN-α. The variability in response also was noted for TNF-α, which has been reported to increase or have no effect on various components of the P-gp system *(78,83)*. A number of studies have examined the effects of cytokines on mdr1 genes in different human colon carcinoma cell lines. In general TNF-α, IFN-γ, and IL-2 appear to increase P-gp expression, specific proteins, and function using rhodamine123 as substrate. As with studies in rodents and hepatocytes, there is a large variation in the response in tumor cell lines that appears to be dependent on the specific cytokine and cell type used.

3.2. Drug Transport Alteration In Vivo

Despite the substantial differences in the studies on cytokine-induced changes to P-gp discussed in Subheading 3.1., there is consistent evidence that, in vivo, the handling of drugs via this transporter protein is disrupted in response to cytokine action. In a model of CNS inflammation that produces high concentrations of cytokines in the peripheral blood, Goralski et al. *(12)* have shown that the distribution and elimination of digoxin is significantly altered. Levels of digoxin were increased in brain, liver, kidney, and blood at times corresponding to the loss in mdr1a expression. The biliary clearance of digoxin was decreased by more than 60% and enhanced by the concomitant administration of the P-gp blocker cyclosporine A. The direct involvement of P-gp was confirmed by the absence of any change to digoxin pharmacokinetics in mice lacking the mdr1a gene. In mice treated with IFN-γ, the elimination of digoxin was decreased, and tissue levels of the drug were elevated *(84)*. Urinary and biliary clearance of digoxin was depressed by approx 40% in these experiments. The loss in digoxin clearance corresponded to a 20% to 30% loss in P-gp in the liver, kidney, and intestine. The bioavailability of digoxin was increased significantly following the administration of a human recombinant interleukin-2 (rIL2) in mice *(85)*. The change to digoxin pharmacokinetics occurred only when the drug was given orally, suggesting that the effect of rIL2 was primarily at the level of intestinal P-gp. Paclitaxel clearance was decreased, and volume of distribution increased by rIL2 in mice and this correlated with changes in P-gp in intestine and brain *(86)*. These studies using clinically useful drugs that are classic substrates for P-gp show that major changes in pharmacokinetics occur in response to cytokines. Although in most instances these changes will result in unwanted complications of therapy in any condition that involves an inflammatory component, there is the possibility of using cytokines to intentionally modify the absorption, distribution, and elimination of therapeutic agents that are substrates for P-gp. For example, cytokine treatment could be used to depress the levels of P-gp in the blood brain barrier thereby increasing the penetration of the anti-neoplastic agents and anti-AIDS drugs that are P-gp substrates.

4. CYTOKINES MODIFY HUMAN DRUG DISPOSITION AND PHARMACOKINETICS

In humans, the first incidence of altered drug pharmacokinetics that resulted from host defense activation was reported in 1978 by Chang et al. *(3)*, who showed that the clearance of theophylline was significantly impaired during upper respiratory tract infections caused by influenza or adenovirus. In the next few years, a number of reports indicted that the use of theophylline in children with infections often resulted in the accumulation of the drug to dangerous levels *(4,5,87–91)*. Although cytokines were not implicated in these early reports, it is most likely that the loss in drug metabolism capacity resulted from the production of cytokines during those infections. In children with organ failure caused by sepsis, the clearance of a test dose of antipyrine diminished to 50%, and the degree of change in drug clearance was correlated with IL-6 and nitrite in the blood and to the number of organs involved in failure *(92)*. The reduced clearance of theophylline, antipyrine, and hexobarbital all have been attributed to altered drug metabolism during septic shock *(93–95)*. A factor in serum (likely cytokines) produced in critically ill patients can depress the biotransformation of midazolam by

CYP3A4 *(96)*. A short-lived reduction in the clearance of midazolam occurs from a combination of changes to hepatic blood flow and cytochrome P450 activity in critically ill patients with infections *(97)*. Also, factors obtained from the serum of humans with viral infections can depress cytochrome P450 when added to cultures of rabbit hepatocytes *(98,99)*. These factors have proved to be Il-1β, IL-6, IFN-γ, and TNF-α and contribute to the downregulation of CYP1A at the transcriptional level and CYP3A6 at both pre- and post-transcriptional levels *(100)*. In *Helibacter pylori* infections, cytochrome P450 activity was significantly lower but did not correlate with TNF-α blood levels *(101)*.

A few studies indicate that the direct administration of cytokines to humans alters the capacity of the liver to handle drugs. The administration of recombinant α-interferons depresses a number of cytochrome P450-dependent drug biotransformation pathways in humans *(31–33,36)*. Antipyrine clearance decreased to a similar degree in patients given α-interferon or β-interferon compared with individuals infected with influenza *(91)*. Direct evidence of alterations in drug-metabolizing pathways was obtained by Okuna et al. *(35)*, who demonstrated that 7-methoxy-coumarin (7-MC) O-demethylase and 7-ethoxycoumarin (7-EC) O-deethylase in liver biopsy samples was depressed in 12 individuals treated with interferon. The effect of interferons on cytochromes P450 in humans is not, however, universal because high-dose interferon (IFN-α-2b) treatment significantly impairs the function of CYP1A2 but has little effect on CYP2E1 in the same subjects *(37)*. IFN-β has no apparent effect on CYP2C19 or CYP2D6 in patients with multiple sclerosis *(102)*.

Some interesting studies have recently emerged linking changes in drug clearance to cytokine production in certain disease states. In a series of surgical patients it has been shown that inflammatory responses have the ability to alter CYP3A4 activity as measured by the erythromycin breath test *(103)*. In this carefully controlled study in which subjects acted as their own controls, the changes in cytochrome P450 correlated with the levels and the area under the curve for IL-6 in blood. The alteration in enzyme activity in these subjects indicated that the presence of acute inflammation and resulting cytokine production after elective surgery is of a magnitude impact on the metabolism and clearance of a large group of commonly used drugs. Another recent study that has potentially important considerations is the finding that, in patients with congestive heart failure who were given a test cocktail of caffeine, mephenytoin, dextromethorphan, and chloroxazone, the activities of CYP1A2 and CYP2C19 but not CYP2E or CYP2D6 were negatively correlated with levels of IL-6 and TNF-α *(104)*. The production of cytokines and loss of cytochrome P450 in patients with congestive heart failure could, in part, be to the result of hypoxia, which recently has been reported to evoke a cytokine-dependent loss in CYP1A1, CYP1A2 in rabbits *(105)*. An inflammatory response to this common cardiac condition, however, likely contributes to the variability in drug response often observed in these patients. Aminopyrine metabolism and CYP1A2 and CYP2E1 activity decreased during rejection of transplanted livers and this correlated with an upregulation of IFN-γ and nitric oxide (NO)-synthase expression, suggesting that intra-graft production of cytokines and NO leads to a marked decrease in the capacity of the liver to metabolize drugs during allograft rejection *(106)*. In a group of subjects immunized against influenza, overall changes in CYP3A4 (as measured by the erythromycin breath test) was not significantly altered by the immunization *(107)*. However, individual changes in CYP3A4 were significantly correlated with the ability of the individual's lymphocytes to produce IFN-γ when challenged with influenza antigen. This study suggests that drug-elimination capacity changes only in the individuals who can mount a significant cytokine release in response to the immunization.

Individuals in the advanced stages of neoplastic disease often activate host defense mechanisms, inflammatory responses, and cytokine production. In a recent review, Slaviero et al. *(108)* provides evidence that the inflammatory response and changes to pharmacokinetics contributes to the interindividual variability of drug response and toxicity. The erythromycin breath test which is indicative of CYP3A activity is compromised in cancer patients with significant acute phase responses *(109,110)*. IL-1β and IL-6 often exist at high concentrations in breast cancer tissue and modify several steps in steroid production in that tissue *(111)*. One study has suggested that intratumoral production of IL-6 explains the low or even absent expression of CYP2C family members in breast cancer cells, as 10 different tumor samples had IL-6 levels and IL-6 receptors but correspondingly low CYP2C8 and CYP2C9 and absent CYP2C18 and CYP2C19 mRNA *(112)*. Because Schmidt *(113)* has suggested that the turnover of ifosphomide in breast cancer tissue may be influenced by the levels of cytochrome P450 forms in breast tumors, it is interesting to speculate that the success or failure of chemotherapy may depend on the ability of a tumor to produce cytokines and the resultant levels of drug biotransformation. When cancer patients received interferons, the metabolism of cyclophosphamide was altered, a response that indicates that CYP3A4 was decreased *(114)*. These studies suggest that cytokine production and changes in cytochrome P450 in advanced cancer patients may not only influence the pharmacokinetics of antineoplastic drugs but also may modulate the production of endogenous steroids within the neoplasm that are essential to tumor growth.

5. CONCLUSIONS AND SIGNIFICANCE

From the view point of therapeutics, the identification of the factors that alter drug biotransformation and/or transport increase the safety and efficacy of a large number of drugs. Understanding the changes to these processes that occur during any disease state that involves an inflammatory component explains many untoward drug responses reported in these conditions and provides a predictive framework to aid their avoidance. Unfortunately, the overall response is complex and the diversity of the changes to cytochrome P450 forms and P-gp in responses to the major proinflammatory and anti-inflammatory cytokines makes it impossible to make blanket predictions. The information presented in this review indicates that the metabolism or transport of most drugs can potentially be altered whenever a cytokine or cascade of cytokines is produced. It would be a major advance to identify the specific conditions that place an individual at risk for aberrant drug handling during an inflammatory response. Although it is impossible to guess the teleological basis for the interaction between drug disposition pathways and cytokines it is important to recognize that the capacity of an individual to metabolize and distribute drugs is likely to be altered, providing a setting for adverse drug responses and drug interactions, during any disease state with an inflammatory component.

ACKNOWLEDGMENTS

This work was supported by grants from the Canadian Foundation for Health Research and the Nova Scotia Foundation for Health Research.

REFERENCES

1. Renton KW, Mannering GJ. Depression of the hepatic cytochrome P-450 monooxygenase system by administered tilorone (2,7-bis(2-(diethylamino)ethoxy) fluoren-9-one dihydrochloride). Drug Metab Dispos 1976;4:223–31.
2. Renton KW, Mannering GJ. Depression of hepatic cytochrome P-450-dependent monooxygenase systems with administered interferon inducing agents. Biochem Biophys Res Commun 1976;73:343–348.
3. Chang KC, Lauer BA, Bell TD, Chai H. Altered theophylline pharmacokinetics during acute respiratory viral illness. Lancet 1978;1:1132–1133.
4. Kraemer MJ, Furukawa C, Koup JP, Shapiro G. Altered theophylline clearance during an influenza outbreak. Pediatrics 1982;69:476–480.
5. Woo OF, Koup JR, Kraemer M, Robertson WO. Acute intoxication with theophylline while on chronic therapy. Vet and Human Toxicol 1980;22:48–51.
6. Renton KW. Altered theophylline kinetics. Lancet 1978;2:160–161.

7. Morgan ET. Regulation of cytochromes P450 during inflammation and infection. Drug Metab Rev 1997;29:1129–1188.

8. Morgan ET. Regulation of Cytochrome P450 by Inflammatory Mediators: Why and How? Drug Metab Dispos 2001;29:207–212.

9. Renton KW. Hepatic drug metabolism and immunostimulation. Toxicology 2000;142:173–178.

10. Renton KW. Cytochrome p450 regulation and drug biotransformation during inflammation and infection. Curr Drug Metab 2004;5:235–243.

11. Piquette-Miller M, Pak A, Kim H, Anari R, Shahzammi A. Decreased expression and activity of P-glycoprotein in rat liver during acute inflammation. Pharmaceutical Res 1998;15:706–711.

12. Goralski KB, Hartmann G, Piquette-Miller M, Renton KW. Downregulation of mdr1a expression in the brain and liver during CNS inflammation alters the in vivo disposition of digoxin. Br J Pharmacol 2003;139:35–48.

13. Nebert DW, Russell DW. Clinical importance of the cytochromes P450. Lancet 2002;360:1155–1162.

14. Gottesman MM, Pastan I. Biochemistry of multidrug resistance mediated by the multidrug transporter. Annu Rev Biochem 1993;62:385–427.

15. Renton KW. Alteration of drug biotransformation and elimination during infection and inflammation. Pharmacol Ther 2001;9:147–163.

16. Leeson GA, Biedenback SA, Chan KY, Gibson JP, Wright GJ. Decrease in activity of the drug metabolizing enzymes of the rat following the administration of tilorone hydrochloride. Drug Metabolism Dispos 1976;4:232–238.

17. Singh G, Renton KW. Interferon-mediated depression of cytochrome P-450-dependent drug biotransformation. Mol Pharmacol 1981;20:681–684.

18. Parkinson A, Lasker J, Kramer MJ, et al. Effects of three recombinant human leukocyte interferons on drug metabolism in mice. Drug Metab Dispos 1982;10:579–585.

19. Singh G, Renton KW, Stebbing N. Homogeneous interferon from *E. coli* depresses hepatic cytochrome P-450 and drug biotransformation. Biochem Biophys Res Commun 1982;106:1256–1261.

20. Calleja C, Eeckhoutte C, Dacasto M, et al. Comparative effects of cytokines on constitutive and inducible expression of the gene encoding for the cytochrome P450 3A6 isoenzyme in cultured rabbit hepatocytes: consequences on progesterone 6beta- hydroxylation. Biochem Pharmacol 1998;56:1279–1285.

21. Carelli M, Porras MC, Rizzardini M, Cantoni L. Modulation of constitutive and inducible hepatic cytochrome(s) P-450 by interferon beta in mice. J Hepatol 1996;24:230–237.

22. Craig PI, Mehta I, Murray M, et al. Interferon down regulates the male-specific cytochrome P450IIIA2 in rat liver. Mol Pharmacol 1990;38:313–318.

23. Morgan ET, Norman CA. Pretranslational suppression of cytochrome P-450h (IIC11) gene expression in rat liver after administration of interferon inducers. Drug Metab Dispos Biol Fate Chem 1990;18:649–653.

24. Morgan ET. Suppression of P450IIC12 gene expression and elevation of actin messenger. Biochem Pharmacol 1991;42:51–57.

25. Sakai H, Okamoto T, Kikkawa Y. Suppression of hepatic drug metabolism by the interferon inducer. J Pharmacol Exp Ther 1992;263:381–386.

26. Stanley LA, Adams DJ, Balkwill FR, Griffin D, Wolf CR. Differential effects of recombinant interferon alpha on constitutive and inducible cytochrome P450 isozymes in mouse liver. Biochem Pharmacol 1991;42:311–320.

27. Cribb AE, Delaporte E, Kim SG, Novak RF, Renton KW. Regulation of cytochrome P-4501A and cytochrome P-4502E induction in the rat. J Pharmacol Exp Ther 1994;268:487–494.

28. Anari MR, Cribb AE, Renton KW. The duration of induction and species influences the downregulation of cytochrome P450 by the interferon inducer polyinosinic acid-polycytidylic acid. Drug Metab Dispos 1995;23:536–541.

29. Delaporte E, Cribb AE, Renton KW. Modulation of rat hepatic CYP3A1 induction by the interferon inducer polyinosinic acid-polycytidylic acid (polyic). Drug Metab Dispos Biol Fate Chem 1993;21:520–523.

30. Moochhala SM, Renton KW, Stebbing N. Induction and depression of cytochrome P-450-dependent mixed- function. Biochem Pharmacol 1989;38:439–447.

31. Williams SJ, Farrell GC. Inhibition of antipyrine metabolism by interferon. Br J Clin Pharmacol 1986;22:610–612.

32. Williams SJ, Baird-Lambert JA, Farrell GC. Inhibition of theophylline metabolism by interferon. Lancet 1987;2:939–941.

33. Jonkman JHG, Nicholson KG, Farrow PR. Effects of interferon on theophylline pharmacokinetics and metabolism. Br J Clin Pharmacol 1989;27:795–802.

34. Echizen H, Ohta Y, Shirataki H, et al. Effects of subchronic treatment with natural human interferons on antipyrine clearance and liver function in patients with chronic hepatitis. Clin Pharmacol 1990;30:562–567.

35. Okuno H, Shiozaki Y, Kiato Y, Kuneida K, Seki T, Sameshima Y. Depression of drug metabolizing activity in human liver by interferon-a. Eur J Clin Pharmacol 1990;39:365–367.

36. Craig PI, Tapner M, Farrell GC. Interferon suppresses erythromycin metabolism in rats and human subjects. Hepatology 1993;17:230–235.

37. Islam M, Frye RF, Richards TJ, et al. Differential effect of IFNalpha-2b on the cytochrome P450 enzyme system: a potential basis of IFN toxicity and its modulation by other drugs. Clin Cancer Res 2002;8:2480–2487.

38. Fukuda Y, Sassa S. Suppression of cytochrome P450IA1 by interleukin-6 in human HepG2 hepatoma cells. Biochem Pharmacol 1994;47:1187–1195.

39. Fukuda Y, Ishida N, Noguchi T, Kappas A, Sassa S. Interleukin-6 down regulates the expression of transcripts encoding cytochrome P450 1A1, 1A2 and III3A in human hepatoma cells. Biochem Biophys Res Comm 1992;184:960–965.

40. Clark MA, Bing BA, Gottschall PE, Williams JF. Differential effect of cytokines on the phenobarbital or 3- methylcholanthrene induction of P450 mediated monooxygenase activity in cultured rat hepatocytes. Biochem Pharmacol 1995;49:97–104.

41. Sanne JL, Krueger KE. Expression of cytochrome P450 side-chain cleavage enzyme and 3 beta-hydroxysteroid dehydrogenase in the rat central nervous system: a study by polymerase chain reaction and in situ hybridization. J Neurochem 1995;65:528–536.

42. Barker CW, Fagan JB, Pasco DS. Interleukin-1 beta suppresses the induction of P4501A1 and P4501A2 mRNAs in isolated hepatocytes. J Biol Chem 1992; 267:8050–8055.

43. Wright K, Morgan ET. Regulation of cytochrome P450IIC12 expression by interleukin-1 alpha, interleukin-6, and dexamethasone. Mol Pharmacol 1991; 39:468–474.

44. Nadin L, Butler AM, Farrell GC, Murray M. Pretranslational down-regulation of cytochromes P450 2C11 and 3A2 in male rat liver by tumor necrosis factor a. Gastroenterology 1995;109:198–205.

45. Tinel M, Robin MA, Doostzadeh J, et al. The interleukin-2 receptor downregulates the expression of cytochrome P450 in cultured rat hepatocytes. Gastroenterology 1995;109:1589–1599.

46. Herrmann M, Scholmerich J, Straub RH. Influence of cytokines and growth factors on distinct steroidogenic enzymes in vitro: a short tabular data collection. Ann N Y Acad Sci 2002;966:166–186.

47. Kurokohchi K, Yoneyama H, Matsuo Y, Nishioka M, Ichikawa Y. Effects of interleukin 1 alpha on the activities and gene expressions of the cytochrome P450IID subfamily. Biochem Pharmacol 1992;44:1669–1674.

48. Morgan ET, Thomas KB, Swanson R, Vales T, Hwang J, Wright K. Selective suppression of cytochrome P-450 gene expression by interleukins 1 and 6 in rat liver. Biochim Biophys Acta 1994;1219:475–483.

49. Monshouwer M, McLellan RA, Delaporte E, Witkamp RF, Van Miert A, Renton KW. Differential effect of pentoxifylline on lipopolysaccharide-induced downregulation of cytochrome P450. Biochem Pharmacol 1996;52:1195–1200.

50. Chen YL, Florentin I, Batt AM, Ferrari L, Giroud JP, Chauvelot-Moachon L. Effects of interleukin-6 on cytochrome P450-dependent mixed-function oxidases in the rat. Biochem Pharmacol 1992;44:137–148.

51. Chen YL, Le VV, Leneveu A, et al. Acute-phase response, interleukin-6, and alteration of cyclosporine pharmacokinetics. Clin Pharmacol Ther 1994;55: 649–660.

52. Warren GW, van Ess PJ, Watson AM, Mattson MP, Blouin RA. Cytochrome P450 and antioxidant activity in interleukin-6 knockout mice after induction of the acute-phase response. J Interferon Cytokine Res 2001;21:821–826.

53. Siewert E, Bort R, Kluge R, Heinrich PC, Castell J, Jover R. Hepatic cytochrome P450 down-regulation during aseptic inflammation in the mouse is interleukin 6 dependent. Hepatology 2000;32:49–55.

54. Mileva M, Bakalova R, Tancheva L, Galabov S. Effect of immobilization, cold and cold-restraint stress on liver monooxygenase activity and lipid peroxidation of influenza virus-infected mice. Arch Toxicol 2002;76:96–103.

55. Renton KW, Dibb S, Levatte TL. Lipopolysaccharide evokes the modulation of brain cytochrome P4501A in the rat. Brain Res 1999;842:139–147.

56. Stern EL, Quan N, Proescholdt MG, Herkenham M. Spatiotemporal induction patterns of cytokine and related immune signal molecule mRNAs in response to intrastriatal injection of lipopolysaccharide. J Neuroimmunol 2000;106:114–129.

57. Nicholson TE, Renton KW. Role of cytokines in the lipopolysaccharide-evoked depression of cytochrome P450 in the brain and liver. Biochem Pharmacol 2001;62:1709–1717.

58. Nicholson TE, Renton KW. The role of cytokines in the depression of CYP1A activity using cultured astrocytes as an in vitro model of inflammation in the central nervous system. Drug Metab Disposition 2002;30:42–46.

59. Morel Y, Barouki R. Down-regulation of cytochrome P450 1A1 gene promoter by oxidative stress - Critical contribution of nuclear factor 1. J Biol Chem 1998; 273:26969–26976.

60. Iber H, Chen Q, Cheng PY, Morgan ET. Suppression of CYP2C11 gene transcription by interleukin-1 mediated by NF-kappaB binding at the transcription start site. Arch Biochem Biophys 2000;377:187–194.

61. Jover R, Bort R, Gomez-Lechon MJ, Castell JV. Down-regulation of human CYP3A4 by the inflammatory signal interleukin-6: molecular mechanism and transcription factors involved. FASEB J 2002;16:1799–1801.

62. Tinel M, Elkahwaji J, Robin MA, et al. Interleukin-2 overexpresses c-myc and down-regulates cytochrome P-450 in rat hepatocytes. J Pharmacol Exp Ther 1999;289:649–655.

63. Van Ess PJ, Mattson MP, Blouin RA. Enhanced induction of cytochrome P450 enzymes and CAR binding in TNF (p55(-/-)/p75(-/-)) double receptor knockout mice following phenobarbital treatment. J Pharmacol Exp Ther 2002;300: 824–830.

64. Sachdeva K, Yan B, Chichester CO. Lipopolysaccharide and cecal ligation/ puncture differentially affect the subcellular distribution of the pregnane X receptor but consistently cause suppression of its target genes CYP3A. Shock 2003;19:469–474.

65. Beigneux AP, Moser AH, Shigenaga JK, Grunfeld C, Feingold KR. Reduction in cytochrome P-450 enzyme expression is associated with repression of CAR (constitutive androstane receptor) and PXR (pregnane X receptor) in mouse liver during the acute phase response. Biochem Biophys Res Commun 2002;293:145–149.

66. Memon RA, Moser AH, Shigenaga JK, Grunfeld C, Feingold KR. In vivo and in vitro regulation of sterol 27-hydroxylase in the liver during the acute phase response. potential role of hepatocyte nuclear factor-1. J Biol Chem 2001; 276:30118–30126.

67. Hakkola J, Hu Y, Ingelman-Sundberg M. Mechanisms of down-regulation of CYP2E1 expression by inflammatory cytokines in rat hepatoma cells. J Pharmacol Exp Ther 2003;304:1048–1054.

68. Wang B, Cai SR, Gao C, Sladek FM, Ponder KP. Lipopolysaccharide results in a marked decrease in hepatocyte nuclear factor 4 alpha in rat liver. Hepatology 2001;34:979–989.

69. Vossen C, Erard M. Down-regulation of nuclear receptor DNA-binding activity by nitric oxide—HNF4 as a model system. Med Sci Monit 2002;8(10): RA217–RA220.

70. Ingelman-Sundberg M, Ronis MJ, Lindros KO, Eliasson E, Zhukov A. Ethanol-inducible cytochrome P4502E1: regulation, enzymology and molecular biology. Alcohol Alcohol Suppl 1994;2:131–139.

71. Abdel-Razzak Z, Corcos L, Fautrel A, Guillouzo A. Interleukin-1 beta antagonizes phenobarbital induction of several major cytochromes P450 in adult rat hepatocytes in primary culture. FEBS Lett 1995;366–:159–164.

72. Abdel-Razzak Z, Loyer P, Fautrel A, et al. Cytokines down-regulate expression of major cytochrome P-450 enzymes in adult human hepatocytes in primary culture. Mol Pharmacol 1993;44:707–715.

73. Abdel-Razzak Z, Garlatti M, Aggerbeck M, Barouki R. Determination of interleukin-4-responsive region in the human cytochrome P450 2E1 gene promoter. Biochem Pharmacol 2004;68:1371–1381.

74. Morgan ET. Suppression of constitutive cytochrome P-450 gene expression in livers of rats undergoing an acute phase response to endotoxin. Mol Pharmacol 1989;36:699–707.

75. Delaporte E, Renton KW. Cytochrome P4501A1 and cytochrome P4501A2 are downregulated at both transcriptional and post-transcriptional levels by conditions resulting in interferon-alpha/beta induction. Life Sci 1997;60:787–796.

76. Vos TA, Hooiveld GJ, Koning H, et al. Up-regulation of the multidrug resistance genes, Mrp1 and Mdr1b, and down-regulation of the organic anion transporter, Mrp2, and the bile salt transporter, Spgp, in endotoxemic rat liver. Hepatology 1998;28:1637–1644.

77. Sukhai M, Yong A, Kalitsky J, Piquette-Miller M. Inflammation and interleukin-6 mediate reductions in the hepatic expression and transcription of the mdr1a and mdr1b Genes. Mol Cell Biol Res Commun 2000;4:248–256.

78. Hartmann G, Kim H, Piquette-Miller M. Regulation of the hepatic multidrug resistance gene expression by endotoxin and inflammatory cytokines in mice. Int Immunopharmacol 2001;1:189–199.

79. Sukhai M, Yong A, Pak A, Piquette-Miller M. Decreased expression of P-glycoprotein in interleukin-1beta and interleukin-6 treated rat hepatocytes. Inflamm Res 2001;50:362–370.

80. Sukhai M, Yong A, Pak A, Piquette-Miller M. Decreased expression of P-glycoprotein in interleukin-1beta and interleukin-6 treated rat hepatocytes. Inflamm Res 2001;50:362–370.

81. Akazawa Y, Kawaguchi H, Funahashi M, et al. Effect of interferons on P-glycoprotein-mediated rhodamine-123 efflux in cultured rat hepatocytes. J Pharm Sci 2002;91:2110–2115.

82. Kang Y, Perry RR. Effect of alpha-interferon on P-glycoprotein expression and function and on verapamil modulation of doxorubicin resistance. Cancer Res 1994;54:2952–2958.

83. Hirsch-Ernst KI, Ziemann C, Foth H, Kozian D, Schmitz-Salue C, Kahl GF. Induction of mdr1b mRNA and P-glycoprotein expression by tumor necrosis factor alpha in primary rat hepatocyte cultures. J Cell Physiol 1998;176:506–515.

84. Kawaguchi H, Matsui Y, Watanabe Y, Takakura Y. Effect of interferon-gamma on the pharmacokinetics of digoxin, a P-glycoprotein substrate, intravenously injected into the mouse. J Pharmacol Exp Ther 2004;308:91–96.

85. Castagne V, Bonhomme-Faivre L, Urien S, et al. Effect of recombinant interleukin-2 pretreatment on oral and intravenous digoxin pharmacokinetics and P-glycoprotein activity in mice. Drug Metab Dispos 2004;32:168–171.

86. Bonhomme-Faivre L, Pelloquin A, Tardivel S, et al. Recombinant interleukin-2 treatment decreases P-glycoprotein activity and paclitaxel metabolism in mice. Anticancer Drugs 2002;13:51–57.

87. Fleetham JA, Nakatsu K, Munt PW. Theophylline pharmacokinteics and respiratory infections. Lancet 1978;2:898.

88. Clarke CJ, Boyd G. Theophylline pharmacokinetics during respiratory viral infection. Lancet 1979;1:492.

89. Walker SB, Middlekamp JN. Theophylline toxicity and viral infection. Pediatrics 1982;70:508.

90. Greenwald M, Koren G. Viral induced changes in theophylline handling in children. Am J Asthma Allergy Pediatr 1990;3:162.

91. Brockmeyer NH, Barthel B, Mertins L, Goos M. Changes of antipyrine pharmacokinetics during influenza and after administration of interferon-alpha and -beta. Int J Clin Pharmacol Ther 1998;36:309–311.

92. Carcillo JA, Doughty L, Kofos D, et al. Cytochrome P450 mediated-drug metabolism is reduced in children with sepsis-induced multiple organ failure. Intensive Care Med 2003;29:980–984.

93. Toft P, Heslet L, Hansen M, Klitgaard NA. Theophylline and ethylenediamine pharmacokinetics following administration of aminophylline to septic patients with multiorgan failure. Intensive Care Med 1991;17:465–468.

94. Shedlofsky SI, Israel BC, McClain CJ, Hill DB, Blouin RA. Endotoxin administration to humans inhibits hepatic cytochrome P450-mediated drug metabolism. J Clin Invest 1994;94:2209–2214.

95. Shedlofsky SI, Israel BC, Tosheva R, Blouin RA. Endotoxin depresses hepatic cytochrome P450-mediated drug metabolism in women. Br J Clin Pharmacol 1997;43:627–632.

96. Park GR, Miller E, Navapurkar V. What changes drug metabolism in critically ill patients?—II Serum inhibits the metabolism of midazolam in human microsomes. Anaesthesia 1996;51:11–15.

97. Shelly MP, Mendel L, Park GR. Failure of critically ill patients to metabolise midazolam. Anaesthesia 1987;42:619–626.

98. Bleau AM, Fradette C, El-Kadi AOS, Cote MC, Du Souich P. Cytochrome P450 down-regulation by serum from humans with a viral infection and from rabbits with an inflammatory reaction. Drug Metab Dispos 2001;29:1007–1012.

99. Bleau AM, Levitchi MC, Maurice H, Du Souich P. Cytochrome P450 inactivation by serum from humans with a viral infection and serum from rabbits with a turpentine-induced inflammation: the role of cytokines. Br J Pharmacol 2000;130:1777–1784.

100. Bleau AM, Maurel P, Pichette V, Leblond F, du Souich P. Interleukin-1beta, interleukin-6, tumour necrosis factor-alpha and interferon-gamma released by a viral infection and an aseptic inflammation reduce CYP1A1, 1A2 and 3A6 expression in rabbit hepatocytes. Eur J Pharmacol 2003;473–:197–206.

101. Giannini E, Fasoli A, Borro P, et al. Impairment of cytochrome P-450-dependent liver activity in cirrhotic patients with *Helicobacter pylori* infection. Aliment Pharmacol Ther 2001;15:1967–1973.

102. Hellman K, Roos E, Osterlund A, et al. Interferon-beta treatment in patients with multiple sclerosis does not alter CYP2C19 or CYP2D6 activity. Br J Clin Pharmacol 2003;56:337–340.

103. Haas CE, Kaufman DC, Jones CE, Burstein AH, Reiss W. Cytochrome P450 3A4 activity after surgical stress. Crit Care Med 2003;31:1338–1346.

104. Frye RF, Schneider VM, Frye CS, Feldman AM. Plasma levels of TNF-alpha and IL-6 are inversely related to cytochrome P450-dependent drug metabolism in patients with congestive heart failure. J Card Fail 2002;8:315–319.

105. Fradette C, Bleau AM, Pichette V, Chauret N, Du Souich P. Hypoxia-induced down-regulation of CYP1A1/1A2 and up-regulation of CYP3A6 involves serum mediators. Br J Pharmacol 2002;137:881–891.

106. Westerholt A, Himpel S, Hager-Gensch B, et al. Intragraft iNOS induction during human liver allograft rejection depresses cytochrome p450 activity. Transpl Int 2004;17:370–378.

107. Hayney MS, Muller D. Effect of influenza immunization on CYP3A4 activity in vivo. J Clin Pharmacol 2003;43:1377–1381.

108. Slaviero KA, Clarke SJ, Rivory LP. Inflammatory response: an unrecognised source of variability in the pharmacokinetics and pharmacodynamics of cancer chemotherapy. Lancet Oncol 2003;4:224–232.

109. Rivory LP, Slaviero K, Seale JP, et al. Optimizing the erythromycin breath test for use in cancer patients. Clin Cancer Res 2000;6:3480–3485.

110. Rivory LP, Slaviero KA, Clarke SJ. Hepatic cytochrome P450 3A drug metabolism is reduced in cancer patients who have an acute-phase response. Br J Cancer 2002;87:277–280.

111. Honma S, Shimodaira K, Shimizu Y, et al. The influence of inflammatory cytokines on estrogen production and cell proliferation in human breast cancer cells. Endocr J 2002;49:371–377.

112. Knupfer H, Schmidt R, Stanitz D, Brauckhoff M, Schonfelder M, Preiss R. CYP2C and IL-6 expression in breast cancer. Breast 2004;13:28–34.

113. Schmidt R, Baumann F, Knupfer H, et al. CYP3A4, CYP2C9 and CYP2B6 expression and ifosfamide turnover in breast cancer tissue microsomes. Br J Cancer 2004;90:911–916.
114. Ibach B, Appel K, Gebicke-Haerter P, et al. Effect of phenytoin on cytochrome P450 2B mRNA expression in primary rat astrocyte cultures. J Neurosci Res 1998;54:402–411.

15

Evaluation of the Immunological Effects of Cytokines Administered to Patients With Cancer

Michael J. Robertson

SUMMARY

The goal of cancer immunotherapy is to induce effective immune responses to malignant tumors. Immunostimulatory cytokines can promote antitumor immunity by augmenting immune responses to cancer cells and by reversing anergy and/or tolerance of immune effector cells. The clinical development of cytokines for cancer therapy has been hampered by difficulty in determining the optimal dose and schedule of these molecules. The successful development of cytokine-based therapy requires assays that can measure appropriate surrogate endpoints for the induction of effective antitumor immunity in vivo. Several in vitro assays that have been useful for monitoring the immunological effects of cytokine therapy in humans will be discussed in this chapter. Each technique has advantages and limitations and no single assay has been found to reliably predict antitumor efficacy during clinical immunotherapy.

Key Words: Immunotherapy; cancer; vaccine; interleukin; tumor immunology.

1. INTRODUCTION

Cytokine-based immunotherapy for cancer has advanced rapidly since its inception in the early 1980s. Several cytokines, including interferon (IFN)-α, IFN-β, IFN-γ, tumor necrosis factor (TNF), interleukin (IL)-1, IL-2, IL-4, IL-12, IL-18, and IL-21 have been given to cancer patients on clinical trials *(1,2)*. Although objective tumor responses have been observed in patients receiving some of these cytokines, only IL-2 and the type I IFNs have thus

From: *Methods in Pharmacology and Toxicology: Cytokines in Human Health:*
Immunotoxicology, Pathology, and Therapeutic Applications
Edited by: R. V. House and J. Descotes © Humana Press Inc., Totowa, NJ

far been approved for clinical use. IL-12 as a single agent has shown activity in melanoma, renal cell carcinoma, and lymphoma *(3–8)*. Nevertheless, the objective response rate to IL-12 alone has been disappointing in patients with advanced cancer. IL-18 and IL-21 currently are being evaluated in phase I studies in patients with advanced solid tumors and lymphoma *(9–12)*.

Clinical development of cytokines for cancer therapy has been hampered by difficulty in determining the optimal dose and schedule of these molecules. New therapeutic agents for cancer have traditionally been tested in a series of phase I, II, and III clinical trials *(13)*. Phase I dose-escalation studies are performed to assess toxicity and determine the maximum tolerated dose of a novel agent. In phase II studies, the agent is given at the maximum tolerated dose to patients with a specific type of tumor to determine the objective response rate. If an agent shows promising activity in phase II studies, approval for nonexperimental use often requires large randomized phase III studies in which the new agent is compared to established treatments.

This paradigm is most apt for development of cytotoxic chemotherapy drugs. It has been generally acknowledged that this paradigm is not suitable for the development of molecularly targeted or biological agents *(13,14)*. The appropriate goal of phase I/II studies of immunostimulatory cytokines or cancer vaccines is to identify the optimal biological dose, rather than the maximum tolerated dose. Therefore, successful development of cytokine-based therapy requires assays that can measure appropriate surrogate endpoints for the induction of effective antitumor immunity in vivo. Unfortunately, there is no assay that has been widely accepted for this purpose. Several in vitro assays that have been proposed to be useful for monitoring the immunological effects of cytokine therapy in humans will be summarized in this chapter. The focus of the discussion will be cytokines developed as primary therapeutic agents in cancer. Use of cytokines (such as erythropoietin, granulocyte colony-stimulating factor, granulocyte-macrophage colony-stimulating factor, and IL-11) for supportive care is relatively straightforward and has been the subject of other reviews *(15–17)*.

2. INNATE AND ADAPTIVE IMMUNE RESPONSES

Assessment of the potential utility of assays for immunological monitoring of cytokine therapy requires a basic understanding of the immune response in general and of antitumor immunity in particular. In vertebrate species, both innate and adaptive effector cells participate in the immune response. Innate effector cells include natural killer (NK) cells, macrophages, and polymorphonuclear phagocytes *(18–20)*. These cells mediate rapid responses that are not antigen specific and are triggered by invariant receptors encoded by germ-line genes. In contrast, adaptive immune responses are mediated by T- and B-lymphocytes *(21,22)*. T- and B-cells express antigen-specific receptors that are generated by somatic recombination of germ-line T-cell receptor and B-cell immunoglobulin gene sequences, respectively. Thus, each mature T- or B-cell expresses a unique, clonotypic antigen receptor. Adaptive immune responses to a novel antigen require the activation, proliferation, and clonal expansion of naive T- and B-cells. Therefore, several days to a week or longer generally are required for full maturation of adaptive immune responses.

There is cooperation and coordination between effectors of innate and adaptive immunity. Innate immune effectors may limit the burden of pathogen during the early phases of the immune response, while adaptive responses are being developed. Moreover, the innate immune response to foreign molecules may affect the character of the subsequent adaptive immune response *(23)*. For example, the activation of NK cells and macrophages by foreign macromolecules can stimulate IFN-γ production, which supports the differentiation of antigen-stimulated CD4 T-cells into helper effector cells of the Th1 phenotype. Th1 cells produce cytokines, such as IL-2, TNF, and IFN-γ, that stimulate NK cell cytolytic activity and macrophage phagocytic activity *(21,24)*. Moreover, Th1 cells promote B-cell switching to immunoglobulin isotypes that optimally stimulate antibody-dependent cell-mediated cytotoxicity by NK cells and macrophages. Th1 cytokines also promote the differentiation of activated CD8 T-cells into functional cytotoxic T-lymphocytes (CTLs).

NK cells and CTLs are complementary mediators of cytotoxicity *(19,25)*. Both cell types possess similar cytolytic effector molecules, including perforin, granzymes, and Fas ligand. However, the triggering of cytolysis by CTL and NK cells involves distinct receptors. CTLs are triggered by engagement of the antigen-specific T-cell receptor, which recognizes short peptides bound to MHC class I molecules. Thus, conventional CTLs cannot kill target cells that do not express MHC class I molecules. In contrast, NK cells generally do not lyse target cells that express high levels of self MHC class I molecules because NK cells express receptors that, when engaged by MHC class I molecules, deliver dominant negative signals that inhibit cytolysis *(19)*. Engagement of several NK cell-activating receptors triggers lysis of cells that lack expression of MHC class I molecules *(19)*. Virus-infected and malignant cells often express abnormally low levels of class I MHC molecules, which renders them relatively resistant to the cytolytic activity of CTLs but more sensitive to lysis by NK cells.

3. IMMUNE RESPONSES TO MALIGNANT TUMORS AND CANCER IMMUNOTHERAPY

There is direct experimental evidence that the mammalian immune system can recognize and eliminate syngeneic malignant tumors in vivo *(2,26,27)*. In most animal models, durable antitumor immunity is dependent on antigen-specific T-cells. Participation of both CD4 and CD8 T-cells may be required for optimal antitumor immune responses in vivo. NK cells also are critical for effective antitumor immunity in several experimental systems. In animal models, elimination of MHC class I-positive tumor cells requires CTL, whereas the elimination of MHC class I negative tumor cells requires NK cells *(28,29)*. Because the expression of MHC class I molecules on primary human tumors can be very heterogeneous, effective antitumor immunity in cancer patients is likely to require the participation of both T-cells and NK cells *(30)*. Moreover, in many preclinical models, IFN-γ is critical for optimal antitumor immune responses *(31–33)*. The predominant producers of IFN-γ are NK cells and Th1 cells, further indicating the importance of these effector cells in the elimination of tumor cells during immune surveillance and cancer immunotherapy.

Although it has been unequivocally shown that the immune system can eradicate tumors in vivo, most patients with cancer ultimately develop progressive disease. Failure to control tumor cells in these patients may reflect several defects in the immune response to cancer. Malignant tumor cells are known to be genetically unstable and to acquire multiple somatic mutations during their development. Loss of immundominant antigens by a subset of tumor cells could lead to expansion of nonimmunogenic tumors after the elimination of tumor cells that express such antigens *(34)*. Furthermore, the downregulation of MHC class I antigens, which frequently is observed in human tumor cells *(35)*, may abrogate effective CTL responses. However, this could potentially render tumors more susceptible to NK cell lysis. Tumor cells also may induce immune tolerance *(36)*. Stimulation of T-cells by specific antigen in the absence of costimulatory signals can produce long-lasting anergy, whereby subsequent stimulation with both antigen and costimulatory molecules fails to activate anergic T-cells. Finally, tumor cells may produce various factors, such IL-10, TGF-β, and prostaglandins, that can inhibit antitumor immune responses *(37)*.

The goal of cancer immunotherapy is to elicit effective immune responses to malignant tumors. Passive immunotherapy involves the administration of agents that can directly target tumor cells and/or recruit host effector cells, but that do not provoke durable antitumor immune responses in vivo. Active immunotherapy aims to induce durable antitumor immunity. A selective list

Table 1
Approaches Used for Cancer Immunotherapy

Passive immunotherapy	Examples
Unconjugated monoclonal antibodies	Rituximab, alemtuzumab
Monoclonal antibodies conjugated to toxins	Gemtuzumab ozogamicin
Monoclonal antibodies conjugated to radio- isotopes	^{131}I-Tositumomab, ^{90}Y-Ibritumomab tiuxetan
Cytokine/toxin fusion protein	Denileukin diftitox

Active immunotherapy	Examples
Immunostimulatory cytokines	IL-2, IL-12, IL-18, IL-21
Cancer vaccines	Idiotype vaccines, peptide vaccines
Adoptive cell transfer (allogeneic)	Donor leukocyte infusion

of different approaches for cancer immunotherapy is given in Table 1. Immunostimulatory cytokines can promote antitumor immunity by augmenting pre-existing but ineffective responses to immunogenic cancer cells, inducing a response to poorly immunogenic cancer cells, and reversing anergy and/or tolerance of immune effector cells. Cytokines that can activate antitumor activity of both T- and NK cells may be the most promising agents for cytokine-based immunotherapy of cancer.

4. MONITORING IMMUNE RESPONSES DURING CANCER IMMUNOTHERAPY

This chapter will discuss several in vitro methods that can be used to monitor the effects of cytokines administered to patients with cancer (Table 2). Many of the methods have been more extensively applied to immune monitoring of cancer vaccine therapy and responses to viral infection (38–42), but the same assays are relevant for monitoring cytokine-based immunotherapy. This discussion is selective rather than comprehensive. Techniques most widely used for immunological monitoring will be discussed, focusing on the advantages and limitations of each technique.

The source of cells for in vitro analysis must be selected based on the type of immunological monitoring that is required. Peripheral blood mononuclear cells (PMBCs), tumor-infiltrating lymphocytes, cells isolated from skin biopsies at the site of cytokine or vaccine injection, and cells isolated from lymph nodes draining tumor or injection sites have been used for in vitro assays of immune function. Despite the known limitations of using PBMCs to assess in vivo immune responses (43), their ready accessibility ensures that PBMCs

Table 2
Methodologies Used to Monitor the Immunological Effects of Cytokines

Methodology	Examples of application
Flow cytometry (cell surface antigens)	Enumeration of lymphocyte subsets; detection of activated cells; detection of cytokine receptors
Flow cytometry (intracellular antigens)	Detection of cytokine production by specific cell subsets; signaling studies; assessment of cell cycle status
ELISA	Measurement of cytokine levels
Quantitative PCR assays	Quantification of gene expression
Cytotoxicity assays	Quantification of target cell lysis by effector cells
Proliferation assays	Assessment of cell proliferation
Limiting dilution analysis	Assesement of CTL or helper T-cell precursor frequency
ELISPOT assays	Quantification of antigen-specific CD4 or CD8 T-cells
MHC tetramer assays	Quantification of antigen-specific CD4 or CD8 T-cells

ELISA, enzyme-linked immunoabsorbance assay; ELISPOT, enzyme-linked immunospot; PCR, polymerase chain reaction.

will remain the cells used most frequently for immunological monitoring. Assays can be performed with PBMCs immediately after isolation or with previously cryopreserved PBMCs. Cryopreservation and thawing are known to affect functional responses of human PBMCs. For instance, cytolytic activity of human NK cells is partially inhibited by cryopreservation (44,45). However, the well-known day-to-day variability in many functional assays dictates that batched assays using cryopreserved PBMCs from sequential time-points may often be preferable to real-time assays using freshly isolated PBMCs.

Careful consideration must be given to the time-points selected for collection of blood samples during immunological monitoring. Several immunostimulatory cytokines, including TNF, IL-2, IL-12, and IL-18, cause transient, profound lymphopenia (5,10,46–48). This lymphopenia most likely reflects the in vivo activation of lymphocytes, followed by their margination and/or extravasation into tissue spaces (48,49). A practical implication of this phenomenon is that informative assays cannot be performed using PBMCs obtained shortly after the administration of some immunostimulatory cytokines. For example, NK-cell cytotolytic activity of PBMCs has been reported to decline during IL-12 therapy *(50)*. However, this observation was based on results of cytotoxicity assays performed with PBMCs obtained 24 h after injections of IL-12 *(50)*, at a time when NK cells are virtually absent from peripheral blood *(5,48,51)*. During the recovery phase after IL-12-induced lymphopenia, augmented NK cell cytolytic activity of PBMCs has been detected *(48)*. Moreover, because the peak activation of lymphocytes is very likely to occur during the period of lymphopenia after cytokine administration, assays performed using recovery phase PBMCs may underestimate the level of immune stimulation that is achieved in vivo during cytokine-based therapy.

5. ASSAYS USED FOR IMMUNOLOGICAL MONITORING OF CYTOKINE THERAPY

5.1. Multiparameter Flow Cytometry

Multiparameter flow cytometry is a powerful and versatile technique with several applications for immunological monitoring during cytokine therapy *(39,41,42,52)*.

5.1.1. Enumeration of Lymphocyte Subsets

Flow cytometric analysis can be used to evaluate the percentages of major lymphocyte subsets (total T-cells, CD4 T-cells, CD8 T-cells, B cells, and NK cells) as well as other subpopulations of cells (e.g., CD45RA+ CD4 T-cells, CD45RO+ CD4 T-cells, CD5+ B cells, CD56^{bright} NK cells) that are present in the blood during cytokine therapy. In conjunction with simultaneous complete blood counts, evaluation of the absolute numbers of various lymphocyte subsets is feasible. Such studies have revealed the selective in vivo expansion of NK cells (and particularly of CD56^{bright} NK cells) during prolonged intravenous or subcutaneous administration of low-dose IL-2 *(53–56)*. In contrast, the intravenous bolus administration of high doses of IL-2 causes rapid lymphopenia, followed several days later by rebound lymphocytosis due to expansion of both T and NK cells *(46,57)*. Administration of IL-12 or IL-18 by bolus intravenous injection also causes a transient, profound lymphopenia; it is generally followed by a recovery of PBMC subsets to pre-treatment baseline levels without rebound lymphocytosis *(3,5,9,10, 48,51)*. Nevertheless, the expansion of major lymphocyte subsets has been seen during prolonged systemic administration of IL-12 to patients with hematological malignancies who have undergone autologous stem cell transplantation *(58)*. A rare subset of CD8 T-cells expressing high levels of LFA-1 is also expanded during IL-12 therapy for advanced solid tumors *(59)*.

Such observations have helped to elucidate the biology of cytokines in humans and have facilitated design of additional clinical immunotherapy studies.

5.1.2. Detection of Cytokine Receptor Expression

Flow cytometry also can be used to assess surface expression of cytokine receptors on different lymphocyte subsets and can provide information that is crucial for interpreting biological responses during cytokine administration. For example, it has been observed that in vivo production of IFN-γ during IL-12 therapy is profoundly deficient in cancer patients who have undergone autologous hematopoietic stem cell transplantation *(58,60)*. Furthermore, post-transplant patient PBMCs directly stimulated in vitro with IL-12 fail to produce significant levels of IFN-γ *(58,60)*. Failure of PBMCs to express either subunit of the IL-12 receptor (IL-12R β1 or IL-12R β2) could have been responsible for defective IL-12-induced IFN-γ production in the post-transplant setting. Nevertheless, flow cytometry studies have demonstrated that post-transplant patient PBMCs express IL-12 receptor subunits at levels that are equivalent to or greater than those expressed by control PBMCs *(60)*. Thus, the defect in IL-12-induced IFN-γ production must be distal to the interaction of IL-12 with its cell surface receptors. Indeed, acquired profound STAT4 protein deficiency has been shown to be responsible for defective IL-12-induced IFN-γ production after autologous transplantation *(60)*.

The limitations of the technique must be taken into consideration when assessing cytokine receptor expression by conventional flow cytometry. Failure to detect cytokine receptors on the cell surface by flow cytometry does not exclude the expression of functional receptor complexes. Cytokine receptor complexes present on the cell surface at levels below the limit of detection by routine flow cytometry (approx 800–1000 receptors per cell) may be sufficient to mediate potent cytokine-induced responses *(61,62)*. Furthermore, cell surface receptors may not be detected because they have been rapidly internalized after binding to exogenous cytokine *(63)*. Thus, negative results for cytokine receptor expression by routine flow cytometry must be interpreted with caution.

5.1.3. Detection of Cellular Activation

Flow cytometry has also been used widely to detect activation of lymphocytes or monocytes during the administration of cytokines. Most unstimulated T-cells and NK cells do not express CD25 (the α chain of the IL-2 receptor), CD69 (a homodimeric C-type lectin *[64–66]*), or MHC class II antigens *(67)* on the cell surface. These antigens are expressed after activation of T-cells and NK cells either in vitro or in vivo *(65,66,68)*. Three- and four-color analysis by flow cytometry can detect expression of activation antigens on specific lymphocyte subsets (e.g., CD8+ CD3+ T-cells or CD3-negative CD56+ NK cells). This approach has confirmed that T-cells, NK cells, and monocytes are activated in vivo during treatment with IL-2 or IL-18 *(10,69)*. Upregulation of adhesion molecules and other functional structures on PBMCs during cytokine therapy has also been detected by this technique *(48,70,71)*.

5.1.4. Assessment of Signal Transduction

Binding of cytokines to their cell surface receptors activates intracellular signaling pathways associated with characteristic post-translational modifications (e.g., tyrosine and serine phosphorylation) of specific proteins *(72,73)*. Immunoblot analysis (Western blotting) has been extensively used to detect such post-translational modifications of signaling molecules. However, detection of a band by immunoblotting does not permit one to distinguish between weak activation of a substrate in many cells vs strong activation in a few cells within a sample. The phosphorylation of relevant tyrosine and serine residues in signaling proteins also can be detected by flow cytometry after intracellular staining with specific fluorochrome-conjugated monoclonal antibodies *(74)*. The simultaneous assessment of intracellular phosphoprotein expression and surface expression of lineage-associated antigens allows one to detect the specific cellular subsets that are responding to cytokine stimulation *(75)*.

5.1.5. Detection of Lymphocyte Proliferation

Intracellular staining and flow cytometry also can be used to detect cellular proliferation during cytokine administration. 5-bromo-2'-deoxyuridine (BrdU) is incorporated in the place of thymidine during deoxyribonucleic acid (DNA) synthesis in proliferating cells *(76)*. Thus, cellular proliferation can be quantified by flow cytometry after staining permeabilized cells with fluorochrome-conjugated antibodies recognizing BrdU. Unlike standard tritiated thymidine incorporation assays (discussed herein), the assessment of cell proliferation by flow cytometry does not require use of radioactive reagents. Moreover, the simultaneous assessment of surface antigens can allow one to discriminate the precise cell subsets that are proliferating in response to cytokine stimulation *(77,78)*. The feasibility of infusing BrdU to cancer patients for in vivo labeling of tumor cells has been demonstrated *(79,80)*. A similar approach could be taken to assess in vivo lymphocyte proliferation during cytokine-based immunotherapy of cancer.

5.2. Enzyme-Linked Immunosorbent Assay

The enzyme-linked immunosorbent assay (ELISA) technique is the basis of most assays used to measure serum cytokine levels for pharmacokinetic analysis after cytokine administration. Most immunostimulatory cytokines that have been given to humans have been found to be rapidly cleared from the circulation. For example, the half-life of IL-2 or TNF after bolus intravenous injection is less than 20 min (46,81–83). In contrast, the elimination half-life of IL-12 is approx 5 to 10 h after intravenous bolus injection (3) and 8 to 24 h after subcutaneous injection (4–6). Moreover, the elimination half-life of IL-18 after bolus intravenous injection is nearly 40 h, a result that was not anticipated based on preclinical animal studies of IL-18 (9,84). The observation of sustained blood levels of IL-12 and IL-18 after the administration of a single dose has obvious implications for designing the optimal schedule of administration for these cytokines.

ELISA techniques also have been used to detect secondary cytokines, such as IFN-γ, TNF, and IL-10, that can be produced in vivo following administration of immunostimulatory cytokines (3,5,10,51,85–87). These secondary cytokines may contribute to both the toxic and beneficial effects of the administered therapeutic cytokines. Sustained in vivo production of IFN-γ is associated with clinical antitumor response during IL-12 therapy for melanoma and renal cell cancer (7). However, excessively high serum IFN-γ levels may also contribute to unacceptable toxicity during IL-12-based immunotherapy (88,89). Similarly, in vivo production of TNF has been associated with both toxicity and efficacy in some studies of high-dose IL-2 therapy (85,90).

5.3. Quantitative Polymerase Chain Reaction Technique

As an alternative to ELISA or cytokine flow cytometry, quantitative polymerase chain reaction (Q-PCR) methods have been used to measure cytokine levels during immunotherapy (91). Complementary DNA primers complementary to messenger ribonucleic acid (mRNA) sequences of interest are generated by reverse transcription and specific sequences are amplified by the PCR technique. An oligoucleotide probe, designed to anneal downstream of one of the primers, is labeled with a fluororescent dye at the 5' end and quenching reagent at the 3' end. During PCR amplification, the fluorophore is released from the probe by the nuclease activity of Taq polymerase; separated from the quencher, the fluorophore produces detectable fluorescence. By reference to a standard curve generated from serial dilution of a control template, measurement of fluorescence emission during PCR amplification can be used to calculate the amount of mRNA in the original sample.

Because the PCR technique involves logarithmic amplification of cDNA, mRNA levels can be measured using a very small number of cells (91,92). Moreover, the extreme sensitivity of the technique permits detection of gene expression by small subpopulations of cells within a heterogeneous sample.

There are also potential disadvantages to Q-PCR. This methodology does not permit the identification of specific subsets of cells within a sample that are expressing the gene of interest. Furthermore, detection of mRNA by Q-PCR does not prove that biologicalally active protein has been produced and/or secreted.

5.4. Cytotoxicity Assays

Both CTL and NK cells are capable of cell-mediated cytotoxicity, resulting in lysis of target cells *(25)*. Chromium release assays are a standard in vitro method for measuring cytotoxicity *(93,94)*. Many cell types take up sodium chromate, and reduced chromate species are retained in viable cells. Disruption of plasma membrane integrity that occurs during cytolysis causes the release of chromium species into the supernatant of lysed cells. Thus, chromium release can be used to quantify the lysis of target cells. In a standard chromium release assay, target cells are labeled by incubation with ^{51}Chromium (^{51}Cr) and admixed with effector cells in wells of microtiter plates at several effector-to-target cell ratios. After incubation for various periods to time at 37°C, supernatants from wells are collected and presence of ^{51}Cr measured using a gamma counter. Spontaneous and maximum release of ^{51}Cr from target cells is determined from wells containing, respectively, no effector cells or membrane-permeabilizing agents. Specific cytotoxicity can be calculated using a standard formula *(94)*.

An alternative method for quantifying cytotoxicity is the JAM test *(95)*. This assay detects the DNA fragmentation that accompanies cytolysis by CTL or NK cells. Target cells are incubated with tritiated thymidine, which is incorporated into the DNA of dividing cells. Labeled target cells are incubated with effector cells as for a chromium release assay. In contrast to the latter, however, after incubation of target and effector cells for the desired period of time, the cells (not cell-free supernatants) are harvested onto fiberglass filters, washed, and tritiated thymidine retained on the filters is measured by liquid scintillation counting. The amount of radioactivity in a sample is inversely proportional to the degree of cytolysis, as unfragmented DNA from viable cells is trapped in fiberglass filters, whereas fragmented DNA from lysed cells are washed through the filters. Advantages of the JAM test include lack of need for shielding from gamma irradiation (required for work with ^{51}Cr) and the much longer half-life of tritiated thymidine compared to ^{51}Cr (allowing much less frequent ordering of radioactive supplies).

Colorimetric or fluorometric techniques also can be used to measure cytotoxicity *(96,97)*. These assays depend on the ability of certain substrates to diffuse into viable cells, where they are hydrolyzed by intracellular enzymes to produce fluorescent or pigmented compounds. Intensity of fluorescence or coloration is directly proportional to the numbers of viable target cells in these assays. Colorimetric and fluorometric assays avoid the use of radio-isotopes and are less cumbersome to perform than chromium release assays. However, the former require use of a microplate fluorimeter or similar detection system.

5.5. Proliferation Assays

Naive T- and B-cells are quiescent until activated by engagement of their antigen receptors. Proliferation of T- and B-cells can therefore be measured as an indication of lymphocyte activation. Tritiated thymidine incorporation assays have been commonly used to detect cellular proliferation (98,99). Unfractionated cells (e.g., PBMCs) or defined cell subsets (e.g., sorted CD4 T-cells or T-cell clones) are generally incubated in the wells of microtiter plates in the presence of specific antigens or non-specific mitogens for several days; tritiated thymidine is added for the last 4 to 18 h of culture. Using automated methods, cells are harvested onto fiberglass filters, which physically trap DNA. Tritiated thymidine incorporated into cellular DNA is detected by liquid scintillation counting. The measured beta particle counts per minute are determined by the amount of tritiated thymidine incorporated into cellular DNA, which in turn reflects the rate of proliferation of activated cells.

A limitation of using tritiated thymidine incorporation assays to measure lymphocyte proliferation in heterogeneous cell samples is that this method does not provide information regarding which subset of cells is responding. As noted above, multiparameter flow cytometry can be used to detect BrdU accumulation in specific cell subsets.

5.6. Limiting Dilution Analysis

Limiting dilution analysis can be used to assess antigen-specific T-cell responses (40,100–103). For limiting dilution analysis, graded numbers of responder cells are incubated for several days to weeks in the presence of antigen and the number of antigen-specific responder cells is calculated by established statistical models. Helper T-cell precursor frequencies have been assessed using IL-2 production as the read out for activated CD4 T-cells and CTL precursor frequency using lysis of antigen-specific target cells in cytotoxicity assays. Frequency of antigen-specific precursor cells also can be determined by limiting dilution analysis by measuring proliferation of responder cells in tritiated thymidine incorporation assays. The latter method appears to preferentially detect helper T-cell precursors as compared to CTL precursors (41).

There are major limitations associated with limiting dilution analyses. These assays are labor intensive and highly operator dependent. Most methods require at least several days of in vitro culture, and the assay may not accurately reflect the true precursor frequency. Furthermore, these assays are usually performed using heterogeneous cell samples (e.g., PBMCs), and the precise subsets of cells responding to antigen cannot be readily identified. Because of these limitations, alternative methods have been developed to detect antigen-specific responder cells, including enzyme-linked immunospot (ELISPOT) and MHC-peptide tetramer assays.

5.7. Enzyme-Linked Immunospot Assays

Enzyme-linked immunospot or ELISPOT assays can detect antigen-specific cells *(100,103,104)*. For ELISPOT assays, antibody specific for a cytokine of interest (typically IFN-γ) is incubated in the wells of a microtiter plate. Antigen and responder cells are added to the wells for 1 to 2 d, responder cells are removed from the wells by washing, and secreted cytokine is detected by ELISA methods. The read-out for ELISPOT assays is a colored spot, corresponding to cytokine released by an antigen-specific T-cell. ELISPOT assays have been used to detect antigen-specific T-cell responses in clinical trials of cancer vaccines and/or immunostimulatory cytokines for cancer treatment *(100,104–106)*.

As an alternative to ELISPOT assays, cytokine-producing responder cells can be detected by multiparameter flow cytometry *(40,101,104,107)*. Simultaneous staining for intracellular cytokine (e.g., IFN-γ) and cell surface antigens can be used to detect the precise subset of cells that is producing cytokine in response to a specific antigen or to cytokine stimulation.

5.8. MHC-Peptide Tetramer Binding Assays

Another powerful technique for detecting antigen-specific T-cells is staining with soluble recombinant MHC-peptide tetramers *(41)*. Recombinant MHC molecules are produced that lack the transmembrane and cytoplasmic domains. Incubation of soluble MHC monomers with β2 microglobulin and specific antigenic peptides yields soluble tetrameric MHC-peptide complexes. Under suitable conditions, soluble MHC-peptide tetramers will bind to T-cells bearing the cognate antigen-specific T-cell receptor. Use of fluorochromes permits the detection of bound cells by flow cytometry *(100,104,105)*. Identification of antigen-specific CD4 T-cells and CD8 T-cells is possible using MHC class II-peptide tetramers or MHC class I-peptide tetramers, respectively. Use of MHC-peptide tetramers permits the detection of rare T-cell populations, with a limit of detection as low as 1 in 10,000. A major limitation of this approach is that MHC-peptide tetramer assays convey no information regarding the functional activity of the cells detected. For example, binding of HLA-A*0201/MART-1 tetramers to CD8 T-cells confirms the specificity of the latter for the MART-1 antigen, but does not reveal whether the detected cells have been previously activated or are capable of effector functions (i.e. cytolysis or cytokine secretion). This is in contrast to the other methods of antigen-specific T-cell detection discussed above, including limiting dilution analysis, ELISPOT assays, and intracellular cytokine staining detected by flow cytometry.

6. CONCLUSION

Several techniques have been used to monitor the immunological effects of cytokines in cancer patients. Studies using these techniques have demonstrated that administration of immunostimulatory cytokines can cause the activation of T-cells, NK cells, and monocytes in vivo, leading to increased effector function and secretion of secondary cytokines. These studies have enhanced our understanding of the biology of cytokines in humans and have facilitated interpretation of clinical results of cytokine-based therapy. Nevertheless, identification of correlative immunological studies that can predict antitumor efficacy has proved elusive. Further investigation is clearly warranted to develop valid surrogate endpoints for effective antitumor immune responses during cancer immunotherapy.

ACKNOWLEDGMENTS

This work was supported in part by NIH grants M01 RR00750–27S3 and MO1 RR750 and by an Immunology and Hematopoiesis Program grant (MJR) from the Indiana University Cancer Center (P30CA82709).

REFERENCES

1. Rosenberg SA. Principles and Practice of the Biological Therapy of Cancer. Philadelphia: Lippincott, 2000.
2. Smyth MJ, Cretney E, Kershaw MH, Hayakawa Y. Cytokines in cancer immunity immunotherapy. Immunol Rev 2004;202:275–293.
3. Atkins MB, Robertson MJ, Gordon M, et al. Phase I evaluation of intravenous recombinant human interleukin 12 in patients with advanced malignancies. Clin Cancer Res 1997;3:409–417.
4. Motzer RJ, Rakhit A, Schwartz LH, et al. Phase I trial of subcutaneous recombinant human interleukin-12 in patients with advanced renal cell carcinoma. Clin Cancer Res 1998;4:1183–1191.

5. Bajetta E, Vecchio MD, Mortarini R, et al. Pilot study of subcutaneous recombinant human interleukin 12 in metastatic melanoma. Clin. Cancer Res 1998;4:75–85.

6. Portielje JEA, Kruit WHJ, Schuler M, et al. Phase I study of subcutaneously administered recombinant human interleukin 12 in patients with advanced renal cell cancer. Clin Cancer Res 1999;5:3983–3989.

7. Gollob JA, Mier JW, Veenstra K, et al. Phase I trial of twice-weekly intravenous interleukin-12 in patients with metastatic renal cell cancer or melanoma: ability to maintain IFN-γ induction is associated with clinical response. Clin Cancer Res 2000;6:1678–1692.

8. Younes A, Robertson MJ, Flinn IW, et al. A phase II study of interleukin-12 in patients with relapsed non-Hodgkin lymphoma Hodgkin disease. Blood 2002;100:364a (abstr 1408).

9. Robertson MJ, Mier J, Weisenbach J, et al. A phase I dose escalation study to assess safety pharmacokinetics of recombinant human IL-18 (rhIL-18) administered as five daily intravenous infusions in adult patients with solid tumors. Proc Am Soc Clin Oncol 2003;22:178 (abstract 713).

10. Robertson MJ, Mier J, Logan T, et al. Tolerability anti-tumor activity of recombinant human IL-18 administered as five daily intravenous infusions in patients with solid tumors. J. Clin. Oncol 2004;22:176s (abstract 2553).

11. Robertson MJ, Kirkwood JM, Logan T, et al. Phase I study of recombinant human interleukin-18 (rhIL-18) administered as five daily intravenous infusions very 28 days in patients with solid tumors. J Clin Oncol 2005;23:169s (abstract 2513).

12. Curti BD, Redman BG, Thompson JA, SIevers EL. Preliminary tolerability anti-tumor activity of intravenous recombinant human interleukin-21 (IL-21) in patients with metastatic melanoma metastatic renal cell carcinoma. J Clin Oncol 2005;23:166s (abstract 2502).

13. Schilsky RL. End points in cancer clinical trials and the drug approval process. Clin Cancer Res 2002;8:935–938.

14. Kelloff GJ, Bast RC Jr, Coffey DS, et al. Biomarkers, surrogate end points, the acceleration of drug development for cancer prevention treatment: an update. Clin Cancer Res 2004;10:3881–3884.

15. Byrne JL, Haynes AP, Russell NH. Use of haematopoietic growth factors: commentary on the ASCO/ECOG guidelines. Blood Rev 1997;11:16–27.

16. Croockewit AJ, Bronchud MH, Aapro MS, et al. A European perspective on haematopoietic growth factors in haemato-oncology: report of an expert meeting of the EORTC. Eur J Cancer 1997;33:1732–1746.

17. Fetscher S, Mertelsmann R Supportive care in hematological malignancies: hematopoietic growth factors, infections, transfusion therapy. Curr Opin Hematol 2000;7:255–260.

18. Staros EB. Innate immunity: new approaches to understanding its clinical significance. Am J Clin Pathol 2005;123:305–312.

19. Lanier LL NK cell recognition. Ann Rev Immunol 2005;23:225–274.
20. Taylor PR, Martinez-Pomares L, Stacey M, Lin H-H, Brown GD, Gordon S. Macrophage receptors and immune recognition. Ann Rev Immunol 2005;23:901–944.
21. Janeway CA Jr, Travers P, Walport M, Shlomchik MJ Immunobiology: the immune system in health disease.New York: Garland Science; 2005.
22. Paul WE. Fundamental Immunology. Philadelphia: Williams and Wilkins; 2003.
23. Medzhitov R, Janeway CA Jr. Innate immunity: impact on the adaptive immune response. Curr Opin Immunol 1997;9:4–9.
24. Abbas AK, Murphy KM, Sher A. Functional diversity of helper T lymphocytes. Nature 1996;383:787–793.
25. Sitkovsky MV Henkart PA. Cytotoxic Cells: Recognition, Effector Function, Generation, and Methods. Boston: Birkhauser; 1993.
26. Roth C, Rochlitz C, Kourilsky P. Immune response against tumors. Adv Immunol 1994;57:281–351.
27. Ostrand-Rosenberg S. Animal models of tumor immunity, immunotherapy, cancer vaccines. Curr Opin Immunol 2004;16:143–150.
28. Smyth MJ, Kelly JM, Baxter AG, Korner H, Sedgwick JD. An essential role for tumor necrosis factor in natural killer cell-mediated tumor rejection in the peritoneum. J Exp Med 1998;188:1611–1619.
29. Levitsky HI, Lazenby A, Hayashi RJ, Pardoll DM. In vivo priming of two distinct antitumor effector populations: the role of MHC class I expression. J Exp Med 1994;179:1215–1224.
30. Kelly JM, Darcy PK, Markby JL, et al. Induction of tumor-specific T cell memory by NK cell-mediated tumor rejection. Nat Immunol 2002;3:83–90.
31. Ikeda H, Old LJ Schreiber RD. The roles of IFN-γ in protection against tumor development and cancer immunoediting. Cytokine Growth Factor Rev 2005;13:95–109.
32. Street SEA, Cretney E, Smyth MJ. Perforin and inteferon-γ independently control tumor initiation, growth, and metastasis. Blood 2001;97:192–197.
33. Street SEA, Trapani JA, MacGregor D, and Smyth MJ. Suppression of lymphoma and epithelial malignancies by inteferon γ. J Exp Med 2002;196:129–134.
34. Fenton RG, Longo DL. Genetic instability tumor cell variation: implications for immunotherapy. J Natl Cancer Inst 1995;87:241–243.
35. Elliott BE, Carlow DA, Rodricks A-M, Wade A. Perspectives on the role of MHC antigens in normal and malignant cell development. Adv Cancer Res 1993; 60:181–245.
36. Mapara MY, Sykes M. Tolerance and cancer: mechanisms of tumor evasion and strategies for breaking tolerance. J Clin Oncol 2004;22:1136–1151.
37. Wojtowicz-Praga S. Reversal of tumor-induced immunosuppression: a new approach to cancer therapy. J. Immunother 1997;20:165–177.
38. Whiteside TL. Monitoring of antigen-specific cytolytic T lymphocytes in cancer patients receiving immunotherapy. Clin Diag Lab Immunol 2000;7:327–332.

39. Clay TM, Hobeika AC, Mosca PJ, Lyerly HK, Morse MA. Assays for monitoring cellular immune responses to active immunotherapy of cancer. Clin Cancer Res 2001;7:1127–1135.
40. Koehne G, Smith KM, Ferguson TL, et al. Quantitation, selection, and functional characterization of Epstein-Barr virus-specific and alloreactive T-cells detected by intracellular interferon-γ production growth of cytotoxic precursors. Blood 2002;99:1730–1740.
41. Keilholz U, Weber J, Finke JH, et al. Immunological monitoring of cancer vaccine therapy: results of a workshop sponsored by the Society for Biological Therapy. J Immunother 2002;25:97–138.
42. Lyerly HK Quantitating cellular immune responses to cancer vaccines. Semin Oncol 2003;30:9–16.
43. Westerman J, Pabst R. Lymphocyte subsets in the blood: a diagnostic window on the lymphoid system? Immunol Today 1990;11:406–410.
44. Kawai H, Komiyama A, Katoh M, Yabuhara A, Miyagawa Y, Akabane T. Induction of lymphokine-activated killer and natural killer cell activities from cryopreserved lymphocytes. Transfusion 1988;28:531–535.
45. Callery CD, Golightly M, Sidell N, Golub SH. Lymphocyte markers and cytotoxicity following cryopreservation. J Immunol Methods 1980;35:213–223.
46. Lotze MT, Matory YL, Ettinghausen SE, et al. In vivo administration of purified human interleukin 2: II. Half-life, immunological effects, and expansion of peripheral lymphoid cells in vivo with recombinant IL 2. J Immunol 1985;135: 2865–2875.
47. Logan TF, Kaplan SS, Bryant JL, et al. Granulocytopenia in cancer patients treated in a phase I trial with recombinant human tumor necrosis factor. J Immunother 1991;10:84–95.
48. Robertson MJ, Cameron C, Atkins MB, et al. Immunologic effects of interleukin 12 administered by bolus intravenous injection to patients with cancer. Clin Cancer Res 1999;5:9–16.
49. Gately MK, Warrier RR, Honasoge S, et al. Administration of recombinant IL-12 to normal mice enhances cytolytic lymphocyte activity induces production of IFN-γ in vivo. Int Immunol 1994;6:157–167.
50. Kohl S, Sigaroudinia M, Charlebois ED, Jacobson MA. Interleukin-12 administered in vivo decreases human NK cell cytotoxicity and antibody-dependent cellular cytotoxicity to human immunodeficiency virus-infected cells. J Infect Dis 1996;174:1105–1108.
51. Portielje JEA, Lamers CHJ, Kruit WHJ, et al. Repeated administrations of interleukin (IL)-12 are associated with persistently elevated plasma levels of IL-10 and declining IFN-γ, tumor necrosis factor-a, IL-6, IL-8 responses. Clin Cancer Res 2003;9:76–83.
52. Wieder ED Real-time monitoring of immune responses. Cytotherapy 2002;4:347–352.
53. Soiffer RJ, Murray C, Cochran K, et al. Clinical and immunological effects of prolonged infusion of low-dose recombinant interleukin-2 after autologous and T-cell-depleted allogeneic bone marrow transplantation. Blood 1992;79:517–526.

54. Caligiuri MA, Murray C, Robertson MJ, et al. Selective modulation of human natural killer cells in vivo after prolonged infusion of low dose recombinant interleukin 2. J Clin Invest 1993;91:123–132.
55. Miller JS, Tessmer-Tuck J, Pierson BA, et al. Low dose subcutaneous interleukin-2 after autologous transplantation generates sustained in vivo natural killer cell activity. Biol Blood Marrow Transplant 1997;3:34–44.
56. Meropol NJ, Porter M, Blumenson LE, et al. Daily subcutaneous injection of low-dose interleukin 2 expands natural killer cells in vivo without significant toxicity. Clin Cancer Res 1996;2:669–677.
57. Phillips JH Lanier LL. Dissection of the lymphokine-activated killer phenomenon: relative contribution of peripheral blood natural killer cells and T lymphocytes to cytolysis. J Exp Med 1986;164:814–825.
58. Pelloso D, Cyran K, Timmons L, Williams BT, Robertson MJ. Immunological consequences of interleukin 12 administration after autologous stem cell transplantation. Clin Cancer Res 2004;10:1935–1942.
59. Gollob JA, Schnipper CP, Orsini E, et al. Characterization of a novel subset of CD8+ T-cells that expands in patients receiving interleukin-12. J Clin Invest 1998;102:561–575.
60. Robertson MJ, Chang H-C, Pelloso D, Kaplan MH. Impaired interferon-γ production as a consequence of STAT4 deficiency after autologous hematopoietic stem cell transplantation for lymphoma. Blood 2005;106:963–970.
61. Nakarai T, Robertson MJ, Streuli M, et al. Interleukin 2 receptor γ chain expression on resting activated lymphoid cells. J Exp Med 1994;180:241–251.
62. Sugamura K, Asao H, Kondo M, et al. The interleukin-2 receptor γ chain: its role in the multiple cytokine receptor complexes and T cell development in XSCID. Annu Rev Immunol 1996;14:179–205.
63. Robertson MJ, Cochran KJ, Cameron C, Le J-M, Tantravahi R, Ritz J. Characterization of a cell line, NKL, derived from an aggressive human natural killer cell leukemia. Exp Hematol 1996;24:406–415.
64. Caligiuri MA, Zmuidzinas A, Manley TJ, Levine H, Smith KA, Ritz J, Functional consequences of interleukin 2 receptor expression on resting human lymphocytes: identification of a novel natural killer cell subset with high affinity receptors. J Exp Med 1990;171:1509–1526.
65. Robertson MJ, Cochran KJ, Ritz J. Characterization of surface antigens expressed by normal and neoplastic human natural killer cells. In: Schlossman SF, Boumsell L, Gilks W, et al, eds. Leucocyte Typing V: White Cell Differentiation Antigens. –Oxford: Oxford University Press; 1995, pp. 1374–1377.
66. Testi R, D'Ambrosio D, DeMaria R, Santoni A. The CD69 receptor: a multipurpose cell surface trigger for hematopoietic cells. Immunol Today 1994;15:479–483.
67. Ortaldo JR, Winkler-Pickett RT, Yagita H, Young HA Comparative studies of CD3- and CD3+ CD56+ cells: examination of morphology, functions, T cell receptor rearrangement, and pore-forming protein expression. Cell Immunol 1991;136: 486–495.

68. Knapp W, Dorken B, Gilks WR, et al. Leucocyte Typing IV: White Cell Differentiation Antigens. Oxford: Oxford University Press, 1989.
69. Phillips JH, Gemlo BT, Myers WW, Rayner AA, Lanier LL. In vivo in vitro activation of natural killer cells in advanced cancer patients undergoing combined recombinant interleukin-2 and LAK cell therapy. J Clin Oncol 1987;5: 1933–1941.
70. Ellis TM, Creekmore SP, McMannis JD, Braun DP, Harris JA, Fisher RI. Appearance and phenotypic characterization of circulating Leu 19+ cells in cancer patients receiving recombinant interleukin 2. Cancer Res 1988;48:6597–6602.
71. Weil-Hillman G, Fisch P, Prieve AF, Sosman JA, Hank JA, Sondel PM. Lymphokine-activated killer activity induced by in vivo interleukin 2 therapy: predominant role for lymphocytes with increased expression of CD2 and Leu19 but negative expression of CD16 antigens. Cancer Res 1989;49:3680–3688.
72. Takeda K, Akira S. STAT family of transcription factors in cytokine-mediated biological responses. Cytokine Growth Factor Rev 2000;11.
73. Ivashkiv LB. Cytokines and STATs: how can signals achieve specificity. Immunity 1995;3:1–4.
74. Krutzig PO, Irish JM, Nolan GP, Perez OD. Analysis of protein phosphorylation and cellular signaling events by flow cytometry: techniques and clinical applications. Clin Immunol 2004;110:206–221.
75. Krutzig PO, Hale MB, Nolan GP. Characterization of the murine immunological signaling network with phosphospecific flow cytometry. J Immunol 2005;175:2366–2373.
76. Bonhoeffer S, Mohri H, Ho D, Perelson AS. Quantification of cell turnover kinetics using 5-bromo-2'-deoxyuridine. J Immunol 2000;164:5049–5054.
77. Plett PA, Frankovitz SM, Orschell-Traycoff CM. In vivo trafficking, cell cycle activity, and engraftment potential of phenotypically defined primitive hematopoietic cells after transplantation into irradiated or nonirradiated recipients. Blood 2002;100:3545–3552.
78. Jamieson AM, Isnard P, Dorfman JR, Coles MC, Raulet DH. Turnover proliferation of NK cells in steady state lymphopenic conditions. J Immunol 2004; 172:864–870.
79. Dobrowsky W, Dobrowksy E, Wilson GD. In vivo cell kinetic measurements in a randomized trial of continuous hyperfractionated accelerated radiotherapy with or without mitomycin C in head and neck cancer. Int J Radiat Oncol Biol Phys 2003;55:576–582.
80. Corvo R, Giaretti W, Sanguineti G, et al. In vivo cell kinetics in head and neck squamous cell carcinomas predicts local control and helps guide radiotherapy regimen. J Clin Oncol 1995;13:1843–1850.
81. Konrad MW, Hemstreet G, Hersh EM, et al. Pharmacokinetics of recombinant interleukin 2 in humans. Cancer Res 1990;50:2009–2017.
82. Blick M, Sherwin SA, Rosenblum M, Gutterman J. Phase I study of recombinant tumor necrosis factor in cancer patients. Cancer Res 1987;47:2986–2989.

83. Chapman PB, Lester TJ, Casper ES, et al. Clinical pharmacology of recombinant human tumor necrosis factor in patients with advanced cancer. J Clin Oncol 1987;5:1942–1951.

84. Koch KM, Roman JR, Jaworski D, et al. PK and PD of recombinant human IL-18 (rhIL-18) administered IV in repeated cycles to patients with solid tumors. J Clin Oncol 2005;23:174s (abstract 2535).

85. Mier JW, Vachino G, VerMeer JWM, et al. Induction of circulating tumor necrosis factor (TNF α) as the mechanism for the febrile response to Interleukin-2 (IL-2) in cancer patients. J Clin Immunol 1988;8:426–436.

86. Heslop HE, Gottlieb DJ, Bianchi ACM, et al. In vivo induction of gamma interferon and tumor necrosis factor by interleukin-2 infusion following intensive chemotherapy or autologous bone marrow transplantation. Blood 1989;74:1374–1380.

87. Robertson MJ, Pelloso D, Abonour R, et al. Interleukin-12 immunotherapy after autologous stem cell transplantation for hematologic malignancies. Clin Cancer Res 2002;8:3383–3393.

88. Rakhit A, Yeon MM, Ferrante J, et al. Down-regulation of the pharmacokinetic-pharmacodynamic response to interleukin-12 during long-term administration to patients with renal cell carcinoma and evaluation of the mechanism of this "adaptive response" in mice. Clin Pharmacol Ther 1999;65:615–629.

89. Leonard JP, Sherman ML, Fisher GL, et al. Effects of single-dose interleukin-12 exposure on interleukin-12-associated toxicity and interferon-γ production. Blood 1997;90:2541–2548.

90. Blay J-Y, Favrot MC, Negrier S, et al. Correlation between clinical response to interleukin 2 therapy and sustained production of tumor necrosis factor. Cancer Res 1990;50:2371–2374.

91. Kammula US, Lee K-H., Riker AI, et al. Functional analysis of antigen-specific T lymphocytes by serial measurement of gene expression in peripheral blood mononuclear cells and tumor specimens. J Immunol 1999;163:6867–6875.

92. Blaschke V, Reich K, Blaschke S, Zipprich S Neumann C. Rapid quantitation of proinflammatory and chemoattractant cytokine expression in small tissue samples and monocyte-derived dendritic cells: validation of a new real-time RT-PCR technology. J Immunol Methods 2000;246:79–90.

93. Pross HF, Baines MG, Rubin P, Shragge P, Patterson MS. Spontaneous human lymphocyte-mediated cytotoxicity against tumor target cells. IX. The quantitation of natural killer cell activity. J Clin Immunol 1981;1:51–63.

94. Bryant J, Day R, Whiteside TL, Herberman RB. Calculation of lytic units for the expression of cell-mediated cytotoxicity. J Immunol Methods 1992;146:91–103.

95. Matzinger P. The JAM test: a simple assay for DNA fragmentation and death. J Immunol Methods 1991;145:185–192.

96. Nociari MM, Shalev A, Benias P, Russo C. A novel one-step, highly sensitive fluorometric assay to evaluate cell-mediated cytotoxicity. J Immunol Methods 1998;213:157–167.

97. Virag L, Kerekgyarto C, Fachet J. A simple, rapid and sensitive fluorimetric assay for the measurement of cell-mediated cytotoxicity. J Immunol Methods 1995;185:199–208.

98. Robertson MJ, Manley TJ, Donahue C, Levine H, Ritz J. Costimulatory signals are required for optimal proliferation of human natural killer cells. J Immunol 1993;150:1705–1714.

99. Robertson MJ, Soiffer RJ, Wolf SF, et al. Response of human natural killer (NK) cells to NK cell stimulatory factor (NKSF): cytolytic activity and proliferation of NK cells is differentially regulated by NKSF. J Exp Med 1992; 175:779–788.

100. Pittet M, Valmori D, Dunbar PR, et al. High frequencies of naive Melan-A/MART-1-specific CD8+ T-cells in a large proportion of human histocompatibility leukocyte antigen (HLA)-A2 individuals. J Exp Med 1999;190:705–715.

101. Kuzushima K, Hoshino Y, Fujii K, et al. Rapid determination of Epstein-Barr virus-specific CD8+ T cell frequencies by flow cytometry. Blood 1999;94: 3094–3100.

102. Healey G Schwarer AP. The helper T lymphocyte (HTLp) frequency does not predict outcome after HLA-identical sibling donor G-CSF-mobilised peripheral blood stem cell transplantation. Bone Marrow Transplant 2002;30:341–346.

103. Scheibenbogen C, Romero P, Rivoltini L, et al. Quantitation of antigen-reactive T-cells in peripheral blood by IFN-γ-ELISPOT assay chromium-release assay: a four-centre comparative trial. J Immunol Methods 2000;244:81–89.

104. Whiteside TL, Zhao Y, Tsukishiro T, Elder EM, Gooding W, Baar J. Enzyme-linked immunospot, cytokine flow cytometry, and tetramers in the detection of T-cell responses to a dendritic cell-based multpeptide vaccine in patients with melanoma. Clin Cancer Res 2003;9:641–649.

105. Lee P, Wang F, Kuniyoshi J, et al. Effects of interleukin-12 on the immune response to a multipeptide vaccine for resected metastatic melanoma. J Clin Oncol 2001;19:3836–3847.

106. Peterson AC, Harlin H, Gajewski TF. Immunization with melan-A peptide-pulsed peripheral blood mononuclear cells plus recombinant human interleukin-12 induces clinical activity and T-cell responses in advanced melanoma. J Clin Oncol 2003;21:2342–2348.

107. Pala P, Hussell T, Openshaw PJM. Flow cytometric measurement of intracellular cytokines. J Immunol Methods 2000;243:107–124.

16

Clinical Adverse Effects of Cytokines on the Immune System

Thierry Vial and Jacques Descotes

SUMMARY

Many clinical studies or case reports have focused on the clinical consequences of immunoenhancement or immune dysregulation mediated by therapeutic cytokines. Flu-like reactions, and the facilitation or exacerbation of
inflammatory diseases, are the main consequences of immunoenhancement.
Flu-like symptoms commonly are observed in patients treated with interferons,
interleukin (IL)-1, IL-2, IL-3, or tumor necrosis factor- α. They typically consist of fever and chills, malaise, tachycardia, arthralgias, and myalgias that
develop within a few hours after administration. The mechanism is not clearly
understood but probably involves the acute release of fever-promoting factors
in the hypothalamus. Because various cytokines are directly or indirectly involved in the pathogenesis of immune-mediated inflammatory disorders and
autoimmune diseases, it is not surprising that such disorders develop after the
administration of pharmacological doses of these cytokines. Another adverse
consequence is the development of cytokine-specific antibodies in the sera of
treated patients.

Key Words: Adverse effect; humans; interferons; interleukins; growth
factors; immune diseases.

1. INTERFERONS

The interferons (IFNs) include at least five natural human glycoproteins
(α, β, γ, ω, and τ), of which only the first three types currently are in therapeutic use. They differ both structurally and antigenically. IFN-α and -β
primarily exert antiviral and antiproliferative effects, whereas IFN-γ acts as
an immunoregulatory cytokine.

From: *Methods in Pharmacology and Toxicology: Cytokines in Human Health:
Immunotoxicology, Pathology, and Therapeutic Applications*
Edited by: R. V. House and J. Descotes © Humana Press Inc., Totowa, NJ

1.1. Interferon-α

IFN-α contains purified natural leukocyte or lymphoblastoid human IFN or recombinant products. Attempts to assign the most frequently observed amino acids at each position led to a consensus IFN-α. Because standard IFN-α has a short half-life, pegylated IFNs obtained by the covalent conjugation of monomethoxy polyethylene glycol have been developed. Pegylated IFN-α is assumed to have the same safety profile as standard IFN-α, although it has been suggested to cause more frequent and severe hematotoxic effects *(1)*. IFN-α currently is used in the treatment of chronic hepatitis C and B virus infections in addition to various hematological or solid neoplasias. A wide range of additional viral diseases or cancers may benefit from IFN-α therapy. The pathogenesis of most adverse effects observed with IFN-α therapy is poorly understood, but the commonly postulated mechanisms involve either a direct toxic effect or an indirect immune-mediated effect.

1.1.1. Flu-Like Syndrome

Virtually all patients experience flu-like syndrome during the first days of treatment, but tachyphylaxis usually develops after 1 to 2 wk *(2)*. Conversion to human leukocyte IFN-α is sometimes successful in patients with a poor tolerance to recombinant IFN-α *(3)*. Although severity increases with the dose, the flu-like syndrome is rarely treatment-limiting and can be partly prevented with the administration of paracetamol (acetaminophen). The acute release of fever-promoting factors, such as the eicosanoids, interleukin (IL)-1, and tumor necrosis factor (TNF)-α, is the suggested mechanism.

1.1.2. IFN-α, Autoantibodies, and Autoimmunity

The exact role of IFN-α in the occurrence of autoimmune disorders is difficult to ascertain as the underlying treated disease also can be associated with immunopathological disorders. Indeed, antinuclear, antithyroid, antiparietal cell, antiliver/kidney microsome, and antismooth muscle autoantibodies and the rheumatoid factor frequently are noted before IFN-α therapy, which suggests that patients may be predisposed to autoimmunity. On the other hand, increased titers, or the new occurrence of autoantibodies, has been observed in 4 to 30% of previously autoantibody-negative patients with the disappearance of autoantibodies after treatment discontinuation in approx two-thirds of patients *(4)*. The clinical significance of autoantibodies in IFN-α-treated patients has been debated. They were repeatedly shown not to affect response to IFN-α treatment *(5,6)*. Although it was initially felt that IFN-α might facilitate autoimmune diseases in patients previously positive for specific or nonorgan-specific autoantibodies, the clinical evidence is still limited. Except for thyroiditis, large studies in patients treated for chronic hepatitis C did not show a significant increase in overt autoimmune diseases despite the pre-existence or subsequent positivity of autoantibodies *(4–6)*. By contrast, a relatively high incidence of immune-mediated complications has been found in patients treated for chronic myeloid leukemia, with a strong association with female gender and long IFN-α exposures *(7,8)*. As a result, there is no clear consensus about the management of patients previously positive for nonorgan-specific autoantibodies. It is, however, usually considered that low autoantibody titers or the absence of concomitant symptoms suggestive of autoimmune disease is not a contraindication to IFN-α therapy.

Because the spectrum of IFN-α-induced autoimmune diseases is extremely wide, including both organ-specific and systemic autoimmune diseases, only few examples will be detailed in this chapter. Most of these disorders correspond to the unmasking of disease in potentially predisposed patients.

1.1.3. Nonorgan-Specific Autoimmune Disorders

The suspicion for an unexpectedly greater incidence of rheumatoid and lupus-like symptoms first arose in patients treated with IFN-α alone or combined with IFN-γ for myeloproliferative disorders *(9)*. However, only a minority of patients fulfilled the diagnostic criteria for systemic lupus erythematosus (SLE), and it is unknown whether this complication is coincidental or truly related to treatment. Additional studies showed that systemic autoimmune diseases are genuine but are very rare complications of IFN-α in chronic hepatitis C. Confirmed cases of SLE have indeed been reported only occasionally *(10)*. The predominance of young patients and female gender, the presence of renal or skin involvement, positive autoantibodies to double-stranded deoxyribonucleic acid (DNA) and decreased serum complement levels, and the rapid onset after starting treatment, as well as the persistence of symptoms after IFN-α withdrawal in most reported cases, are more in keeping with the unmasking of idiopathic SLE by IFN-α than with the drug-induced lupus syndrome. The reactivation or appearance of clinical or biological symptoms consistent with rheumatoid arthritis, lupus-like polyarthritis, or systemic sclerosis has sometimes been reported, and most patients were found to have underlying rheumatoid disease, increased levels of rheumatoid factor, or positive titers of antinuclear antibodies before treatment *(11–13)*.

1.1.4. Organ-Specific Autoimmune Diseases

Whereas the prevalence of pancreatic autoantibodies increased during IFN-α treatment in patients with chronic hepatitis *(14)*, diabetes mellitus was found in only 10 of 11,241 patients treated for chronic hepatitis C *(15)*. However, IFN-α-associated diabetes mellitus is probably more than coincidental, as suggested by several reports of acute onset or worsening of diabetes mellitus shortly after IFN-α initiation, with subsequent improvement or complete recovery after IFN-α withdrawal *(14,16)*. An autoimmune mechanism is suggested by the presence of the HLA-DR4 haplotype and/or islet cell antibody (ICA) positivity at the time of diagnosis in several patients. Although the induction of ICA antibodies occasionally has been reported, they do not predict for the development of diabetes *(17)*. Others suggested that repetitive treatment with IFN-α could increase the risk of type 1 diabetes in patients previously positive for islet auto-antibodies *(18)*. The triggering rather than the induction of a latent autoimmune phenomenon in patients with genetic susceptibility is the most probable mechanism, but a direct interference with glucose metabolism cannot be excluded.

Although a myelosuppressive effect accounts for most of the hematological toxicities associated with IFN-α, reports of immune-mediated thrombocytopenia, immune hemolytic anemia, or asymptomatic positive direct Coombs' test *(19–21)* indicate that IFN-α also can mediate immune blood cell destruction. A mechanism close to that observed with α-methyldopa is thought to be involved in autoimmune hemolytic anemias *(22)*. IFN-α-induced immune-mediated thrombocytopenias share many features with idiopathic thrombocytopenic purpura, but the recurrence of thrombocytopenia upon IFN-α readministration strongly supports the causal role of IFN-α *(21)*. By contrast, IFN-α was not considered to be harmful in patients with chronic hepatitis C who were previously positive for platelet associated IgG *(23)*. IFN-α also has been reported to induce multiple antibody formation to transfused blood cell antigens with subsequent massive hemolysis *(24)* and pernicious anemia with positive anti-intrinsic factor antibodies *(25)*. IFN-α also has been associated with the development of anti-factor VIII autoantibodies in very few patients *(26)*, or with the production of antiphospholipid antibodies, potentially increasing the risk of venous thrombosis *(27)*.

The possible acute exacerbation of latent autoimmune hepatitis emerged as a therapeutic dilemma in patients treated for chronic hepatitis C because of the possible simultaneous presence of unequivocal serological evidence of chronic hepatitis C and serological markers of autoimmune hepatitis *(28,29)*. The therapeutic management is therefore difficult because the distinction between both diseases cannot readily be made. The systematic detection of specific autoantibodies was unable to predict the risk of autoimmune hepatitis *(28)*. In this situation, a possible increase in viremia can be observed with glucocorticosteroids, whereas an acute exacerbation of the latent autoimmune liver disease may be expected with IFN-α. As a result, several investigators advocated the use of glucocorticosteroids as a first-line treatment in patients with high antibody titers, whereas others considered IFN-α to be more appropriate when autoantibody titers are low and IFN-α is expected to be effective *(30)*. Although prospective studies usually failed to evidence the induction of autoantibodies specifically linked to autoimmune liver disease, de novo induction rather than exacerbation of autoimmune hepatitis is possible *(31)*.

Thyroid disorders are well-established complications of IFN-α therapy, with a spectrum of effects ranging from the asymptomatic presence of antithyroid autoantibodies to severe clinical features of hypothyroidism, hyperthyroidism, or acute biphasic thyroiditis *(32)*. In large prospective studies, the incidence of thyroid abnormalities was 5% to 8% in patients treated for chronic hepatitis C, and only 1% to 3% in patients treated for chronic hepatitis B *(4,32,33)*. The incidence reached 12% in patients treated for cancer *(4)*. The reversibility of thyroid disorders or the decline of thyroid antibody levels after IFN-α withdrawal is in accordance with a causal relationship. Although spontaneous resolution is expected in most patients, sustained hypothyroidism requiring long-term substitutive therapy occurred in patients with initially severe hypothyroidism and elevated thyroid antibody titers. Among the many potential susceptibility factors examined, only female gender and pretreatment positivity or development of thyroid autoantibodies during treatment have been consistently associated with the occurrence of thyroid dysfunction, whereas the duration of treatment, the dose or the type of IFN-α, natural or recombinant, was not *(32,34)*. The HLA-A2 haplotype also was suggested to be a predisposing factor *(35)*. An autoimmune reaction or immune dysregulation leading to the induction or exacerbation of pre-existing latent thyroid autoimmunity is therefore the most attractive hypothesis, in accordance with the relatively frequent occurrence of other autoantibodies or clinical autoimmune disorders in patients who develop thyroid disorders. It is not yet proven, however, that autoimmunity is the universal mechanism and a direct effect of IFN-α on the thyroid functions has been considered *(36)*. Finally, the pattern of thyroid autoantibodies in patients who developed thyroid dysfunction during IFN-α treatment differed significantly from that of patients with spontaneous autoimmune thyroid disease *(37)*. Other organ-specific autoimmune diseases have been attributed to IFN-α, including unmasking myasthenia gravis and celiac disease in predisposed patients *(38,39)*.

1.1.5. Other Adverse Effects Possibly Involving the Immune System

Beside autoimmune reactions, a number of adverse effects in treated patients may reflect the immunomodulatory properties of IFN-α, or an enhanced T-cell-mediated reaction. IFN-α is a probable triggering factor for the reactivation or new occurrence of cutaneous or generalized sarcoidosis *(40)*. In a number of patients, interstitial pneumonitis or bronchiolitis obliterans that may be triggered by T-cell activation have been associated with IFN-α *(41)*. A similar mechanism has been discussed in IFN-α-induced nephrotoxicity, such as acute tubulointerstitial nephritis, the nephrotic syndrome with severe glomerular changes, membranoproliferative glomerulonephritis, extracapillary glomerulonephritis, and focal segmental glomerulosclerosis *(42)*.

Possible immune-mediated neurological complications include demyelinating events, such as chronic inflammatory demyelinating polyneuropathy, multiple sclerosis-like disease, or the Guillain Barré syndrome *(43–45)*. Although IFN-α initially was thought to have some benefit in multiple sclerosis, the long-term use of IFN-α actually caused a more rapid aggravation of the disease *(46)*.

A large range of skin lesions, including injection site reactions and generalized cutaneous reactions, has been attributed to IFN-α. IFN-α can undoubtedly induce psoriasis in patients without a history of psoriasis, or worsen pre-existing psoriasis *(2)*, which is in accordance with IFN-α-induced skewing toward a Th1 response. The occurrence or exacerbation of lichen planus has been the matter of considerable debate because most cases were observed in patients with chronic hepatitis C, a disease that is controversially associated with a spontaneously higher incidence of lichen planus. The recurrence of lesions after the readministration of IFN-α or reports of lichen planus in treated cancer patients argues strongly for a direct causal link *(47)*. The induction of bullous lesions with circulating pemphigus-like autoantibodies or pemphigus foliaceus with the new occurrence of anti-intercellular IgG antibodies has been observed *(4)*. Other dermatological complications of IFN-α possibly involving an immune-mediated reaction include alopecia areata, cutaneous vasculitis, vitiligo, eczema-like lesions, or lichenoid eruptions *(48,49)*. More recently, pegylated IFN-α has been suggested to be associated with more frequent or more severe dermatological adverse effects with histological features resembling contact dermatitis *(50)*. Of interest, positive cutaneous tests to pegylated IFN-α, but not to standard IFN-α have been documented in three patients who had experienced severe rash while receiving pegylated IFN-α *(51)*.

1.1.6. Immediate Hypersensitivity

No IgE-mediated reactions to IFN-α have been documented, and reports of urticaria or angioedema are very scarce. In one patient with urticaria after treatment with IFN-α-2b, anti-IFN-α IgG, but not IgE, was identified *(52)*. A recurrent anaphylactoid reaction, possibly the result of mast cell degranulation, has been described in a patient with mastocytosis *(53)*.

1.1.7. Immunosuppression

Possible immunosuppressive effects of IFN-α have been suggested after reports of exacerbation of latent parasitic infections *(54)*, unexpectedly severe abscesses *(55)*, or opportunistic infections *(56,57)* in previously immuno-competent patients. The available evidence, however, is very limited, so that no conclusion can be drawn on a possible association between IFN-α therapy and deleterious effects related to immunosuppression.

1.1.8. IFN-α Antibodies

Both binding and neutralizing anti-IFN-α antibodies have been detected in treated patients *(4,58,59)*. The rate of anti-IFN-α antibody formation ranged from zero to more than 50% of patients, but comparison between studies is difficult because the underlying treated disease, the studied population, the type of interferon used, the route of administration, the dosing regimen, the duration of treatment, and the method of assay differed across studies. Higher rates of anti-IFN-α antibodies were noted in patients on long-term maintenance treatment, low-dose IFN-α, and subcutaneous rather than intravenous administration. Using the same antibody assay, a greater frequency of anti-IFN-α-2a antibodies was found compared with other recombinant or natural IFN-α *(58)*. The mechanism accounting for the difference in immunogenicity is speculative. The role of the single amino acid substitution, the lack of the IFN-α-2a gene in the Caucasian population, or the absence of glycosylation sites on recombinant IFN-α has been proposed.

The clinical significance of binding antibodies appears to be limited to changes in IFN-α pharmacokinetics. By contrast, neutralizing antibodies can theoretically reduce clinical response, but this has been strongly debated *(4,58)*. Whereas several studies could not detect a loss in therapeutic response, others noted response failure, or breakthrough hepatitis or viremia concomitantly to the appearance of neutralizing antibodies. Crossreactivity between antibodies to the various recombinant forms of IFN-α has consistently been demonstrated in vitro. By contrast, the ability of antirecombinant IFN-α antibodies for neutralizing the antiviral or antiproliferative activity of natural IFN-α was not or seldom documented. That natural IFN-α can overcome the neutralizing activity of antibodies to recombinant IFN-α was further demonstrated: a change to natural IFN-α proved successful in restoring the response in some patients who had ceased to respond after they had developed anti-recombinant IFN-α antibodies *(58,59)*. This discovery led researchers to suggest that the formation of neutralizing antibodies results from a specific immune response toward recombinant preparations and that natural IFN-α can overcome the neutralizing activity of antibodies to recombinant IFN-α. Finally, neutralizing antibodies were not associated with immune complex-associated diseases or hypersensitivity reactions, and exerted no influence on IFN-α-associated adverse effects.

1.2. Interferon-β

IFN-β is available as the natural fibroblast or recombinant form (β-1a or -1b). It exerts antiviral and antiproliferative properties similar to those of IFN-α. Although it is mostly used in the treatment of multiple sclerosis, it also has been investigated in chronic viral hepatitis. The general toxicity of IFN-β is very similar to that of IFN-α, with no marked differences between the two recombinant forms *(60)*. Fatigue and a transient flu-like syndrome are observed in approx 60% of patients at the initiation of treatment, and tachyphylaxis usually develops after several doses *(61)*.

1.2.1. Autoimmune Disorders

In contrast to IFN-α, the evaluation of autoimmune effects associated with IFN-β therapy is limited. A 6-mo course of IFN-β was not associated with the appearance of autoantibodies or increased autoantibody titers, and no clinical features of autoimmune disease were observed *(62,63)*. The incidence of clinically overt thyroid disorders is far lower in patients treated with IFN-β than IFN-α *(64,65)*. Antithyroid autoantibodies usually are found in these patients. Other possible autoimmune complications, such as reversible autoimmune hemolytic anemia *(66)*, transient autoimmune hepatitis *(67)*, subcutaneous lupus erythematosus *(68)*, myasthenia gravis *(69)*, and acquired hemophilia A *(70)*, were described in isolated case reports.

1.2.2. Other Adverse Effects of IFN-β Possibly Involving the Immune System

Injection-site reactions after subcutaneous IFN-β-1b are more frequent than with any other available interferons *(71)*. They mostly consisted of benign inflammatory reactions, but deep cutaneous ulcers with skin necrosis also are possible. The mechanism may involve a local vascular inflammatory process or platelet-dependant thrombosis. However, positive intracutaneous tests to IFN-β in several patients suggest the involvement of an immunological reaction *(72)*. Rare isolated reports of dermatological adverse effects possibly involving the immune system include psoriasis exacerbation, sarcoidosis or vasculitis *(73–75)*.

Hypersensitivity reactions to IFN-β have been rarely reported *(76)*. In one patient, a positive intradermal test to IFN-β-1b, but not to IFN-β-1a or the diluents, suggested a specific hypersensitivity type I reaction *(77)*. Allergic contact dermatitis after the use of IFN-β eye-drops has been reported once *(78)*.

1.2.3. Anti-IFN-β Antibody Formation

Neutralizing antibodies toward recombinant IFN-β have been noted in 12 to 38% of patients treated for 2 to 3 yr at a higher frequency with subcutaneous IFN-β-1b compared with subcutaneous IFN-β-1a or intramuscular IFN-β-1b *(79,80)*. Neutralizing antibodies against recombinant IFN-β were found to crossreact in both binding and biological in vitro assays *(81)*. Whether neutralizing antibodies to IFN-β are associated with adverse clinical consequences is still debated *(82)*. Early studies in patients receiving IFN-β-1b found decreased clinical efficacy *(83)*, but more recent studies that sequentially assessed neutralizing antibodies in a large number of patients treated with various IFN-β preparations showed that neutralizing antibodies may be associated with a significantly higher relapse rate or shortened time to first relapse, but did not affect overall disease progression *(79,80)*. Although the impact of neutralizing antibodies on the relapse rate was inconsistent, there was an increased relapse rate during periods of high neutralizing antibody titers. No predictors of antibody formation were identified. Other recent studies with long-term follow-up indicated that the simultaneous presence of high titers of both binding and neutralizing antibodies, or their persistence, is correlated with greater levels of disease activity and worsening *(84,85)*. These results, however, were again strongly disputed and faced with discordant results, most investigators agree that treatment decisions in patients with positive neutralizing antibodies should rather be based on individual clinical outcome and when possible, on the unequivocal demonstration of neutralizing antibodies.

1.3. Interferon-γ

Recombinant IFN-γ-1b is used in the treatment of chronic granulomatous disease. Its immunoregulatory potential is under investigation in other diseases. The clinical experience is still limited, and most common adverse effects include flu-like symptoms and moderate injection-site reactions. Although the therapeutic use of IFN-γ is mainly based on its immunoregulatory properties, the potential for immune-mediated adverse clinical consequences has rarely been investigated. Most patients treated for chronic hepatitis B developed autoantibodies, but none had clinical evidence of autoimmune disease *(86,87)*. By contrast, no change in antinuclear antibodies was reported in one study of rheumatoid arthritis patients *(88)*, whereas other studies found increased titers or new antinuclear antibodies associated with the development of SLE-like symptoms, or clinical exacerbation of rheumatoid arthritis *(89,90)*. IFN-γ was involved in the unexpected exacerbation of multiple sclerosis *(91)* and in one case of autoimmune thrombocytopenia *(92)*. Altogether, these findings suggest a possible, but seemingly limited potential of IFN-γ to cause deleterious immune-mediated effects. An anaphylactoid reaction with severe bronchospasm was reported in one patient after the first IFN-γ injection *(93)*, but no data substantiated an immunological mechanism. Finally, neutralizing anti-IFN-γ antibodies have been found exceptionally, and their clinical significance is unknown *(94)*.

2. INTERLEUKINS

2.1. Interleukin-1

IL-1α and IL-1β act through the same receptor and share similar biological properties in vitro. IL-1 has modest antitumor activity and limited hematopoietic effects. Both forms of IL-1 have been investigated in humans, and they produce a wide and very similar spectrum of adverse effects *(95)*. Whatever the dose, the flu-like syndrome is quite universal, but only occasionally treatment-limiting. Tachyphylaxis develops during prolonged administration. Because IL-1 induces a dose-limiting hypotension with clinical features of septic shock resulting from a capillary leak phenomenon, its use is considerably limited.

2.2. Interleukin-2

Recombinant IL-2 is approved for the treatment of metastatic renal cell carcinoma. Its potential benefits also have been investigated in other malignant neoplasias and in HIV-infected patients *(96)*. Because high-dose IL-2 is associated with quite universal constitutional symptoms and various severe dose-dependent and limiting toxicities *(97)*, low-dose subcutaneous or continuous intravenous administrations are preferred *(98,99)*.

2.2.1. Autoimmune-Like Adverse Reactions

Experimental data suggest that IL-2 may activate autoreactive lymphocytes, facilitate *de novo* immune response, reactivate quiescent autoimmunity, or exacerbate inflammatory diseases. A number of studies have indeed reported thyroid dysfunction, which usually consisted of moderate and reversible hypothyroidism in patients receiving IL-2 alone or in combination with LAK cells, IFN-α, IFN-β, or TNF-α *(100,101)*. Patients receiving IL-2 plus IFN-α more commonly developed biphasic thyroiditis with subsequent hypothyroidism or hyperthyroidism. In a survey of 281 cancer patients receiving IL-2, up to 41% of previously euthyroid patients developed thyroid dysfunction *(102)*. Combined immunotherapy with IL-2 and IFN-α produced more frequent thyroid disorders with an incidence of laboratory thyroid dysfunction up to 100% *(103)*. Female gender and the presence of antithyroid antibodies correlated significantly with the development of thyroid disease *(101)*. These findings together with the strong expression of HLA-DR antigens on thyrocytes or the presence of mononuclear cell infiltrates in the thyroid gland make an autoimmune phenomenon likely *(100)*.

However, a possible direct effect on thyroid hormonal function has also been suggested in patients with no detectable thyroid antibodies.

IL-2 rarely exacerbated other latent autoimmune diseases, including diabetes mellitus, myasthenia gravis and rheumatoid arthritis *(104–106)*. Immunostimulation induced by IL-2 was thought to break tolerance to self-antigens and enhance latent autoimmunity in patients with a genetic predisposition and/or several autoantibodies before treatment.

2.2.2. Other Adverse Reactions Possibly Involving the Immune System

One of the most severe adverse effects of IL-2 is the vascular leak syndrome, the severity of which is dose-related. It is characterized by damage to endothelial cells with extravasation of plasma proteins and fluid from capillaries into the extravascular space. The clinical consequence is a third-space syndrome with severe hypotension, weight gain, generalized edema, pulmonary congestion, ascites, and cardiovascular and renal complications *(99)*. The increased vascular permeability has been suggested to result from IL-2-induced suppression of endothelin-1 secretion by endothelial cells, or activation of the complement cascade, or TNF-α release from IL-2-activated T cells with subsequent activation of polymorphonuclear neutrophils *(107,108)*. Although the hemodynamic and cardiac complications of high-dose IL-2 mostly result from a reduction in systemic vascular resistance and left ventricular ejection fraction, clinical and histological findings of eosinophilic, lymphocytic, or mixed lymphocytic-eosinophilic myocarditis have occasionally been observed, which suggests that an immune-mediated reaction may also occur *(109,110)*.

Immunostimulation caused by IL-2 may have played an important role in the occasional exacerbation of Crohn's disease, rheumatoid arthritis, or IgA glomerulonephritis, or the development of acute interstitial nephritis with a predominant T-lymphocyte kidney infiltration *(4,100)*. A number of dermatological adverse effects in IL-2-treated patients are also probably the consequence of an aberrant immune response. Cutaneous reactions generally comprise pruritus, flushing, mild-to-moderate erythematous macular and desquamative eruptions and, rarely, generalized erythroderma *(111)*. Histological and immunopathological examination of the skin showed mild infiltrates of activated T-helper lymphocytes, increased expression of HLA-DR, and intercellular adhesion molecule-1 on keratinocytes and endothelial cells. A possible role of IFN-γ has been suggested *(111,112)*. Unmasking of erythema nodosum, linear IgA bullous dermatosis, extensive bullous skin eruption, toxic epidermal necrolysis, dermatitis exfoliativa, recurrence of pemphigus vulgaris, exacerbation of localized or widespread psoriasis, acute reactivation of eczema, rapid progression of scleroderma with myositis, and leukocytoclastic vasculitis, in isolated case reports suggest that immunostimulation caused by IL-2 can induce or exacerbate cutaneous reactions *(113)*.

2.2.3. Hypersensitivity Reactions

Although IL-2 can induce sustained eosinophilia possibly mediated by IL-5 or GM-CSF, this was not associated with allergic reactions *(114)*. So far, no hypersensitivity reactions directly related to IL-2 have indeed been described, and only two case reports of angioneurotic edema and urticaria questioned the role of IL-2 in hypersensitivity reactions *(115)*. IL-2-induced antigen-independent T-cell activation, however, can increase the risk of drug-induced hypersensitivity reactions. A three to four times greater incidence of hypersensitivity reactions to iodinated and nonionic contrast media injection was observed when radiological examination was performed several weeks after IL-2 withdrawal in patients who had previously well tolerated contrast media injection *(113,116)*. The reactions usually developed within 1 to 4 h after contrast media injection, and delayed reactions up to 24 h were sometimes noted. Enhancement of the immune response to contrast media after IL-2 was suggested as the likely mechanism. An unexpectedly high incidence of immediate allergic reactions to cisplatin and dacarbazine also was observed in patients who had received a combination of IL-2 and IFN-α *(117)*. Successive episodes of multifocal fixed drug eruption in response to chemically unrelated drugs (acetaminophen, ondansetron, and tropisetron) were described in one patient *(118)*.

2.2.4. Infectious Complications

Clinically relevant bacterial infections not associated with severe neutropenia occurred with an incidence of 10% to 40% during the first intravenous course of IL-2 therapy *(119)*. The mechanism is not understood. Impaired cell-mediated or humoral immune responses after high-dose IL-2, reduced neutrophil chemotaxis, superoxide production, and/or neutrophil Fc receptor expression have been suggested to be involved.

2.2.5. Interleukin-2 Antibodies

Recombinant IL-2 binding antibodies were detected in one half of 205 patients with metastatic cancer, but there were neutralizing in only 7% *(120)*. No significant difference in incidence was found between subcutaneous and continuous intravenous administration. In another study, no patients receiving IL-2 alone developed neutralizing antibodies *(121)*. Anti-IL-2 antibodies have been shown to recognize both the recombinant and natural cytokine, and patients developing neutralizing antibodies had significantly lower serum soluble IL-2 receptor than patients without antibody. However, the clinical relevance of neutralizing antibodies has not been evaluated.

2.3. Denileukin Diftitox

Denileukin diftitox is a fusion protein formed by the binding of human IL-2 to the cytotoxic A chain of diphtheria toxin, which by binding to the IL-2 receptor and inhibiting protein synthesis, results in T-cell death. It has been approved for the treatment of persistent or recurrent cutaneous T-cell lymphoma and is being evaluated in severe psoriasis. A dose-related flu-like reaction is very frequent and approx 60% of patients experienced dyspnea, back pain, hypotension, and chest pain or tightness within 24 h of its infusion *(122)*. Skin reactions compatible with delayed hypersensitivity reactions were noted in several patients treated for severe psoriasis, including one case of exfoliative dermatitis *(123)*.

2.4. Interleukin-3

IL-3 produced by activated T lymphocytes acts as a colony-stimulating factor. Because IL-3 given alone has only limited clinical effects, a genetically engineered GM-CSF/IL-3 fusion protein (PIXY321) has been developed. However, neither product had demonstrable advantage over conventional growth factors. The most frequent adverse effects of subcutaneous recombinant human IL-3 in healthy volunteers were flu-like symptoms *(124)*, and preliminary clinical trials confirmed that patients subsequently develop tachyphylaxis *(95)*. A similar safety profile was reported in patients receiving PIXY321 *(125)*. Some of these adverse effects were supposedly the result of a dose-dependent increase in IL-6 and acute phase protein production. Minor erythematous reactions at the injection site were also consistently described. Mild-to-severe skin rashes or urticaria were sometimes observed, and one patient had histological features of allergic vasculitis resembling those reported with GM-CSF *(126)*. In 185 patients with ovarian cancer, the most frequent adverse effects were allergic-type reactions (50% *[127]*). Histamine release from circulating basophils was suggested as the possible mechanism of an anaphylactoid reaction to recombinant human IL-3 *(128)*.

2.5. Interleukin-4

IL-4 is a pleiotropic cytokine, mostly produced by activated T-cells, that acts on the proliferation and differentiation of B- and T-lymphocytes, and enhances the function of NK cells, eosinophils, and mast cells. It has been investigated for potential antitumoral and hemopoietic actions. Flu-like symptoms frequently were observed at all doses and by all routes of administration, but they were more severe and prolonged at high doses *(95)*. Periorbital, facial, and peripheral edema also were noted. Frequent discomfort caused by severe and resistant nasal congestion supposedly caused by edema and vascular engorgement caused by histamine release was sometimes dose-limiting. A putative antibody-mediated mechanism has been suggested in several rIL-4-treated patients with transient acantholytic dermatosis *(129)*. Reversible Coomb's positive hemolytic anemia, as yet not clearly related to IL-4, has been described. The vascular leak syndrome was observed after bolus or continuous intravenous administration, but a moderate syndrome was also noted at lower subcutaneous doses *(130)*. Cardiac toxicity consistent with myocardial infarction was observed in three of seven cancer patients treated with bolus high dose IL-4 *(131)*. A unique pattern of myocarditis with predominant polymorphonuclear, eosinophil, and mast cell infiltration was the possible cause of death in one patient, which suggests a possibly allergic myocardial process.

2.6. Interleukin-6

IL-6 produced by T-cells, monocytes, fibroblasts, endothelial cells, and keratinocytes regulates pleiotropic biological functions. Recombinant IL-6 has been evaluated for thrombopoietic and antitumoral properties. During clinical trials, IL-6 consistently produced moderate fever and flu-like symptoms *(95)*. Moderate injection-site reactions were seen after subcutaneous administration, and a diffuse maculopapular erythema was sometimes treatment-limiting. IL-6 has not been associated with the vascular leak syndrome or hypotension. Neutralizing antibodies to IL-6 were rarely evidenced.

2.7. Interleukin-10

IL-10 is a potent anti-inflammatory and immunosuppressive cytokine with beneficial effects expected in a wide range of diseases. In healthy volunteers, adverse effects mostly consisted of flu-like symptoms at the highest dose *(132)*. Because of its immunomodulating properties, potential adverse immunological effects, namely an increased risk of infections, autoimmune disorders, or B-cell lymphoproliferative disorders, can be anticipated.

2.8. Interleukin-11

IL-11 has thrombopoietic activity and is approved to prevent severe thrombopenia and reduce the need for platelet transfusion after myelosuppressive chemotherapy. Common adverse effects included myalgia and arthralgias, fatigue, headache, and conjunctival injection *(133)*. So far, no systemic immune adverse effects have been reported.

2.9. Interleukin-12

IL-12 is an immunomodulatory cytokine with potential therapeutic activity in several cancerous and infectious diseases. Severe and sometimes-fatal multiple organ adverse effects have been described in early studies, but this unexpected profile of toxicity was later shown to be schedule-dependent *(134)*. IL-12 has been involved in autoimmune disorders, but only one report described the acute exacerbation of severe rheumatoid arthritis after each course of IL-12 in one patient with a previously stable disease *(135)*. A recent controlled trial of IL-12 in patients with chronic hepatitis C was halted because the treatment was ineffective and poorly tolerated with one case of immune thrombocytopenic purpura *(136)*.

3. GRANULOCYTE MACROPHAGE COLONY-STIMULATING FACTOR AND GRANULOCYTE COLONY-STIMULATING FACTOR

Although a number of myeloid hemopoietic growth factors or colony-stimulating factors (CSFs) have been purified, most clinical studies involved granulocyte-macrophage CSF (GM-CSF) or granulocyte CSF (G-CSF). GM-CSF causes a dose-related increase in peripheral neutrophil numbers and functions, and a delayed increase in circulating monocytes and eosinophils, whereas the effect of G-CSF appears to be more restricted to neutrophils by stimulating the proliferation of committed myeloid precursors *(95)*.

Both G-CSF and GM-CSF have been extensively investigated for the treatment of chemotherapy-induced neutropenia, the reduction of neutropenia duration after bone marrow transplantation, or the mobilization of peripheral blood progenitor cells after myelosuppressive chemotherapy *(137)*. CSFs also are used in severe chronic neutropenic diseases and in healthy donors to mobilize blood progenitor cells.

Recombinant human forms of G-CSF include filgrastim, lenograstim, nargrastim, and pegfilgrastim, a pegylated derivative of filgrastim. Recombinant human forms of GM-CSF include molgramostim and sargramostim. Active recombinant proteins are glycosylated, and glycosylation may be clinically relevant with regard to efficacy and antigenicity. Although G-CSF was considered to be better tolerated than GM-CSF *(138)*, there were no major differences in the safety profile and severity of adverse effects in the few studies that compared G-CSF and GM-CSF *(139)*. Overall, GM-CSF produced more frequent noninfectious fever, fatigue, diarrhea, injection-site reactions, edema, and skin rash, whereas skeletal pains were more frequent with G-CSF. Mild to moderate flu-like symptoms are mostly observed with GM-CSF and probably result from the release of cytokines, such as TNF-α and IL-1 *(95)*.

3.1. Exacerbation of Autoimmune Diseases

Isolated case reports suggested that GM-CSF may exacerbate underlying autoimmune thyroiditis (140,141). By contrast, no influence on thyroid function or autoimmunity was observed in cancer patients treated with G-CSF (142). Worsening of rheumatoid symptoms has been reported in patients with neutropenia because of Felty's syndrome receiving G-CSF or GM-CSF (95). Although there was concern on the short-term safety of CSFs in these patients or in those with rheumatoid arthritis, other investigators felt that G-CSF can be used for a prolonged period of time without a flare-up of rheumatoid symptoms (143). G-CSF has also been associated with possible exacerbation in neutropenic patients with severe SLE (144). The mechanism of flare-up is unclear and a localized neutrophil activation or acute IL-6 release with an increase in acute phase proteins were thought to be involved.

3.2. Complications Related to CSF-Induced Hematopoietic Activation

Asymptomatic but marked increases in spleen volume concomitant with neutrophilia have been reported in patients or healthy donors after receiving G-CSF or GM-CSF (145,146). Splenomegaly with extramedullary hematopoiesis was thought to result from the mobilization of early hemopoietic progenitors from the bone marrow to the spleen (147). Spontaneous splenic rupture associated with G-CSF has been reported (148,149).

G-CSF and GM-CSF-induced increased production and functions of neutrophils, or activation of monocytes/macrophages can play a critical role in the occurrence of neutrophilic dermatoses and a wide range of skin disorders. Although the ability of G-CSF to induce acute neutrophilic dermatitis (Sweet's syndrome) has been disputed, recurrence of the lesions after G-CSF re-administration has been noted (150–152). GM-CSF or G-CSF-induced neutrophilic dermatoses also include neutrophilic abscesses or panniculitis, bullous pyoderma gangrenosum and neutrophilic eccrine hidradenitis (95,150). Disseminated vesiculopustular lesions, generalized and indurated erythematous papules or plaques, severe exacerbation of acne or palmoplantar pustulosis, erythema multiforme or erythema nodosum, and leukocytoclastic vasculitis were mentioned as possible consequences of acutely increased neutrophil count after G-CSF administration (153–155). Finally, G-CSF or GM-CSF has been convincingly associated with acute exacerbations of psoriasis or psoriatic arthritis (156,157). The accumulation of activated neutrophils in the epidermis and dermis may play an important role in the occurrence or worsening of these complications.

Eosinophil activation and the subsequent release of eosinophil-derived toxic products were supposedly involved in GM-CSF-induced maculopapular, exfoliative, and urticarial eruptions with perivascular infiltration by lymphocytes, neutrophils, and eosinophils *(158)*. A similar phenomenon may account for atopic dermatitis-like eruptions with elevated serum IgE levels *(159)*. By contrast, dose-related and sometimes marked eosinophilia after GM-CSF is usually not associated with systemic symptoms, and Loeffler's endocarditis has been only exceptionally described *(160)*. Extensive and persistent bone marrow histiocytosis was suggested to result from GM-CSF-induced proliferation and activation of monocytes/macrophages *(161)*.

A dose-dependent vascular leak syndrome was consistently described with high-dose GM-CSF, but low doses also induce clinically relevant symptoms *(161)*. Endothelial cell damage with an increase in the transcapillary escape rate of albumin and the possible role of IL-1 and TNF-α production by GM-CSF-activated monocytes were suggested as possible mechanisms. By contrast, a typical capillary leak syndrome has been anecdotally reported after G-CSF administration *(162)*.

Whether G-CSF or GM-CSF can cause direct pulmonary toxicity or enhance chemotherapy-induced pulmonary toxicity is a matter of continuing debate. Conflicting data suggested a possibly increased risk of interstitial pneumonia in patients treated with anticancer drugs combined with G-CSF or GM-CSF *(163,164)*, whereas others did not *(165)*. G-CSF also can cause severe pulmonary toxicity in patients not receiving concomitant chemotherapy *(166)*, or play a role in the development or worsening of the adult respiratory distress syndrome *(167)*. Anyway, G-CSF should be regarded as a possible cause of pulmonary complications in treated cancer patients. The abrupt increase in activated neutrophils after G-CSF may account for exacerbation of latent chemotherapy-induced pulmonary damage. Endothelial damage subsequent to increased neutrophil activity (i.e., enhanced superoxide release and increased adhesion molecule expression and adherence), or the release of cytokines (IL-1, IL-6, TNF-α) have been suggested to be involved.

3.3. Immediate Hypersensitivity Reactions

Although specific IgE antibodies have not been detected, G-CSF and GM-CSF are undoubtedly associated with immediate hypersensitivity reactions, although rare, including systemic anaphylaxis, bronchospasm, urticaria, and angioedema, with positive skin tests in a few patients *(168–170)*. Crossreactivity between recombinant forms of GM-CSF, or between filgrastim and other products derived from *Escherichia coli* have been documented *(171)*. Patients can, however, subsequently tolerate the alternative growth factor and possible crossreactivity between G-CSF and GM-CSF with successful desensitization has been reported *(172)*.

3.4. Antibodies to G-CSF or GM-CSF

Antibodies to recombinant G-CSF have not been reported, whereas antibodies to recombinant GM-CSF has been detected in 31% of patients treated with sagramostim and in 95% of patients treated with molgramostim *(173,174)*. The clinical relevance of these findings is uncertain, but a significant modification of exogenous GM-CSF pharmacokinetics, a reduction in the rise of leukocyte counts, and a reduction in the frequency of GM-CSF-associated adverse effects are possible consequences. Subcutaneous and repeated administrations have been considered to increase the likelihood of antibody formation. The fact that most patients receiving CSFs are likely to be immunocompromised as a result of intensive chemotherapy may also account for discrepancies between the widespread use of growth factors and the paucity of reports on antibodies against CSFs.

4. TUMOR NECROSIS FACTOR-α

TNF-α is naturally produced by activated macrophages and monocytes and exerts pleiotropic effects on normal and malignant cells. The systemic administration of TNF-α as a single therapeutic agent gave disappointing results with severe dose-limiting hypotension or neurotoxicity, and no significant clinical antitumor effect *(175)*. Other frequent adverse effects include fever, chills and rigors, myalgias, diarrhea, nausea or vomiting, and local reactions at the injection sites. Exacerbation of hypothyroidism was noted in one patient with chronic thyroiditis *(176)*. Anaphylactic-like reactions, dyspnea, or acute bronchospasm have been attributed to TNF-α in patients also treated with IL-2 *(177)*.

REFERENCES

1. Manns MP, McHutchison JG, Gordon SC, et al. Peginterferon alfa-2b plus ribavirin compared with interferon alfa-2b plus ribavirin for initial treatment of chronic hepatitis C: a randomised trial. Lancet 2001;358:958–965.
2. Vial T, Descotes J Clinical toxicity of the interferons. Drug Safety 1994;10:115–150.
3. Tripi S, Soresi M, Di Gaetano G, et al. Leucocyte interferon-alpha for patients with chronic hepatitis C intolerant to other alpha interferons. BioDrugs 2003;17:201–205.
4. Vial T, Descotes J. Immune-mediated side-effects of cytokines in humans. Toxicology 1995;105:31–57.

5. Wada M, Kang KB, Kinugasa A, et al. Does the presence of serum autoantibodies influence the responsiveness to interferon-α2a treatment in chronic hepatitis C. Intern Med 1997;36:248–254.

6. Cassani F, Cataleta M, Valentini P, et al. Serum autoantibodies in chronic hepatitis C: comparison with autoimmune hepatitis and impact on the disease profile. Hepatology 1997;26:561–566.

7. Steegmann JL, Requena MJ, Martin-Regueira P, et al. High incidence of autoimmune alterations in chronic myeloid leukemia patients treated with interferon-alpha. Am J Hematol 2003;72:170–176.

8. Tothova E, Kafkova A, Stecova N, et al. Immune-mediated complications during interferon alpha therapy in chronic myelogenous leukemia. Neoplasma 2002;49:91–94.

9. Wandl UB, Nagel-Hiemke M, May D, et al. Lupus-like autoimmune disease induced by interferon therapy for myeloproliferative disorders. Clin Immunol Immunopathol 1992;65:70–74.

10. Fukuyama S, Kajiwara E, Suzuki N, Miyazaki N, Sadoshima S, Onoyama K. Systemic lupus erythematosus after alpha-interferon therapy for chronic hepatitis C: a case report and review of the literature. Am J Gastroenterol 2000;95: 310–312.

11. Jumbou O, Berthelot JM, French N, Bureau B, Litoux P, Dréno B. Polyarthritis during interferon alpha therapy: 3 cases and a review of the literature. Eur J Dermatol 1995;5:581–584.

12. Nesher G, Ruchlemer R Alpha-interferon-induced arthritis: clinical presentation, treatment, and prevention. Semin Arthritis Rheum 1998;27:360–365.

13. Solans R, Bosch JA, Esteban I, Vilardell M. Systemic sclerosis developing in association with the use of interferon alpha therapy for chronic viral hepatitis. Clin Exp Rheumatol 2004;22:625–628.

14. Fabris P, Floreani A, Tositti G, Vergani D, De Lalla F, Betterle C. Type 1 diabetes mellitus in patients with chronic hepatitis C before and after interferon therapy. Aliment Pharmacol Ther 2003;18:549–558.

15. Fattovich G, Giustina G, Favarato S, Ruol A. A survey of adverse events in 11,241 patients with chronic viral hepatitis treated with alfa interferon. J Hepatol 1996;24:38–47.

16. Mofredj A, Howaizi M, Grasset D, et al. Diabetes mellitus during interferon therapy for chronic viral hepatitis. Dig Dis Sci 2002;47:1649–1654.

17. Wesche B, Jaeckel E, Trautwein C, et al. Induction of autoantibodies to the adrenal cortex and pancreatic islet cells by interferon alpha therapy for chronic hepatitis C. Gut 2001;48:378–383.

18. Schories M, Peters T, Rasenack J, Reincke M. Autoantikörper gegen Inselzellantigene und Diabetes mellitus Typ 1 unter Interferon alpha-Kombinationtherapie. Dtsch Med Wochenschr 2004;129:1120–1124.

19. Khan HA, Khawaja FI, Mahrous ARS. Life-threatening severe immune thrombocytopenia after alpha-interferon therapy for chronic hepatitis C infection. Am J Gastroenterol 1996;91:821–822.

20. Landau A, Castera L, Buffet C, Tertian G, Tchernia G. Acute autoimmune hemolytic anemia during interferon-alpha-therapy for chronic hepatitis C. Dig Dis Sci 1999;44:1366–1377.
21. Zuffa E, Vianelli N, Martinelli G, Tazzari P, Cavo M, Tura S. Autoimmune mediated thrombocytopenia associated with the use of interferon-α in chronic myeloid leukemia. Haematologica 1996;81:533–535.
22. Barbolla L, Paniagua C, Outeirino J, Prieto E, Sanchez Favos J. Haemolytic anaemia to the alpha-interferon treatment: a proposed mechanism. Vox Sanguinis 1993;65:156–157.
23. Taliani G, Duca F, Clementi C, De Bac C. Platelet-associated immunoglobulin G, thrombocytopenia and response to interferon treatment in chronic hepatitis C. J Hepatol 1996;25:999.
24. McNair ANB, Jacyna MR, Thomas HC. Severe haemolytic transfusion reaction occurring during alpha-interferon therapy for chronic hepatitis. Eur J Gastroenterol Hepatol 1991;3:193–194.
25. Borgia G, Reynaud L, Gentile I, et al. Pernicious anemia during IFN-alpha treatment for chronic hepatitis C. J Interferon Cytokine Res 2003;23:11–12.
26. English KE, Brien WF, Howson-Jan K, Kovacs MJ. Acquired factor VII inhibitor in a patient with chronic myelogenous leukemia receiving interferon-alpha therapy. Ann Pharmacother 2000;34:737–739.
27. Becker JC, Winkler B, Klingert S, Bröcker EB. Antiphospholipid syndrome associated with immunotherapy for patients with melanoma. Cancer 1994;73: 1621–1624.
28. Garcia-Buey L, Garcia-Monzon C, Rodriguez S, et al. Latent autoimmune hepatitis triggered during interferon therapy in patients with chronic hepatitis C. Gastroenterology 1995;108:1770–1777.
29. Iorio R, Giannattasio A, Vespere G, Vegnente A. LKM1 antibody and interferon therapy in children with chronic hepatitis C. J Hepatol 2001;35:685–687.
30. Sezaki H, Arase Y, Tsubota A, et al. Type C-chronic hepatitis patients who had autoimmune phenomenon and developed jaundice during interferon therapy. J Gastroenterol 2003;38:493–500.
31. Steegmann JL, Requena MJ, Garcia-Buey ML, et al. Severe autoimmune hepatitis in a chronic myeloid leukemia patient treated with interferon alpha and with complete genetic response. Am J Hematol 1998;59:95–97.
32. Monzani F, Caraccio N, Dardano A, Ferrannini E. Thyroid autoimmunity and dysfunction associated with type I interferon therapy. Clin Exp Med 2004;3: 199–210.
33. Preziati D, La Rosa L, Covini G, et al. Autoimmunity and thyroid dysfunction in patients with chronic active hepatitis treated with recombinant interferon alpha-2a. Eur J Endocrinol 1995;132:587–593.
34. Watanabe U, Hashimoto E, Hishamitsu T, Obata H, Hayashi N. The risk factors for develoment of thyroid disease during interferon-α therapy for chronic hepatitis C. Am J Gastroenterol 1994;89:399–403.

35. Kakizaki S, Takagi H, Murakami M, Takayama H, Mori M. HLA antigens in patients with interferon-α-induced autoimmune thyroid disorders in chronic hepatitis C. J Hepatol 1999;30:794–800.
36. Roti E, Minelli R, Guiberti T, et al. Multiple changes in thyroid function in patients with chronic active HCV hepatitis treated with recombinant interferon-alpha. Am J Med 1996;172:482–487.
37. Schuppert F, Rambusch E, Kirchner H, Atzpodien J, Kohn LD, von zur Muhlen A. Patients treated with interferon-α, interferon-β, and interleukin-2 have a different thyroid autoantibody pattern than patients suffering from endogenous autoimmune thyroid disease. Thyroid 1997;7:837–842.
38. Borgia G, Reynaud L, Gentile I, et al. Myasthenia gravis during low-dose IFN-alpha therapy for chronic hepatitis C. J Interferon Cytokine Res 2001;21:469–470.
39. Durante-Mangoni E, Iardino P, Resse M, et al. Silent celiac disease in chronic hepatitis C: impact of interferon treatment on the disease onset and clinical outcome. J Clin Gastroenterol 2004;38:901–905.
40. Tahan V, Ozseker F, Guneylioglu D, et al. Sarcoidosis after use of interferon for chronic hepatitis C: report of a case and review of the literature. Dig Dis Sci 2003;48:169–173.
41. Midturi J, Sierra-Hoffman M, Hurley D, Winn R, Beissner R, Carpenter J. Spectrum of pulmonary toxicity associated with the use of interferon therapy for hepatitis C: case report and review of the literature. Clin Infect Dis 2004;39:1724–1729.
42. Bremer CT, Lastrapes A, Alper AB, Mudad R. Interferon-alpha-induced focal segmental glomerulosclerosis in chronic myelogenous leukemia: a case report and review of the literature. Am J Clin Oncol 2003;26:262–264.
43. Anthoney DA, Bone I, Evans TR. Inflammatory demyelinating polyneuropathy: a complication of immunotherapy in malignant melanoma. Ann Oncol 2000;11:1197–1200.
44. Bachmann T, Koetter KP, Muhler J, Fuhrmeister U, Seidel G. Guillain-Barre syndrome after simultaneous therapy with suramin and interferon-alpha. Eur J Neurol 2003;10:599.
45. Kataoka I, Shinagawa K, Shiro Y, et al. Multiple sclerosis associated with interferon-alpha therapy for chronic myelogenous leukemia. Am J Hematol 2002;70: 149–153.
46. Kinnunen E, Timonen T, Pirtilla T, et al. Effect of recombinant alfa-2b interferon therapy in patients with progressive MS. Acta Neurol Scand 1993;87: 457–460.
47. Barreca T, Corsini G, Franceschini R, Gambini C, Garibaldi A, Rolandi E. Lichen planus induced by interferon-alpha-2a therapy for chronic active hepatitis C. Eur J Gastroenterol Hepatol 1995;7:367–368.
48. Guillot B, Blazquez L, Bessis D, Dereure O, Guilhou JJ. A prospective study of cutaneous adverse events induced by low-dose alpha-interferon treatment for malignant melanoma. Dermatology 2004;208:49–54.

49. Kerl K, Negro F, Lübbe J. Cutaneous side-effects of treatment of chronic hepatitis C by interferon alfa and ribavirin. Br J Dermatol 2003;149:656.
50. Cottoni F, Bolognini S, Deplano A, Garrucciu et al. Skin reaction in antiviral therapy for chronic hepatitis C: a role for polyethylene glycol interferon? Acta Dermatol Venereol 2004;84:120–123.
51. Jessner W, Kinaciyan T, Formann E, Steindl-Munda P, Ferenci P. Severe skin reactions during therapy for chronic hepatitis C associated with delayed hypersensitivity to pegylated interferons. Hepatology 2002;36:361:2002
52. Beckman DB, Mathisen TL, Harris KE, Boxer MB, Grammer LC. Hypersensitivity to IFN-alpha. Allergy 2001;56:806–807.
53. Pardini S, Bosincu L, Bonfilgi S, Dore F, Longinotti M. Anaphylactic-like syndrom in systemic mastocytosis treated with alpha-2-interferon. Acta Haematol 1991;85:220.
54. Parana R, Portugal M, Vitvitski L, Cotrim H, Lyra L, Trepo C. Severe strongyloidiasis during interferon plus ribavirin therapy for chronic HCV infection. Eur J Gastroenterol Hepatol 2000;12:245–246.
55. Gogos CA, Starakis JK, Bassaris HP, Skoutelis AT. Remote abscess formation during interferon-alpha therapy for viral hepatitis. Clin Microbiol Infect 2003;9:540–542.
56. Pesce A, Taillan B, Rosenthal E, et al. Opportunistic infections and CD4 lymphocytopenia with interferon treatment in HIV-1 infected patients. Lancet 1993;341:1597.
57. Soriano V, Bravo R, Samaniego JG, et al, and the HIV-Hepatitis Spanish Study Group. CD4 T-lymphocytopenia in HIV-infected patients receiving interferon therapy for chronic hepatitis C. AIDS 1994;8:1621–1622.
58. Antonelli G. In vivo development of antibody to interferons: an update to 1996. J Interferon Cytokine Res 1997;17(Suppl 1):39–46.
59. Hanley JP, Haydon GH. The biology of interferon-alfa and the clinical significance of anti-interferon antibodies. Leuk Lymph 1998;29:257–268.
60. Weinstock–Guttman B, Rudick RA. Prescribing recommendations for interferon-beta in multiple sclerosis. CNS Drugs 1997;8:102–112.
61. Walther EU, Hohlfeld R. Multiple sclerosis. Side effects of interferon beta and their management. Neurology 1999;53:1622–1627.
62. Colosimo C, Pozzilli C, Frontini M, et al. No increase of serum autoantibodies during therapy with recombinant human interferon-β1a in relapsing-remitting multiple sclerosis. Acta Neurol Scand 1997;96:372–374.
63. Kivisäkk P, Lundahl J, von Heigl Z, Fredrikson S. No evidence for increased frequency of autoantibodies during interferon-β 1b treatment of multiple sclerosis. Acta Neurol Scand 1998;97:320–323.
64. Kreisler A, de Seze J, Stojkovic T, et al. Multiple sclerosis, interferon beta and clinical thyroid dysfunction. Acta Neurol Scand 2003;107:154–157.
65. Rotondi M, Oliviero A, Profice P, et al. Occurrence of thyroid autoimmunity and dysfunction throughout a nine-month follow-up in patients undergoing interferon-β therapy for multiple sclerosis. J Endocrinol Invest 1998:21:748–752.

66. Kazuta Y, Watanabe N, Sagawa K, et al. A case of autoimmune hemolytic anemia induced by IFN-beta therapy for type-C chronic hepatitis. Fukushima J Med Sci 1995;41:43–49.
67. Wallack EM, Callon R. Liver injury associated with the beta-interferons for MS. Neurology 2004;63:1142–1143.
68. Nousari HC, Kimyai-Asadi A, Tausk FA. Subacute cutaneous lupus erythematosus associated with interferon beta-1a. Lancet 1998;352:1825–1826.
69. Blake G, Murphy S. Onset of myasthenia gravis in a patient with multiple sclerosis during interferon-1b. Neurology 1997;49:1747–1748.
70. Kaloyannidis P, Sakellari I, Fassas A, et al. Acquired hemophilia-A in a patient with multiple sclerosis treated with autologous hematopoietic stem cell transplantation and interferon beta-1a. Bone Marrow Transplant 2004;34:187–188.
71. Gaines AR, Varricchio F. Interferon beta-1b injection site reactions and necroses. Multiple Sclerosis 1998:4:70–73.
72. Feldmann R, Löw-Weiser H, Duschet P, Gschnait F Necrotizing cutaneous lesions caused by interferon beta injections in a patient with multiple sclerosis. Dermatology, 195:52–53:1997.
73. Bobbio-Pallavicini E, Valsecchi C, Tacconi F, Moroni M, Porta C. Sarcoidosis following beta-interferon therapy for multiple myeloma. Sarcoidosis 1995;12: 140–142.
74. Debat Zoguereh D, Boucraut J, Beau-Salinas F, Bodiguel E, Lechapois D, Pomet E. Vascularite cutanée avec atteinte rénale compliquant un traitement par interféron-beta 1a pour une sclérose en plaques. Rev Neurol 2004;160:1081–1084.
75. Webster GF, Knobler RL, Lublin FD, Kramer EM, Hochman LR. Cutaneous ulcerations and pustular psoriasis flare caused by recombinant interferon beta injections in patients with multiple sclerosis. J Am Acad Dermatol 1996;34: 365–367.
76. Corona T, Leon C, Ostrosky-Zeichner L. Severe anaphylaxis with recombinant interferon beta. Neurology 1999;52:425.
77. Brown DL, Login, IS, Borish L, Powers PL. An urticarial Ig-E-mediated reaction to interferon beta-1b. Neurology 2001;56:1416–1417.
78. Pigatto PD, Bigardi A, Legori A, Altomare GF, Riboldi A. Allergic contact dermatitis from beta-interferon in eyedrops. Contact Dermatitis 1991;25:199–200.
79. Polman C, Kappos L, White R, et al. Neutralizing antibodies during treatment of secondary progressive MS with interferon beta-1b. Neurology 2003;60:37–43.
80. Sorensen PS, Ross C, Clemmesen KM, et al. Clinical importance of neutralising antibodies against interferon beta in patients with relapsing-remitting multiple sclerosis. Lancet 2003;362:1184–1191.
81. Khan OA, Dhib-Jalbut SS. Neutralizing antibodies to interferon β-1a and interferon β-1b in MS patients are cross-reactive. Neurology 1998;51:1698–1702.
82. Cross AH, Antel JP. Antibodies to beta-interferons in multiple sclerosis. Can we neutralize the controversy? Neurology 1998;50:1206–1208.

83. Wolinsky JS, Toyka KV, Kappos L, Grossberg SE. Interferon-beta antibodies: implications for the treatment of MS. Lancet Neurol 2003;2:528.
84. Malucchi S, Sala A, Gilli F, et al. Neutralizing antibodies reduce the efficacy of betaIFN during treatment of multiple sclerosis. Neurology 2004:62:2031–2037.
85. Perini P, Calabrese M, Biasi. The clinical impact of interferon beta antibodies in relapsing-remitting MS. J Neurol 2004;251:305–309.
86. Kung AWC, Jones BM, Lai CL. Effects of interferon-gamma therapy on thyroid function, T-lymphocytes subpopulation and induction of autoantobodies. J Clin Endocrinol Metabol 1990;71:1230–1234.
87. Weber P, Wiedmann KH, Klein R, Walter E, Blum HE, Berg PA. Induction of autoimmune phenomena in patients with chronic hepatitis B treated with gamma-interferon. J Hepatol 1994;20:321–328.
88. Cannon GW, Emkey RD, Denes A, et al. Prospective 5-year follow up of recombinant interferon-γ in rheumatoid arthritis. J Rheumatol 1993;20:1867–1873.
89. Machold KP, Smolen JS Interferon gamma induced exacerbation of systemic lupus erythematosus. J Rheumatol 1990;17:831–832.
90. Seitz M, Kranke M, Kirchner H. Induction of antinuclear antibodies in patients with rheumatoid arthritis receiving treatment with recombinant human gamma interferon. Ann Rheum Dis 1988;47:642–644.
91. Panitch HS, Hirsch RL, Schindler J, Johnson KP. Treatment of multiple sclerosis with gamma interferon: exacerbations associated with activation of the immune system. Neurology 1987;37:1097–1102.
92. Aihara Y, Mori M, Katakura S, Yokota S. Recombinant IFN-gamma treatment of a patient with hyperimmunoglobulin E syndrome triggered autoimmune thrombocytopenia. J Interfer Cytokine Res 1998;18:561–563.
93. Mattson K, Niiranen A, Pyrhonen S, Farkkila M, Cantell K. Recombinant interferon gamma treatment in non-small cell lung cancer. Antitumour effect and cardiotoxicity. Acta Oncol 1991;30:607–610.
94. Fiehn C, Prummer O, Gallati H, Heilig B, Hunstein W. Treatment of systemic mastocytosis with interferon-gamma: failure after appearance of anti-IFN-gamma antibodies. Eur J Clin Invest 1995;25:615–618.
95. Vial T, Descotes J. Clinical toxicity of cytokines used as haemopoietic growth factors. Drug Safety 1995;13:371–406.
96. Gaffen SL, Liu KD. Overview of interleukin-2 function, production and clinical applications. Cytokine 2004;28:109–123.
97. Schwartzentruber DJ. Guidelines for the safe administration of high-dose interleukin-2. J Immunother 2001;24:287–293.
98. Stadler WM, Vogelzang NJ. Low-dose interleukin-2 in the treatment of metastatic renal-cell carcinoma. Semin Oncol 1995;22:67–73.
99. Geertsen PF, Gore ME, Negrier S, Tourani JM, von der Maase H. Safety and efficacy of subcutaneous and continuous intravenous infusion rIL-2 in patients with metastatic renal cell carcinoma. Br J Cancer 2004;90:1156–1162.

100. Vial T, Descotes J. Clinical toxicity of interleukin-2. Drug Safety 1992;7: 417–433.

101. Vialettes B, Guillerand MA, Viens P, et al. Incidence rate and risk factors for thyroid dysfunction during recombinant interleukin-2 therapy in advanced malignancies. Acta Endocrinol 1993;129:31–38.

102. Krouse RS, Royal RE, Heywood G, et al. Thyroid dysfunction in 281 patients with metastatic melanoma of renal carcinoma treated with interleukin-2 alone. J Immunother 1995;18:272–278.

103. Jacobs EL, Clare-Salzer MJ, Chopra IJ, Figlin RA. Thyroid function abnormalities associated with the chronic outpatient administration of recombinant interleukin-2 and recombinant interferon-alpha. J Immunother 1991;10:448–455.

104. Soni N, Meropol NJ, Porter M, Caligiuri MA. Diabetes mellitus induced by low-dose interleukin-2. Cancer Immunol Immunother 1996;43:59–62.

105. Fraenkel PG, Rutkove SB, Matheson JK, et al. Induction of myasthenia gravis, myositis, and insulin-dependent diabetes mellitus by high-dose interleukin-2 in a patient with renal cell cancer. J Immunother 2002;25:373–378.

106. Massarotti E, Liu NY, Mier J, Atkins MB. Chronic inflammatory arthritis after treatment with high-dose interleukin-2 for malignancy. Am J Med 1992; 92:693–697.

107. Baars JW, Hack CE, Wagstaff J, et al. The activation of polymorphonuclear neutrophils and the complement system during immunotherapy with recombinant interleukin-2. Br J Cancer 1992;65:96–101.

108. Dubinett SM, Huang M, Lichtenstein A, et al. Tumor necrosis factor-alpha plays a central role in interleukin-2-induced pulmonary vascular leak and lymphocyte accumulation. Cell Immunol 1994;157:170–180.

109. Junghans RP, Manning W, Safar M, Quist W. Biventricular cardiac thrombosis during interleukin-2 infusion. N Engl J Med 2001;344:859–860.

110. Truica CI, Hansen CH, Garvin DF, Meehan KR. Idiopathic giant cell myocarditis after autologous hematopoietic stem cell transplantation and interleukin-2 immunotherapy. A case report. Cancer 1998;83:1231–1236.

111. Wolkenstein P, Chosidow O, Wechsler J, et al. Cutaneous side effects associated with interleukin-2 administration for metastatic melanoma. J Am Acad Dermatol 1993;28:66–70.

112. Blessing K, Park KG, Heys SD, King G, Eremin O. Immunopathological changes in the skin following recombinant interleukin-2 treatment. J Pathol 1992;167:313–319.

113. Asnis LA, Gaspari AA. Cutaneous reactions to recombinant cytokine therapy. J Am Acad Dermatol 1995;33:393–410.

114. MacDonald D, Gordon AA, Kajitani H, Enokihara H, Barrett AJ. Interleukin-2 treatment-associated eosinophilia is mediated by interleukin-5 production. Br J Haematol 1990;76:168–173.

115. Baars JW, Wagstaff J, Hack CE, Wolbink GJ, Eerenberg-Belmer AJ, Pinedo HM. Angioneurotic oedema and urticaria during therapy with interleukin-2. Ann Oncol 1992;3:243–244.

116. Zukiwski AA, David CL, Coan J, Wallace S, Gutterman JU, Mavligit GM. Increased incidence of hypersensitivity to iodine-containing radiographic contrast media after interleukin-2 administration. Cancer 1990;65:1521–1524.

117. Heywood GR, Rosenberg SA, Weber JS. Hypersensitivity reactions to chemotherapy agents in patients receiving chemoimmunotherapy with high-dose interleukin 2. J Natl Cancer Inst 1995;87:915–922.

118. Bernand S, Scheidegger EP, Dummer R, Burg G. Multifocal fixed drug eruption to paracetamol, tropisetron and ondansetron induced by interleukin-2. Dermatology 2000;201:148–150.

119. Pockaj BA, Topalian SL, Steinberg SM, White DE, Rosenberg SA. Infectious complications associated with interleukin-2 administration: a retrospective review of 935 treatment courses. J Clin Oncol 1993;11:136–147.

120. Scharenberg JG, Stam AG, von Blomberg BM, et al. The development of anti-interleukin-2 (IL-2) antibodies in cancer patients treated with recombinant IL-2. Eur J Cancer 1994;30A:1804–1809.

121. Atzpodien J, Lopez Hänninen E, Kirchner H, Knuver-Hopf J, Poliwoda H. Human antibodies to recombinant interleukin-2 in patients with hypernephroma. J Interfer Res 1994;14:177–178.

122. Olsen E, Duvic M, Frankel A, et al. Pivotal phase III trial of two dose levels of denileukin diftitox for the treatment of cutaneous T-cell lymphoma. J Clin Oncol 2001;19:376–388.

123. Martin A, Gutierrez E, Muglia J, et al. A multicenter dose-escalation trial with denileukin diftitox in patients with severe psoriasis. J Am Acad Dermatol 2001;45:871–881.

124. Huhn RD, Yurkow EJ, Kuhn JG, et al. Pharmacodynamics of daily subcutaneous recombinant human interleukin-3 in normal volunteers. Clin Pharmacol Ther 1995;57:32–41.

125. Schuh JCL, Morrissey PJ. Development of a recombinant growth factor and fusion protein: lessons from GM-CSF. Toxicol Pathol 1999;2:72–77.

126. Bridges AG, Helm TN, Bergfeld WF, Lawlor KB, Dijkstra J. Interleukin-3-induced urticaria-like eruption. J Am Acad Dermatol 1996;34:1076–1078.

127. Hofstra LS, Kristensen GB, Willemse PHB, et al. Randomized trial of recombinant human interleukin-3 versus placebo in prevention of bone marrow depression during first-line chemotherapy for ovarian carcinoma. J Clin Oncol 1998;16:3335–3344.

128. Zeidman A, Fradin Z, Menachem Y, Mittelman M. Anaphylactic shock due to recombinant human interleukin-3. Eur J Haematol 1999;62:199–200.

129. Mahler SJ, de Villez RL, Pulitzer DR. Transient acantholytic dermatosis induced by recombinant human interleukin 4. J Am Acad Dermatol 1993;29: 206–209.

130. Prendiville J, Thatcher N, Lind M, et al. Recombinant human interleukin-4 (rhu IL-4) administered by the intravenous and subcutaneous routes in patients with advanced cancer: a phase I toxicity study and pharmacokinetic analysis. Eur J Cancer 1993;29A:1700–1707.

131. Trehu EG, Isner JM, Mier JW, Karp DD, Atkins MB. Possible myocardial toxicity associated with interleukin-4 therapy. J Immunother 1993;14:348–351.
132. Huhn RD, Radwanski E, O'Connell SM, et al. Pharmacokinetics and immunomodulatory properties of intravenously administered recombinant human interleukin-10 in healthy volunteers. Blood 1996;87:699–705.
133. Smith JW. Tolerability and side-effect profile of rhIL-11. Oncology 2000;14 (Suppl 8):41–47.
134. Leonard JP, Sherman ML, Fisher GL, et al. Effects of single-dose interleukin-12 exposure on interleukin-12 associated toxicity and interferon-γ production. Blood 1997;90:2541–2548.
135. Peeva E, Fishman AD, Goddard G, Wadler S, Barland P. Rheumatoid arthritis exacerbations caused by exogenous interleukin-12. Arthr Rheum 2000;43: 461–463.
136. Pockros PJ, Patel K, O'Brien C, et al. A multicenter study of recombinant human interleukin 12 for the treatment of chronic hepatitis C virus infection in patients nonresponsive to previous therapy. Hepatology 2003;37:1368–1374.
137. Hübel K, Engert A. Clinical applications of granulocyte colony-stimulating factor: an update and summary. Ann Hematol 2003;82:207–213.
138. Milkovich G, Moleski J, Reitan JF, et al. Comparative safety of filgrastim versus sargramostim in patients receiving myelosuppressive chemotherapy. Pharmacotherapy 2000;20:1432–1440.
139. Beveridge RA, Miller JA, Kales AN, et al. A comparison of efficacy of sargramostim (Yeast-derived RhuGM-CSF) and filgrastim (bacteria-derived RhuG-CSF) in the therapeutic setting of chemotherapy-induced myelosuppression. Cancer Invest 1998;16:366–373.
140. Hansen PB, Johnsen HE, Hippe E. Autoimmune hypothyroidism and granulocyte-macrophage colony-stimulating factor. Eur J Haematol 1993;50:183–184.
141. Hoekman K, von Blomberg-van der Flier BME, Wagstaff J, Drexhage HA, Pinedo HM. Reversible thyroid dysfunction during treatment with GM-CSF. Lancet 1991;338:541–542..
142. Van Hoef MEH.M., Howell A. Risk of thyroid dyfunction during treatment with G-CSF. Lancet 1992;340:1169–1170.
143. Stanworth SJ, Bhavnani M, Chattopadhya C, Miller H, Swinson DR. Treatment of Felty's syndrome with the haemopoietic growth factor granulocyte colony-stimulating factor (G-CSF). Q J Med 1998;9:49–56.
144. Euler HH, Harten P, Zeuner RA, Schwab UM. Recombinant human granulocyte colony stimulating-factor in patients with systemic lupus erythematosus associated neutropenia and refractory infections. J Rheumatol 1997;24:2153–2157.
145. Lindemann A, Hermann F, Mertelsmann R, Gamm H, Rumpelt HJ. Splenic haematopoiesis following GM-CSF therapy in a patient with hairy cell leukaemia. Leukemia 1990;4:606–607.
146. Picardi M, De Rosa G, Selleri C, et al. Spleen enlargement following recombinant human granulocyte colony-stimulating factor administration for peripheral blood stem cell mobilization. Haematologica 2003;88:794–800.

147. Nakayama T, Kudo H, Suzuki S, Sassa S, Mano Y, Sakamoto S. Splenomegaly induced by recombinant human granulocyte-colony stimulating factor in rats. Life Sci 2001;69:1521–1529.

148. Falzetti F, Aversa F, Minelli O, Tabilio A. Spontaneous rupture of spleen during peripheral blood stem-cell mobilisation in a healthy donor. Lancet 1999;353:555.

149. O'Malley DP, Whalen M, Banks PM. Spontaneous splenic rupture with fatal outcome following G-CSF administration for myelodysplastic syndrome. Am J Hematol 2003;73:294–295.

150. Johnson MML, Grimwood CRE. Leukocyte colony-stimulating factors. A review of associated neutrophilic dermatoses and vasculitides. Arch Dermatol 1994;130:77–81.

151. Kumar G, Bernstein JM, Waibel JS, Baumann MA. Sweet's syndrome associated with sargramostim (granulocyte-macrophage colony stimulating factor) treatment. Am J Hematol 2004;76:283–285.

152. Paydas S, Sahin B, Zorludemir S. Sweet's syndrome accompanying leukaemia: seven cases and review of the literature. Leuk Res 2000;24:83–86.

153. Alvarez-Ruiz S, Penas PF, Fernandez-Herrera J, Sanchez-Perez J, Fraga J, Garcia-Diez A. Maculopapular eruption with enlarged macrophages in eight patients receiving G-CSF or GM-CSF. J Eur Acad Dermatol Venereol 2004; 18:310–313.

154. Jain KK. Cutaneous vasculitis associated with granulocyte colony-stimulating factor. J Am Acad Dermatol 1994;31:213–215.

155. Kurokawa I, Umehara M, Iwai T, Hamanishi S. Exacerbation of palmoplantar pustulosis by granulocyte colony-stimulating factor. Int J Dermatol 2005;44: 529–530.

156. Kelly R, Marsden RA, Bevan D. Exacerbation of psoriasis with GM-CSF therapy. Br J Dermatol 1993;128:468–469.

157. Mossner R, Beckmann I, Hallermann C, Neumann C, Reich K. Granulocyte colony-stimulating-factor-induced psoriasiform dermatitis resembles psoriasis with regard to abnormal cytokine expression and epidermal activation. Exp Dermatol 2004;13:340–346.

158. Mehregan DR, Fransway AF, Edmonson JH, Leiferman KM. Cutaneous reactions to granulocyte-monocyte colony-stimulating factor. Arch Dermatol 1992;128:1055–1059.

159. Yamada H, Tubaki K, Ashida T, et al. Does recombinant granulocyte-macrophage colony-stimulating factor (GM-CSF) play a crucial role in the pathogenesis of atopic dermatitis after bone marrow transplantation (BMT)? Med Sci Res 1991;19:395.

160. Donhuijsen K, Haedicke C, Hattenberger S, Hauswaldt C, Freund M. Granulocyte-macrophage colony-stimulating factor-related eosinophilia and Loeffler's endocarditis. Blood 1992;79:2798.

161. Al-Homaidhi A, Prince HM, Al-Zahrani H, Doucette D, Keating A. Granulocyte-macrophage colony-stimulating factor-associated histiocytosis and cap-

illary-leak syndrome following autologous bone marrow transplantation: two case reports and a review of the literature. Bone Marrow Transplant 1998;21:209–214.

162. Oeda E, Shinohara K, Kamei S, Nomiyama J, Inoue H. Capillary leak syndrome likely the result of granulocyte colony-stimulating factor after high-dose chemotherapy. Intern Med 1994;33:115–119.

163. Couderc LJ, Stelianides S, Frachon I, et al. Pulmonary toxicity of chemotherapy and G/GM-CSF: a report of five cases. Respir Med 1999;93:65–68.

164. Niitsu N, Iki S, Motomura S, et al. Interstitial pneumonia in patients receiving granulocyte colony-stimulating factor during chemotherapy: survey in Japan 1991–96. Br J Cancer 1997;76:1661–1666.

165. Bastion Y, Reyes F, Bosly A, et al. Possible toxicity with the association of G-CSF and bleomycin. Lancet 1994;343:1221–1222.

166. Ruiz-Argüelles G, Arizpe-Bravo D, Sanchez-Sosa S, Rojas-Ortega S, Moreno-Ford V, Ruiz-Argüelles A. Fatal G-CSF-induced pulmonary toxicity. Am J Hematol 1999;60:82–83.

167. Takatsuka H, Takemoto Y, Mori A, Okamoto T, Kanamaru A, Kakishita E. Common features in the onset of ARDS after administration of granulocyte colony-stimulating factor. Chest 2002;121:1716–1720.

168. Jaiyesimi I, Giralt SS, Wood J. Subcutaneous granulocyte colony-stimulating factor and acute anaphylaxis. N Engl J Med 1991:325:587.

169. Keung YK, Suwanvecho S, Cobos E. Anaphylactoid reaction to granulocyte colony-stimulating factor used in mobilization of peripheral blood stem cell. Bone Marrow Transplant 1999;23:200–201.

170. Sasaki O, Yokoyama A, Uemura S, Fujino S, Inoue Y, Kohno N, Hiwada K. Drug eruption caused by recombinant human G-CSF. Intern Med 1994;22: 641–643.

171. Stone HD, DiPiro C, Davis PC, Meyer CF, Wray BB. Hypersensitivity reactions to *Escherichia coli*-derived polyethylene glycolated-asparaginase associated with subsequent immediate skin test reactivity to *E. coli*-derived granulocyte colony-stimulating factor. J Allergy Clin Immunol 1998;101: 429–431.

172. Shahar E, Krivoy N, Pollack S. Effective acute desensitization for immediate-type hypersensitivity to human granulocyte-monocyte colony stimulating factor. Ann Allergy Asthma Immunol 1999;83:543–546.

173. Gribben JG, Devereux S, Thomas NS, et al. Development of antibodies to unprotected glycosylation sites of recombinant human GM-CSF. Lancet 1990; 335:434–437.

174. Ragnhammar P, Friesen HJ, Frodin JE, et al. Induction of anti-recombinant human granulocyte-macrophage colony-stimulating factor (*Escherichia coli*-derived) antibodies and clinical effects in nonimmunocompromised patients. Blood 1994;84:4078–4087.

175. Sidhu RS, Bollon AP. Tumor necrosis factor activities and cancer therapy—a perspective. Pharmacol Ther 1993;57:79–128.

176. Miyakoshi H, Ohsawa K, Yokoyama H, et al. Exacerbation of hypothyroidism following tumor necrosis factor-alpha infusion. Intern Med 1992;31:200–203.
177. Negrier MS, Pourreau CN, Palmer PA, et al. Phase I-trial of recombinant interleukin-2 followed by recombinant tumor necrosis factor in patients with metastatic cancer. J Immunother 1992;11:93–102.

Index